Understanding and improving crop photosynthesis

It is widely recognised that agriculture is a significant contributor to global warming and climate change. Agriculture needs to reduce its environmental impact and adapt to current climate change whilst still feeding a growing population, i.e. become more 'climate-smart'. Burleigh Dodds Science Publishing is playing its part in achieving this by bringing together key research on making the production of the world's most important crops and livestock products more sustainable.

Based on extensive research, our publications specifically target the challenge of climate-smart agriculture. In this way we are using 'smart publishing' to help achieve climate-smart agriculture.

Burleigh Dodds Science Publishing is an independent and innovative publisher delivering high quality customer-focused agricultural science content in both print and online formats for the academic and research communities. Our aim is to build a foundation of knowledge on which researchers can build to meet the challenge of climate-smart agriculture.

For more information about Burleigh Dodds Science Publishing simply call us on +44 (0) 1223 839365, email info@bdspublishing.com or alternatively please visit our website at www.bdspublishing.com.

Related titles:

Achieving sustainable cultivation of maize Volume 1: From improved varieties to local applications
Print (ISBN 978-1-78676-008-1); Online (ISBN 978-1-78676-010-4, 978-1-78676-011-1)

Achieving sustainable cultivation of wheat Volume 1: Breeding, quality traits, pests and diseases
Print (ISBN 978-1-78676-016-6); Online (ISBN 978-1-78676-018-0, 978-1-78676-019-7)

Advances in breeding techniques for cereal crops
Print (ISBN 978-1-78676-244-3); Online (ISBN 978-1-78676-246-7; 978-1-78676-247-4)

Achieving sustainable cultivation of barley
Print (ISBN 978-1-78676-308-2); Online (ISBN 978-1-78676-310-5, 978-1-78676-311-2)

Understanding and improving crop root function
Print (ISBN 978-1-78676-360-0); Online (ISBN 978-1-78676-362-4, 978-1-78676-363-1)

Chapters are available individually from our online bookshop:
https://shop.bdspublishing.com

climate•SMART•publishing
IN AGRICULTURAL SCIENCE

BURLEIGH DODDS SERIES IN AGRICULTURAL SCIENCE
NUMBER 130

Understanding and improving crop photosynthesis

Edited by Dr Robert Sharwood, Western Sydney University, Australia

burleigh dodds
SCIENCE PUBLISHING

Published by Burleigh Dodds Science Publishing Limited
82 High Street, Sawston, Cambridge CB22 3HJ, UK
www.bdspublishing.com

Burleigh Dodds Science Publishing, 1518 Walnut Street, Suite 900, Philadelphia, PA 19102-3406, USA

First published 2023 by Burleigh Dodds Science Publishing Limited

Library of Congress Control Number: 2022949283

British Library Cataloguing in Publication Data
A catalogue record for this book is available from the British Library

ISBN 978-1-80146-129-0 (print)
ISBN 978-1-80146-132-0 (PDF)
ISBN 978-1-80146-131-3 (ePub)
ISSN 2059-6936 (print)
ISSN 2059-6944 (online)

DOI 10.19103/AS.2022.0119

Typeset by Deanta Global Publishing Services, Dublin, Ireland

Contents

Series list

Title	Series number
Achieving sustainable cultivation of maize - Vol 1 From improved varieties to local applications Edited by: Dr Dave Watson, CGIAR Maize Research Program Manager, CIMMYT, Mexico	001
Achieving sustainable cultivation of maize - Vol 2 Cultivation techniques, pest and disease control Edited by: Dr Dave Watson, CGIAR Maize Research Program Manager, CIMMYT, Mexico	002
Achieving sustainable cultivation of rice - Vol 1 Breeding for higher yield and quality Edited by: Prof. Takuji Sasaki, Tokyo University of Agriculture, Japan	003
Achieving sustainable cultivation of rice - Vol 2 Cultivation, pest and disease management Edited by: Prof. Takuji Sasaki, Tokyo University of Agriculture, Japan	004
Achieving sustainable cultivation of wheat - Vol 1 Breeding, quality traits, pests and diseases Edited by: Prof. Peter Langridge, The University of Adelaide, Australia	005
Achieving sustainable cultivation of wheat - Vol 2 Cultivation techniques Edited by: Prof. Peter Langridge, The University of Adelaide, Australia	006
Achieving sustainable cultivation of tomatoes Edited by: Dr Autar Mattoo, USDA-ARS, USA and Prof. Avtar Handa, Purdue University, USA	007
Achieving sustainable production of milk - Vol 1 Milk composition, genetics and breeding Edited by: Dr Nico van Belzen, International Dairy Federation (IDF), Belgium	008
Achieving sustainable production of milk - Vol 2 Safety, quality and sustainability Edited by: Dr Nico van Belzen, International Dairy Federation (IDF), Belgium	009
Achieving sustainable production of milk - Vol 3 Dairy herd management and welfare Edited by: Prof. John Webster, University of Bristol, UK	010
Ensuring safety and quality in the production of beef - Vol 1 Safety Edited by: Prof. Gary Acuff, Texas A&M University, USA and Prof. James Dickson, Iowa State University, USA	011
Ensuring safety and quality in the production of beef - Vol 2 Quality Edited by: Prof. Michael Dikeman, Kansas State University, USA	012
Achieving sustainable production of poultry meat - Vol 1 Safety, quality and sustainability Edited by: Prof. Steven C. Ricke, University of Arkansas, USA	013
Achieving sustainable production of poultry meat - Vol 2 Breeding and nutrition Edited by: Prof. Todd Applegate, University of Georgia, USA	014

Introduction

It is widely recognised that photosynthesis in many important crops is well below its theoretical potential. With crop yields and stability under threat from the impact of climate change, there is now an urgent need to synthesise existing research on best practices for improving C_3 photosynthesis in crops to optimise sustainable crop production and yields. This volume reviews the wealth of current research that addresses this challenge.

The book is split into three parts: Part 1 chapters provide a general overview of C_3 photosynthesis, focusing specifically on its biochemistry and genetics in crop plants. Chapters in Part 2 focus on improving photosynthesis through the process of light harvesting, drawing attention to the relationship between photosynthesis and the circadian clock as well as modifying the photosystem antenna complex to improve light harvesting. Chapters also focus on relaxing non-photochemical quenching to improve photosynthesis, the modification of mesophyll conductance and the modification of leaf and canopy architecture to optimise photosynthesis in crops. Part 3 chapters focus on optimising chloroplast function and light conversion to improve photosynthesis, highlighting how modifying photorespiration, maximising the efficiency of ribulose biphosphate respiration and understanding and modifying protein function in plant chloroplasts can optimise crop performance.

Part 1 General

Part 1 opens with a chapter that focuses on understanding the biochemistry of C_3 photosynthesis in crop plants. Chapter 1 begins by first describing the carboxylation phase of the C_3 cycle, which is then followed by sections on the reduction phase and regeneration phase of the C_3 cycle. The chapter also addresses how C_3 cycle enzymes can be regulated, highlighting methods such as light activation and how analysing the crystal structure of the C_3 cycle enzymes can help to regulate them. It moves on to discuss approaches to determine which enzymes limit the flow of carbon through the C_3 cycle, such as empirical studies and modelling of the C_3 cycle. A section on future opportunities to improve the C_3 cycle is also included.

Moving on from Chapter 1, Chapter 2 draws specific attention to the genetics of C_3 photosynthesis in crop plants. The chapter introduces the photosynthesis-associated core genes encoded either in the nucleus or in the chloroplast and discusses how different internal (e.g. redox state, circadian rhythm) and external (e.g. abiotic stresses, light) signals regulate their transcription, particularly in crop plants. Since the molecular mechanisms underlying the regulation of photosynthesis-associated genes is poorly understood, the chapter also discusses what is known regarding the transcriptional regulation of photosynthesis in C_3 crops, mainly rice and tomato. Among the regulators described, few were shown in the field to have the potential to improve photosynthesis. How the state-of-the-art knowledge can be used for photosynthesis improvement and future work perspectives, including the use of transplastomics, is also discussed.

Part 2 Improving photosynthesis: light harvesting

The first chapter of Part 2 discusses interactions between photosynthesis and the circadian system. Chapter 3 begins by describing how the circadian system regulates photosynthetic processes, as well as how it can regulate photosynthesis in C_4 and Crassulacean Acid Metabolism species. The chapter then draws specific attention to the interactions between photosynthates and the circadian system and highlights how starch metabolism is used as a consistent energy supply during the night. This is then followed by an analysis of the indicators of cellular energy status and its integration into the circadian system. The chapter also discusses the generation of reactive oxygen species during photosynthesis and how reactive oxygen species can also be integrated into circadian timing.

Chapter 4 looks at modifying photosystem antennas to improve light harvesting for photosynthesis in crops. The chapter first examines photopigments and their functions in light-harvesting complexes, then moves on to review several photosynthetic light harvesting protein complexes. A section on photoinhibition is also provided, which is then followed by an analysis of photosynthesis efficiency. Current challenges with modifying photosystem antennas to improve light harvesting are highlighted, which are then followed by the potential areas of research that can be developed in the future.

The next chapter focuses on relaxing non-photochemical quenching to improve photosynthesis in crops. Chapter 5 starts by reviewing light harvesting and photochemistry, then moves on to analyse the process of non-photochemical quenching, focusing specifically on dynamic regulation of light harvesting efficiency in the PSII antenna. The chapter also looks at assessing non-photochemical quenching via fluorescence measurements and how the photosystem II subunit S and zeaxanthin are important factors in controlling non-photochemical quenching formation and relaxation in higher plants. A section on manipulating non-photochemical quenching to improve photosynthetic efficiency is also provided.

The subject of Chapter 6 is modifying mesophyll conductance to optimise photosynthesis in crops. The chapter first describes the points of resistance to diffusion of carbon dioxide in plants, which in turn, influences mesophyll conductance. It then moves on to review the interaction between mesophyll cell anatomy, light and mesophyll conductance. The link between leaf age and mesophyll conductance is also examined, which is followed by an analysis of cell wall diffusion, how cellular membranes can be used to diffuse carbon dioxide and how mesophyll conductance can be improved via the use of aquaporins as carbon dioxide channels. The chapter reviews carbon dioxide solubility in liquids, the improvement of mesophyll conductance through using carbonic anhydrases and how mesophyll conductance can be estimated. Strategies for altering mesophyll conductance are also highlighted in the chapter.

The final chapter of Part 2 looks at modifying canopy architecture to optimise photosynthesis in crops. Chapter 7 begins by reviewing how light is modelled within crop canopies and how breeding has impacted modern crop canopy architecture. It moves on to examine potential targets for canopy improvement, such as optimising leaf area, leaf traits as well as leaf optical properties to improve radiation use efficiency. Finally, the chapter discusses considerations for the adoption of canopies better suited to their environment and management practices.

Part 3 Improving photosynthesis: optimising chloroplast function/light conversion

The first chapter of Part 3 discusses modifying photorespiration to optimise crop performance. Chapter 8 first highlights both the advantages and disadvantages of using photorespiration, focusing on how carbon and energy is lost during the process and how future climates could affect photorespiration. The chapter moves on to examine recent efforts to improve photorespiration, highlighting the importance of keeping things moving to prevent inhibition as well as discussing current photorespiratory bypass approaches. A section on how photorespiration can beat rising temperatures is also included, which is then followed by an overview of how improving photorespiration under non-steady-state conditions could improve carbon and nitrogen budgets.

The subject of Chapter 9 is maximising the efficiency of ribulose bisphosphate regeneration to optimise photosynthesis in crops. It begins by first describing the various component processes of photosynthesis, such as the use of enzymes, photosynthetic responses to light and carbon dioxide as well as optimising the use of resources important for photosynthesis. The chapter then moves on to review ways of optimising ribulose bisphosphate regeneration through resource investment, light use efficiency and interactions between light use and carbon metabolism. Examples of improving ribulose bisphosphate regeneration are also provided, such as sedoheptulose bisphosphatase and sucrose-phosphate synthase.

The final chapter of the book draws attention to understanding and modifying protein function in plant chloroplasts to optimise photosynthesis. Chapter 10 starts by describing a general carbon dioxide concentrating mechanism, focusing on cyanobacterial carbon dioxide and green algal carbon dioxide models first. It then goes on to examine the differences between the cyanobacterial carbon dioxide concentrating mechanism and the Chlamydomonas carbon dioxide concentrating mechanism, as well as how algal carbon dioxide concentrating mechanism components can be put into crops. The chapter dedicates individual sections to the various strategies currently available, specifically the envelope transporter-based, thylakoid-based, protopyrenoid-based, pyrenoid-based and finally the carboxysome-based strategies. It also highlights currently technical challenges associated with targeting inorganic carbon as well as the challenges of replacing and relocating carbonic anhydrases.

Part 1

General

Chapter 1

Understanding the biochemistry of C_3 photosynthesis in crop plants

C. A. Raines, A. P. Cavanagh, C. Afamefule, K. Chibani, H. Gherli, P. Lopez, V. Mengin, B. Moreno-García and S. Wall, The University of Essex, UK

1 Introduction

All photosynthetic organisms use the C_3 (Calvin-Benson) cycle to fix atmospheric CO_2 into organic compounds needed for the development and growth of the plant. In higher plants, the reactions of this pathway in plants take place in the chloroplast stroma of green tissues, predominantly in the leaves. Energy in the form of ATP and NADPH, needed to drive this cycle, is generated by the light reactions of photosynthesis that take place on the thylakoid membranes. The steps in this cycle were first elucidated in the 1950s by Calvin, Benson, Bassham and colleagues (see for a recent historical perspective Sharkey, 2019), and the enzymes involved were identified and catalytic properties determined during the 1970s and 1980s (Leegood, 1990). In photosynthetic eukaryotes, there are 11 different enzymes in the C_3 cycle catalysing 13 reactions, and the cycle can be divided in response to 3 phases: carboxylation, reduction and regeneration. Carbon compounds produced in this cycle are essential for growth and

http://dx.doi.org/10.19103/AS.2022.0119.02

development of the plant and while the majority (five-sixths) of the triose phosphate produced in the C_3 cycle remains within the cycle to regenerate RuBP, one-sixth of the carbon exits the cycle for biosynthesis of a range of compounds (Fig. 1). The balance carbon being delivered to these different biosynthesis pathways is not fixed but is likely to change during development, under different environmental conditions and also to changes in the rate of photosynthesis. However, what is fixed is the need to retain five out of every six carbons fixed in the pathway to prevent the cycle becoming depleted.

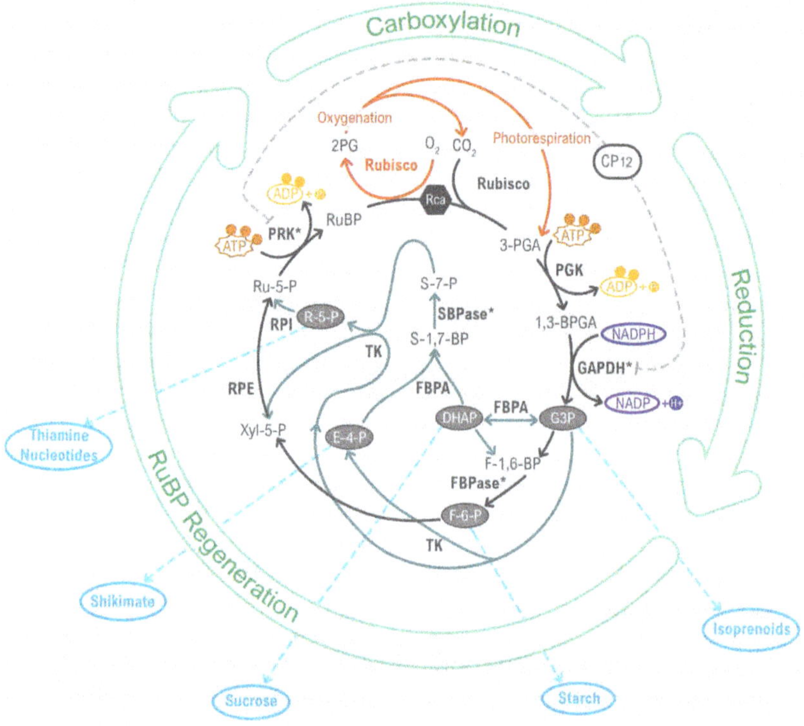

Figure 1 The reactions of the C_3 cycle. The first step in the fixation of atmospheric CO_2 is catalysed by the enzyme Rubisco which results in the formation of 3-PGA, the first stable compound in the cycle. The next two reactions form the reductive phase and are catalysed by PGK, forming BPGA using ATP and GAPDH which forms G-3-P consuming NADPH. The enzyme TPI produced DHAP. G-3-P and DHAP, known as TPs, enter the regenerative phase of the C_3 cycle catalysed by FBPA, forming S1,7-BP and F1,6-BP. SBPase and FBPase then produce S7-P and F6-P which are then converted to 5C compounds in reactions catalysed by TK, RPI and RPE resulting in the formation of Ru5P. The final step in the cycle is catalysed by PRK producing the CO_2 acceptor molecule Ru1,5-BP. The C_3 cycle produces compounds which are exported to a number of biosynthetic pathways for the synthesis of isoprenoids, starch, sucrose, shikimate, thiamine and nucleotides. Rubisco also has an oxygenase reaction which results in the formation of 2-phospoglycerate which enters the photorespiratory pathway.

The first two phases of the C_3 cycle, carboxylation and reduction, form the linear, assimilatory part of the cycle catalysed by the enzymes phosphoribulokinase (PRK), ribulose 1,5-bisphosphate carboxylase/oxygenase (Rubisco), phosphoglycerate kinase (PGK) and glyceraldehyde 1,3-dehydrogenase (GAPDH) resulting in the formation of triose phosphates (TPs). These TPs can be exported from the pathway out to the cytosol for sucrose and isoprenoid synthesis, or they can be retained in the cycle and enter the branched, regeneration phase of the cycle. This regenerative phase results in the formation of the CO_2 acceptor molecule ribulose 1,5-bisphosphate (RuBP), and it also produces a number of different carbon compounds which exit the cycle and form the basis of the biosynthetic pathways for starch (fructose 6-phosphate), thiamine, nucleotides (ribose 5-phosphate) and phenylpropanoids (ertyhrose-4-phosphate) (Fig. 1).

1.1 CO_2 assimilation in C_3 photosynthesis

The initial carboxylation reaction of the C_3 cycle is catalysed by the enzyme Rubisco, in which atmospheric CO_2 is added to the acceptor molecule, ribulose 1,5-bisphosphate (RuBP). This results in the formation of an unstable six-carbon compound, 2-carboxy-3-ketoarabinotol-1,5-bisphosphate, which splits in half forming two molecules of 3-phosphoglycerate (3-PGA). This three-carbon molecule is the first stable product following carboxylation, giving this pathway the name the C_3 cycle. This 3-PGA then enters the reduction phase of the cycle and is converted to glycerate 1,3-bisphosphate (G1,3-BP) using a molecule of ATP. In the second step in this reductive phase, NADPH is consumed and G1,3-BP is converted to glyceraldeyde-3-phosphate (G3-P), which is then converted to dihydroxyacetone phosphate (DHAP) by triose phosphate isomerase (TPI). These two 3C compounds, G1,3-P and DHAP, are then converted to the 6C and 7C compounds fructose 1,6-bisphosphate (F1,6-BP), sedoheptulose 1,7-bisphosphate (S1,7-BP), by fructose 1,6-bisphosphate/sedoheptulose 1,7-bisphosphate aldolase (FBPA). Dephosphorylation of F1,6-BP and S1,7-BP to fructose 6-phosphate (F6-P) and sedoheptulose 7-phosphate (S7-P) is catalysed by fructose 1,6-bisphosphatase (FBPase) and sedoheptulose 1,7-bisphosphate (SBPase). Both of these reactions are essentially irreversible *in vivo* due to the high-energy requirement for phosphorylation, which commits these carbon compounds to the regenerative phase of the C_3 cycle. A series of reactions convert F6-P and S7-P into 4C and 5C carbon compounds resulting in the formation of the CO_2-acceptor molecule ribulose 1,5-bisphosphate (RuBP) (Fig. 1). Three of these reactions are catalysed by transketolase (TK) (1) removes 2C from F6-P, forming erythrose 4-phosphate (E4-P), (2) transfers the 2C to G3-P forming xylulose 5-phosphate (Xyl5-P) and (3) removes to 2C from S7-P forming ribose 5-phosphate (R5-P). The last steps in the branched regenerative phase of the cycle are catalysed by ribulose 5-phosphate epimerase (RPE), which converts Xyl 5-P to ribulose 5-P (Ru5-P) and ribose 5-P isomerase (RPI) catalysing the conversion of ribose 5-phosphate (R5-P) to Ru5-P. The final regeneration reaction is the start of the linear section of the C_3 cycle, and Ru5-P is phosphorylated by PRK to form the CO_2 acceptor RuBP (Fig. 1).

The carbon fixed through the C_3 in the daytime is partitioned between immediate use for biosynthesis, sustaining growth and maintenance costs, and reserves for the night time (concept reviewed in Smith and Stitt, 2007). The regulation of carbon utilisation is a complex process involving the integration of environmental cues, multiple signalling networks, and circadian regulation of carbon availability and demand across various plant tissues (see examples for the model plant Arabidopsis) (Flis et al., 2019; Graf et al., 2010; Mengin et al., 2017; Pilkington et al., 2015; Sulpice et al., 2014; Scialdone et al., 2013). Improving both photosynthesis and downstream physiological processes will be a critical part of strategies aiming to improve the high efficiency of our crops. To this end in recent years a major focus of research on the C_3 cycle has been on identifying bottlenecks to provide strategies for improving crop yield. There have been one or two notable successes from this work but as yet this has mainly been in model species or in controlled environment conditions (López-Calcagno et al., 2020; Simkin et al., 2019; Driever et al., 2017; Raines, 2022). However, given that the C_3 cycle is highly conserved, it is reasonable to predict that lessons learned from model plants will be transferable to crops, and there are now reports on improvements in wheat, tomato and brassica that this is a feasible route to improve yield (Long et al., 2006). This chapter provides an update on the current understanding of the biochemistry and regulation of the C_3 cycle enzymes, their role as limiting factors in the C_3 cycle and strategies to modify the C_3 cycle for crop improvement.

2 The carboxylation phase

The carboxylation phase of C_3 cycle is facilitated by the photosynthetic carbon fixing enzyme Rubisco (EC 4.1.1.39), and its ubiquity among photosynthetic organisms makes Rubisco the most abundant protein on Earth, with an estimated global terrestrial mass of 0.7 gigatons (Bar-On and Milo, 2019). Rubisco is often considered the rate-limiting step for photosynthetic carbon assimilation and is characterised by four biochemical features considered to be evidence of the enzyme's inefficiency. Firstly, Rubisco has a turnover rate of 2–10 s^{-1} orders of magnitude lower than other so-called 'perfect catalysts' which are limited only by the rate of diffusion of the substrate to the catalytic site. Secondly, the Rubisco carboxylation reaction is inhibited by a competing RuBP oxygenation reaction, which catalyses the first step of the energetically demanding photorespiratory pathway. Finally, Rubisco is heavily regulated, requiring both a post-translational carbamylation and a regulatory protein to maximise full catalytic activity. Although reports of Rubisco's inefficiencies may be over-exaggerated, as recently argued by Bathellier et al. (2018), and ignore the significant rate enhancement (i.e. ratio of catalysed to non-spontaneous reactions) of plant Rubisco, improvements in Rubisco carboxylation rate or substrate specificity could improve crop productivity and growth.

Natural variation in the kinetic parameters of Rubisco has been studied with a view to identifying more efficient forms of this enzyme that could be used to improve the efficiency of the C_3 cycle. The full set of Rubisco catalytic properties, including Michaelis-Menten constants for both gaseous substrates, determined in plant species representing crops and undomesticated plants from diverse climates showed the existence of significant variation even between closely related species (Orr et al., 2016; Kubien et al., 2008; Galmés et al., 2016; Hermida-Carrera et al., 2016; Perdomo et al., 2015; Sharwood et al., 2016). Some of these newly characterised Rubiscos were naturally 'better' compared to crops and could provide opportunities for improving photosynthesis in crops. Interestingly, a number of candidate amino acid differences between the catalytic variants were identified (see, for example, Whitney et al., 2011), paving the way for engineering of improved Rubisco in crops using emerging chloroplast gene editing technologies. A more recent study took a systematic data mining approach to identify Rubisco enzyme variants that had faster carboxylation rates, resulting from this over 30 000 unique Rubisco sequences were identified (Davidi et al., 2020). Based on this analysis of similarity groups, 143 sequences were synthesised, and the carboxylation rates were determined for >100 variants. Most variants were active *in vitro* with some showing a twofold faster rate than the fastest rates previously determined; however, the fastest variants did not come from plants or cyanobacteria but from photosynthetic bacteria that use Rubiscos with a different structure. Realising an improvement in plant productivity with these faster Rubisco variants would require a functional carbon-concentrating mechanism due to the poor CO_2 specificity of these enzymes, but this approach is interesting in that it can be adapted to explore other enzymes including those of the C_3 cycle.

Rubisco performance is tightly regulated in all plants species, including crop plants, and full carboxylation capacity depends on a regulatory protein called Rubisco activase (Rca) which removes inhibitory sugar phosphates from Rubisco's catalytic site (Fig. 2). Rubisco can be inhibited when sugar-phosphate molecules such as Xyl5-P, 2-carboxy-D-arabinitol-1-phosphate (CA1P) and even Rubisco's substrate RuBP bind to the catalytic site of an uncarbamylated Rubisco and prevent carboxylation. Rubisco activase binds to Rubisco and, using energy from ATP hydrolysis, induces a conformational change, facilitating the release of these inhibitors and maintaining full Rubisco activity. Rubisco activation plays an important role in regulating Rubisco activity above the thermal optima, as

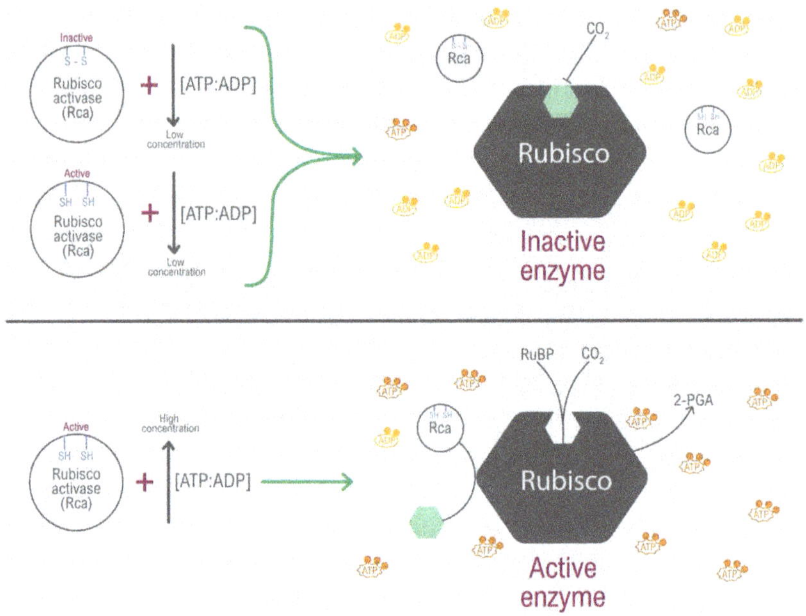

Figure 2 Regulation of carboxylation. Rubisco regulation is mediated through the regulatory protein Rubisco activase (Rca), which removes tightly bound sugar-phosphate inhibitors from the Rubisco catalytic site. Rca activity depends on the [ATP:ADP] ratio in the stroma and is further regulated through the reduction of sulphydryl groups in some species.

Rca is more heat sensitive than Rubisco. Rca activity is dependent on the ratio of ATP:ADP in the chloroplast stroma and can be regulated via TRX f (Zhang and Portis, 1999), coordinating its activity with rates of photosynthetic electron transport, and so the rate of Rubisco activation can be a limiting factor in transitions from low to high light. Many species express two isoforms of Rca, α and β, that are both active in ATP hydrolysis and Rubisco activation. Rca-α isoforms are redox-regulated due to a C-terminal extension containing two conserved redox-sensing cysteine residues that, when oxidised, form a disulphide bond that decreases the affinity of Rca for ATP and increases its susceptibility for ADP inhibition. Some species (including *Solanacea* species such as tobacco, tomato and pepper) express only a single, non-redox-regulated Rca-β isoform (Portis, 2003). Removing the redox regulation of Rca might be a strategy to improve Rubisco activation, and indeed mutant Arabidopsis plants in which Rca-α is absent and only Rca-β is expressed produce more dry weight under fluctuating light conditions (Carmo-Silva and Salvucci, 2013). The expression of Rca-β isoforms at elevated temperatures is associated with improved photosynthesis in many C_3 species (Degen et al., 2020, 2021; Scafaro et al., 2019). However, the less efficient Rca-α isoforms are expressed in response to high temperature in C_4 bioenergy grasses (Kim et al., 2021), and this upregulation is associated with sustained Rubisco activation, suggesting that Rca-α isoforms may play thermoprotective role in carbon fixation in C_4 species.

3 The reduction phase

The reductive phase of the C_3 cycle converts six molecules of 3-phosphoglycerate (3-PGA) resulting from carboxylation into six molecules of glyceraldehyde 3-phosphate (G3-P) using ATP and NADPH produced in the light reactions. Phosphoglycerate kinase (PGK; EC 2.7.2.3) catalyses the reversible reaction of ATP with the carboxyl group of 3-PGA, giving rise to G1,3-BP. This is one of the two ATP-consuming steps in the C_3 cycle and is favoured under conditions of high ATP:ADP ratios. Plant genomes contain three or more PGK genes, including cytosolic and chloroplastic isoforms. For example, in Arabidopsis two of the three PGK genes are chloroplast isoforms, while in maize three of the five PGK genes are predicted to be chloroplastic (Massange-Sánchez et al., 2020; Rosa-Téllez et al., 2017). While Arabidopsis PGK1 is exclusive to photosynthetic tissues, PGK2 is present in photosynthetic and non-photosynthetic cells, and both isoforms have a redundant role in photosynthesis (Li et al., 2019). PGK is a monomeric enzyme which functions as a hinge-bending enzyme with two structural domains linked by a flexible α-helix. The N-terminal domain binds 3-PGA or G1,3-BP and the C-terminal domain binds ATP or ADP. In the closed conformation, the two domains get closer, allowing the substrates to become correctly oriented for the reaction to occur (Szabó et al., 2008). The crystal structure of PGK has been deciphered for a number of prokaryotic organisms, and it is believed that it could be conserved in plants due to the strong sequence similarities between cytosolic and chloroplast isoforms. For example, the chloroplast PGK from *Chlamydomonas* shows sequence identities of 50–60% with human and bacterial PGKs and 65–80% with PGKs from photosynthetic organisms, including land plants (Morisse et al., 2014).

In the second step in the reduction phase of the C_3 cycle, G1,3-BP is converted to G3-P and inorganic phosphate (Pi) by glyceraldehyde 3-phosphate dehydrogenase (GAPDH; EC 1.2.1.13), with the consumption of NAD(P)H. GAPDH is encoded by four genes in land plants: GAPA, GAPB, GAPC and GAPCp (Petersen et al., 2003). GAPA and GAPB encode the C_3 cycle enzymes while GAPC and GAPCp encode the cytosolic and plastidic glycolytic isoforms. The GAPA and GAPB proteins form the tetrameric isoforms A_4-GAPDH and A_2B_2-GAPDH in the chloroplast. The amino acid sequence and protein structure are conserved between these two isoforms and, in particular, the crystal structure of A_4-GAPDH (NAD-bound) has been resolved in Spinach and Arabidopsis (Fermani et al., 2001; Gurrieri et al., 2019). The A and B subunits are structurally similar, but the B subunit includes a C-terminal extension that allows for direct redox regulation of the A_2B_2-GAPDH heterotetramer. Each subunit binds the substrate in the catalytic domain and NAD(H) or NADP(H) in the cofactor-binding domain. The regulation of the GAPDH enzyme also involves a multiprotein complex with PRK and a small protein CP12 (Marri et al., 2004).

4 The regeneration phase

The first enzyme in the regenerative phase of the C_3 cycle is TPI (EC 5.3.1.1), which converts G3-P to DHAP. Plants contain both a chloroplastic and cytosolic isoform of this enzyme, encoded by different nuclear genes. In the C_3 cycle, the chloroplastic TPI catalyses the reversible reaction between G3-P and DHAP and the cytosolic form carries out this reaction in glycolysis and gluconeogenesis (Castro-Torres et al., 2019). The enzyme is composed of eight domains of β-strand and α-helix which fold into a dimer structure named after the enzyme, TIM-barrel, including the active site. It was demonstrated that in Arabidopsis, the chloroplastic TPI can be redox-regulated through post-translational glutathionylation of specific cysteine residues; however, the functional significance *in vivo* has not been elucidated (López-Castillo et al., 2016).

Transketolase (TK; EC 2.2.1.1) is among the few C_3 cycle enzymes to catalyse a reversible reaction. In plants, TK forms a functional dimer, with the active site located in a groove formed by the contact site of each subunit (Gerhardt et al., 2003). There are two highly conserved TK genes found in Arabidopsis, with one predominantly expressed in the chloroplasts of photosynthetic tissue while the other is mainly expressed during embryo development. In the C_3 cycle, TK catalyses the transfer of a two-carbon molecule to G3-P (acceptor) from S7-P, generating Xyl5-P and R5-P, or from F 6-P to produce Xyl5-P and erythrose 4-phosphate (E 4-P). Like many C_3 cycle enzymes, TK also plays a critical role in the oxidative pentose phosphate pathway where it catalyses these reactions in reverse, and its dual substrate specificity is regulated through a post-translational calcium-dependent phosphorylation that acts as a regulatory kinetic switch. Under pH values representing stromal conditions in the light (i.e. pH of 8.0), phosphorylation of TK increases the affinity of the enzyme for F 6-P over that of R5-P, promoting flux through the C_3 cycle. When the enzyme is assayed at a pH representing dark conditions (i.e. pH of 7.2), the opposite reaction is favoured, and TK promotes flux through the oxidative pentose phosphate pathway (Rocha et al., 2014).

In the C_3 cycle, the reactions catalysed by TK are not only essential for the regeneration of RuBP to maintain active photosynthesis but also provide the immediate precursor molecules for the shikimic acid pathway and phenylpropanoid metabolism (E4-P) and may also influence the level of R5-P a substrate for thiamine biosynthesis. TK activity requires the presence of the cofactor thiamine pyrophosphate (TPP), which is synthesised from and is the active form of the vitamin, thiamine. Biosynthesis of thiamine (and TPP) is dependent on intermediates of the C_3 cycle, R5-P, which is a substrates/product of TK. The phosphorylation of TK might represent a regulatory link between flux through the C_3 cycle for the regeneration of RuBP and flux of intermediates to secondary metabolism. Both decreased TK expression via antisense transformation and increased TK expression via transgenic overexpression have a negative impact on photosynthesis and growth, highlighting the integration of the C_3 cycle with wider metabolism (Henkes et al., 2001; Khozaei et al., 2015).

SBP/FBP Aldolase (FBPA EC 4.1.2.13), as the name implies, catalyses two steps in the C_3 cycle forming F1,6 BP from DHAP and G3-P and S1,7-P from E4-P and DHAP. In Arabidopsis, three genes-encoding plastid forms of FBPA have been identified, *AtFBA1–AtFBA3*. Analysis of knockout mutants has shown that FBA2 contributes the majority of the activity in the chloroplast and has partial redundancy with FBA1. This allows both single mutants to survive, but combining both mutations is lethal, indicating a total block of photosynthesis. In contrast, *FBA3* is expressed predominantly in heterotrophic tissues, especially in the leaf and root vasculature, but not in the leaf mesophyll. These data indicate that the two aldolase reactions are catalysed predominantly by the FBP2 isoform and indicate that it is unlikely that there are distinct SBP and FBPA enzymes (Carrera et al., 2021). Overexpression of the plastid FBPA results in a positive effect on photosynthesis and yield under some conditions (Batista-Silva et al., 2020).

The reaction in the C_3 cycle catalysed by SBPase (EC 3.1.3.37) is irreversible under *in vivo* conditions and commits intermediates to the regenerative part of the cycle, placing SBPase at an important junction in carbon metabolism. SBPase functions as a homodimer, is unique to the chloroplast and does not have a cytosolic counterpart. In most species, there is only one SBPase gene present but in some species such as tobacco four SBPase genes are present. SBPase activity limits the C_3 cycle and increasing the levels of this protein in transgenic plants leads to an increase in photosynthesis and yield (Lefebvre et al., 2005) (Driever et al., 2017).

Similar to SBPase, the chloroplastic FBPase (EC 3.1.3.11) catalyses an irreversible reaction producing F6-P which either enters the regenerative phase of the C_3 cycle or can exit the cycle and be used for starch synthesis in the chloroplast. In plants there are three FBPase isozymes, one in the cytosol and two in the chloroplast one of which is involved in the C_3 cycle (Gütle et al., 2017). At the structural level, the tetrameric FBPase is composed of two pairs of dimers. Despite their different ancestors, α-proteobacteria for the FBPase and Archea for the SBPase, protein sequences are clearly related, and a bifunctional version of FBPase/SBPase is present in cyanobacteria (Gütle et al., 2016).

The penultimate intermediate of the regeneration phase of the C_3 cycle is ribulose 5-phosphate (Ru5-P), the product of two reactions. Ribulose phosphate

3-epimerase (RPE; EC 5.1.3.1) catalyses the interconversion of Xyl5-P to Ru5-P. The RPE enzyme is a homodimeric metalloprotein containing a TIM-barrel domain (Akana et al., 2006). Structural and biochemical studies revealed that RPE uses Zn^{2+} in the active site as cofactor for catalysis. In Arabidopsis, one plastid RPE is encoded by a single gene, while two other genes encode cytosolic isoforms. RPI (EC 5.3.1.6) catalyses the conversion of R5-P to Ru5-P. The structure of RPI was resolved from *Chlamydomonas*, and it functions as a homodimer. Although this enzyme has four cysteine residues present in the 3D structure, two are inaccessible and the distance between the second pair of cysteines is 20 Å which is too far apart to allow the formation of a disulphide bridge. These findings suggest that it is highly unlikely that the RPI is regulated by the ferredoxin/thioredoxin (Fd/TRX) system (Le Moigne et al., 2020).

Phosphoribulokinase (PRK, EC 2.7.1.19) catalyses the final step in the regeneration phase, resulting in the formation of ribulose 1,5-bisphosphate, the CO_2 accepting substrate for the Rubisco reaction. This reaction uses one molecule of ATP resulting in the formation of ADP and transfers a phosphate to R5-P. In higher plants this homodimeric enzyme is regulated by TRX, is unique to plastids and therefore does not have a cytosolic counterpart. The 3D structure of PRK was determined from Arabidopsis and *Chlamydomonas* and was the last in the cycle to be characterised at the structural level, revealing the presence of a large central ß-sheet that extends between the two catalytic sites. The TRX-regulated cysteine residues are 13 Å apart and are connected by a flexible loop region (Gurrieri et al., 2019).

5 Regulation of the C_3 cycle enzymes

5.1 Light activation

It has been known for over 50 years that the activity of enzymes in the C_3 cycle is regulated by light. Four C_3 cycle enzymes, PRK, GAPDH, FBPase and SBPase, have canonical regulation through the Fd/TRX system (reviewed in Buchanan, 1980, 1991; Schürmann and Buchanan, 2008; Gurrieri et al., 2021; Lemaire et al., 2007; Schürmann and Jacquot, 2000). Additionally, in many species, Rubisco carboxylation activity is also regulated through an Rca isoform with thioredoxin (TRX)-dependent regulation (Zhang and Portis, 1999). More recently the activity of PGK from *Chlamydomonas* and cyanobacteria has also been shown to be regulated by this system, though this remains to be demonstrated in higher plants (Morisse et al., 2014; Tsukamoto et al., 2013). This light-dependent activation of the C_3 cycle is crucial in ensuring light-responsive control of photosynthetic reactions in the chloroplast and balancing the carbon-fixation reactions of photosynthesis with the light-harvesting reactions. These C_3 cycle enzymes contain a pair of cysteines that can be reversibly oxidised in the dark and then reduced by a chloroplastic TRX following a transition to light, conferring full activity (Chibani et al., 2010; Michelet et al., 2013) (Fig. 3). The Fd/TRX system is composed of three classes of soluble proteins located in the chloroplast stroma: ferredoxin (Fd), Fd/TRX reductase and TRX. TRXs are ubiquitous and small proteins of 10–14 kDa, which belongs to the oxidoreductase family. TRXs can be divided into three main groups consisting of the typical, atypical and the TRX reductase forms. Of these it is the typical TRXs, specifically TRX f, m and z with the canonical WCGPC active site, that transfer the reducing power from the photosynthetic Fd/TRX system to C_3 cycle enzymes by reducing disulphide bridges to thiol groups on the cysteine residues of C_3 cycle enzymes (Chibani et al., 2021) (Table 1).

The light-driven reductive activation of the C_3 cycle enzymes via the TRX system can take between 1 minute and 10 minutes, and this results in a delay in the C_3 cycle enzymes becoming fully active and in turn causes a lag in the time taken for photosynthesis to reach maximum steady-state levels (Thormählen et al., 2017). In the natural environment, such shifts from low to high light can occur many times throughout the day either due to cloud cover or leaf shading.

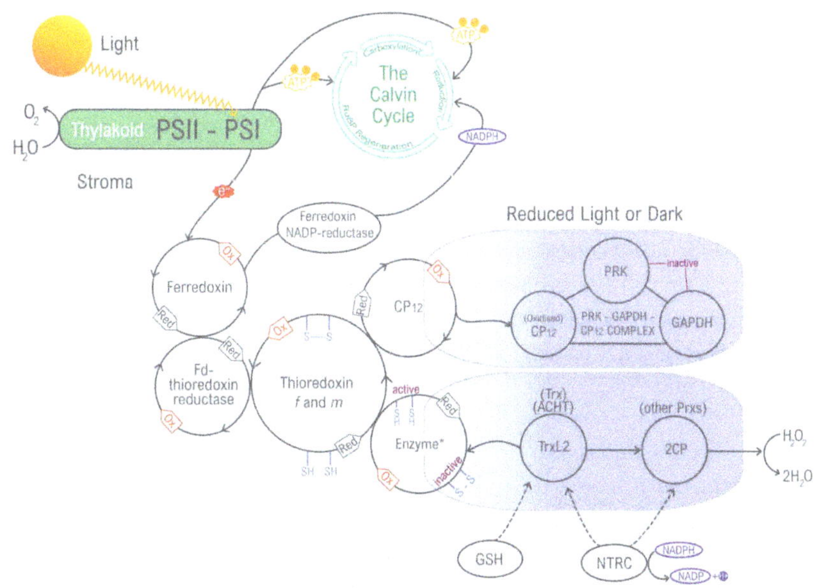

Figure 3 Light regulation of the C_3 cycle. In the light, ferredoxin is reduced by the photosynthetic electron transfer chain through photosystem I (PSI) which initiates a redox relay reducing ferredoxin-thioredoxin reductase which then reduces thioredoxins f, m and z, and they in turn activate the C_3 cycle enzymes SBPase, FBPase, GAPDH and PRK through the conversion of a disulphide bridge to thiol groups. Recently a mechanism to directly deactivate the C_3 cycle enzymes has been proposed to involve a Thioredoxin-like protein (TRL2) and an atypical Cys His-rich TRX (ACHT).

Table 1 Thioredoxin types and target enzymes

	Thioredoxin type	Target C_3 cycle enzymes	References
Typical thioredoxin	TRX f	FBPase, SBPase	Okegawa and Motohashi (2015) Naranjo et al. (2016)
		GAPDH	Sparla et al. (2002)Marri et al. (2009)
	TRX m	FBPase, SBPase	Okegawa and Motohashi (2015)
		GAPDH	Marri et al. (2009)
	TRX z	PRK	Le Moignet et al. (2021)
Thioredoxin reductase	NTRc	FBPase, PRK	Nikkanen et al. (2016)

GAPDH, glyceraldehyde 1,3-dehydrogenase; PRK, phosphoribulokinase; FBPase, fructose 1,6-bisphosphatase; SBPase, sedoheptulose 1,7-bisphosphate.

Cumulatively, this can result in plants failing to achieve the maximum potential rate of photosynthesis over the day. This raises the possibility that improving the rate of activation of the C_3 cycle enzymes could improve photosynthesis and result in an increased yield of crops.

In contrast to light activation of the C_3 cycle enzymes the mechanism of dark deactivation of the TRX-regulated enzymes is not yet well understood. Nevertheless, a recently published study suggests that a stroma-localised atypical TRX from Arabidopsis, designated TRX-like2 (TRXL2), could be responsible for oxidatively deactivating the C_3 cycle enzymes (Fig. 3). This might be a novel target to explore if manipulation of this process can impact photosynthetic efficiency (Yokochi et al., 2021; Yoshida et al., 2018).

There is an additional C_3 cycle regulatory mechanism linked to redox which involves the small 12 kDa chloroplast protein, CP12 (Fig. 3). This mechanism functions in the dark or under low light when the CP12 protein becomes oxidised enabling it to mediate the formation of a supra-molecular complex with GAPDH and PRK, GAPDH/CP12/PRK, resulting in a decrease in the activity of both of these enzymes. The formation and breakdown of the GAPDH/CP12/PRK complex are also dependent on TRX and have been shown to operate *in vivo* in response to changes in light availability, modulating the deactivation and activation of PRK and GAPDH enzymes more rapidly than by the sole action of TRXs in higher plants (Howard et al., 2008). This regulatory mechanism also allows redox regulation of the A4-GAPDH homotetrameric form of GAPDH, which lacks the C-terminal region present in the GAPB isoform that contains the cysteine residues responsible for direct interaction with TRXs (Gurrieri et al., 2021). Antisense and mutant studies in higher plants have shown that loss of the CP12 protein results in lower photosynthetic rates, slow growth, low GAPDH, PRK and NADP-MDH, activities and reduced levels of PRK protein highlighting its importance in normal photosynthetic function (Howard et al., 2011; López-Calcagno et al., 2017; Li et al., 2018).

5.2 Insight from crystal structures of the C_3 cycle enzymes

The crystal structures for all five TRX-regulated enzymes have now been resolved. The cysteine residues responsible for the regulation of these enzymes have been known for some time, and there is no conserved sequence motif or conserved location in the target C_3 cycle enzymes. This is in contrast to the TRX proteins that have a highly conserved redox motif CXXXXC.

In FBPase, the TRX-binding site is far from the conserved sugar-binding domain, while in SBPase, it is present at the dimer interface. When the cysteines in FBPase are reduced, the substrate-binding site adopts a catalytically competent conformation due to the loosening of secondary structures. In A_2B_2-GAPDH, the redox-active cysteines are found in the highly flexible carboxy-terminal extension (CTE), which is characteristic of B subunits. When oxidised, the last residue in the CTE blocks the NADPH-binding site and inactivates the enzyme. The regulatory pair of cysteines in PRK is embedded in the substrate-binding site, and oxidation can be achieved by the presence of the flexible clamp loop. From these studies, it has been proposed that TRX regulation has been acquired through the introduction of regions in the target protons that facilitate flexibility.

6 Approaches to determine which enzymes limit the flow of carbon through the C_3 cycle

6.1 Empirical studies

The *in vitro* determined kinetic properties of individual C_3 enzymes led to the identification of a number of so-called 'key' enzymes in the C_3 cycle such as Rubisco, SBPase, FBPase and PRK because the reactions each catalysed were essentially irreversible, and their activity was tightly regulated by light, stromal pH and [Mg^{2+}] (Lawlor, 2001). The conclusion from this biochemical analysis was that these 'key' enzymes were likely to have the greatest importance in controlling the rate of CO_2 fixation. However, these studies did not provide any information on the extent to which any individual enzyme in the C_3 cycle controlled the rate of carbon dioxide fixation *in vivo*. Metabolic control analysis looks more holistically at the whole pathway and can be used to provide a quantitative measure of the relative importance of an individual enzyme in controlling the flow of intermediates through a pathway (Fell, 1997).

To undertake a metabolic control analysis of a pathway, it is necessary to be able to reduce specifically the amount of an individual enzyme in that pathway; the effect of this reduction on the flow of intermediates can then be compared to the control with the normal level of enzyme activity (for a fuller description of this approach, see Fell, 1997). The control exerted by an individual enzyme can be defined mathematically:

$$C = \frac{\partial J / J}{\partial E / E}$$

where C is the flux control coefficient; *J* is the original flux through the pathway; ∂J is the change in flux; *E* is the original enzyme activity; and ∂E is the change in enzyme activity. The flux control coefficient of an enzyme can vary from zero to one; an enzyme exerting no control being 0 and 1 denoting an enzyme with total control. An important point to note is that the flux control value for any single enzyme is not a constant and can change depending on the conditions under which the analysis is carried out. However, the sum of the flux control coefficients for all of the enzymes in a single pathway should equal 1. Therefore, in order that this be the case, when the control value for one enzyme in the system increases, then the control values for one or more of the other enzymes in the system must decrease. One fundamental difference between this approach and that of earlier studies based on the kinetics of individual enzymes is that metabolic control analysis allows for all enzymes in a pathway to share control of flux in that pathway. The contribution that individual enzymes make to the control of carbon flux through the C_3 cycle has been investigated using antisense plants with reduced levels of individual enzymes. The application of metabolic control analysis to address this problem was initiated by Stitt and co-workers using transgenic antisense Rubisco plants (Rodermel et al., 1988; Stitt and Schulze, 1994). Antisense plants for 8 of the 11 enzymes in the C_3 cycle have been analysed using this approach (Raines, 2003).

Overexpression of individual enzymes in the C_3 cycle has resulted in some positive effects. Increasing the levels of SBPase FBPA, TPI and FBPase have demonstrated that increases in the activity of these enzymes increase photosynthesis and growth in controlled conditions. Interestingly combining overexpression of these enzymes with each other or with components of the electron transport chain have shown further positive effect on growth and yield (Simkin et al., 2019). In the case of GAPDH, overexpression did not result in improved photosynthesis but did result in drought tolerance (Kappachery et al., 2015).

6.2 Modelling of the C_3 cycle

The C_3 cycle is complex and a number of models have been developed to link the biochemical function of this cycle to *in vivo* photosynthesis rates. A well-established and widely used model of CO_2 uptake, developed by Farquhar and colleagues, partitioned the limitation of photosynthesis between Rubisco activity and the capacity for RuBP regeneration under steady-state conditions (von Caemmerer, 2000; von Caemmerer and Farquhar, 1981). Under light saturating, CO_2 limiting conditions, Rubisco activity is the major determinant of carbon fixation but as CO_2 increases limitation shifts to regeneration capacity. Application of this model demonstrated that the balance of co-limitation of photosynthetic carbon fixation is poised between carboxylation and regeneration, under a wide range of environmental conditions in a large number of plant species (Von Caemmerer, 2000; Wullschleger, 1993). However, this model does not provide information on which biochemical steps in the regenerative part of the cycle are limiting or how limitations shift between the different enzymes under conditions that plants experience in the dynamic external environment.

A kinetic model of the C_3 cycle, developed by Fell and Poolman, has also predicted that the control of photosynthetic carbon flux is co-limited by Rubisco and the regenerative capacity of the C_3 cycle, and they went on to identify the enzyme SBPase as being the major controlling enzyme in this phase of the cycle. Interestingly, this model also predicted that the share of control between Rubisco and SBPase was balanced and that this co-limitation could be shifted dependent on conditions within the cell (Poolman et al., 2001).

A complete dynamic model of photosynthetic carbon metabolism was developed by extending existing metabolic models to include the photorespiratory pathway, starch and sucrose metabolism (Zhu et al., 2007). This model consists of linked differential equations, each representing the change of concentration of one metabolite, and this model has been shown to be able to mimic established physiological responses. Coupling this model to an evolutionary algorithm, in which a fixed amount of protein-nitrogen was shared between enzymes and based on the outputs following 1500 iterations, indicated that there was an over-investment in the photorespiratory enzymes and an under-investment in Rubisco, SBPase and FBPA, together with sink capacity, such as the increase in ADP-glucose pyrophosphorylase. These results suggest that a multigene manipulation strategy targeting these enzymes could substantially improve photosynthesis.

Extension of this model has led to the development of a more comprehensive model of leaf photosynthesis, *e-photosynthesis* (Zhu et al., 2013). This dynamic model linked the detailed reactions of C_3 carbon metabolism, photorespiration and downstream carbon metabolism pathways (Zhu et al., 2007), a fluorescence induction model (Zhu et al., 2005) and the main regulatory processes of photosynthesis, namely Rubisco activation by Rubisco activase, non-photochemical quenching and the light-dependent regulation of enzyme activities via the Fd/TRX system. *In silico* parameter perturbation experiments showed that *e-photosynthesis* provided realistic simulations of photosynthesis.

Recently, the C_3 carbon metabolism module of *e-photosynthesis* (Zhu et al., 2013) was used to examine the impact of the activity of individual C_3 cycle enzymes on the rate of photosynthesis (Zhao et al., 2020). It was observed that CO_2 assimilation rates responded differently when changing the maximum reaction rates of the enzymes. For most enzymes, assimilation rates increased with enzyme capacity until reaching a plateau, in a mono-phasic manner. Other enzymes were found to show a biphasic response, namely TK, FBPase, FBPA and PGK, with assimilation rates increasing initially before decreasing in response to increasing enzyme capacity. It was observed that these biphasic enzymes control the distribution of substrates in sub-cycles within the C_3 cycle, consuming a shared intermediate and producing a substrate used by another enzyme, e.g. PGK and PRK both consume ATP, and TPs are used as a common substrate by FBPA, TK and SBPase. The existence of enzymes with biphasic responses may help explain why the overexpression of C_3 cycle enzymes does not always lead to improvements in photosynthetic capacity, as these could create imbalances between sub-cycles and in the overall flux through the C_3 cycle. In particular, the feedback inhibition of the biochemical regulators ADP and F6-P was found to be of particular importance in the coordination of fluxes between sub-cycles. The simulations suggest that a balanced investment in biphasic enzymes is needed to maximise flux through the C_3 cycle by maintaining metabolite concentrations and flux balances between sub-cycles.

7 Future opportunities to improve the C_3 cycle

Although the C_3 cycle is ubiquitous and highly conserved between species, there remain a number of significant gaps in our knowledge including:

- The extent of natural variation that exists in the kinetic parameters and regulation of individual enzymes is unknown with the exception of Rubisco.
- The regulation of the allocation of carbon from the C_3 cycle to other pathways during development and in response to changes in photosynthetic rates has not been addressed holistically.
- The response of the C_3 cycle across species to environmental variables has not been widely investigated.
- The genetic factors regulating the co-ordinated expression of the C_3 cycle genes are not known in full for even one individual species.

Addressing these questions will require the application and integration of information from new technologies/approaches, and some examples are presented in this section.

It has been known for some time that the RNA abundance (levels of mRNA) of individual C_3 cycle enzymes changes during plant development, in response to changes in light conditions and accumulation of downstream feedback signals such as sugars. However, the detailed molecular mechanisms underlying the regulation of individual genes have not been elucidated, and our understanding of the coordination of expression of the pathway as a whole is even more limited. Many studies examine photosynthetic gene expression and regulation between C_3 and C_4 species, to take advantage of the striking difference in expression pattern between the two photosynthetic types. Notably, the regulatory motifs and associated transcription factors identified are not limited to C_4 or C_3 species, and there is much to learn from the methods used.

Approaches to identify new regulatory elements could be classified into two categories: those based on transcriptomics data (i.e. comparative transcriptomics) and those based on genomic data (i.e. promoter motif scanning, chromatin accessibility evaluation). These methods are complementary and have been used in combination to further our understanding of these processes. For example, comparative transcriptomics between maize and rice RNAseq data to identify motifs regulating the expression of photosynthetic genes in both C_3 and C_4 species (Xu et al., 2016). With the decrease in the cost of sequencing and the availability of sequencing platforms, genomic data is readily available, and many plant species now have fully sequenced and annotated genomes, containing a wealth of information about regulatory regions. After annotation of genes based on closely related species, scanning genomic sequences of putative regulatory regions (i.e. promoters) for common motifs has identified conserved sequences that can function as gene repressors or enhancers (Afamefule and Raines, 2020; Wang et al., 2019). However, chromatin accessibility adds another layer of complexity to the identification of regulatory motifs. Regulatory motifs are recognised by transcription factors that, when bound to the sequence, will trigger a response in the expression level, either an increase or a decrease depending on whether the transcription factor acts as an enhancer or a repressor. Despite the critical importance of the motif sequence, it is also required that the specific sequence is accessible to the transcription factor. Thus, several experiments have aimed to evaluate the chromatin accessibility of photosynthetic promoters, in an effort to identify regions that could be regulating the expression level of the downstream genes. A good example of this approach is Burgess et al. (2019), where they used DNAse-seq (a method to identify open chromatin) to identify putative conserved motifs across C_3 cycle genes in both C_3 and C_4 species.

Individual C_3 cycle enzymes from different C_3 species can exhibit considerable diversity in their primary protein sequences. With the exception of Rubisco, the functional implication of this diversity has not been studied systematically, and to date detailed data on the catalytic properties of any individual enzyme of the cycle is available for only a small number of species. Extending our knowledge of the natural diversity of these enzymes will allow a better understanding of the relationship between the structure/function of the C_3 enzymes and the specific catalytic roles of the conserved and non-conserved amino acids. To achieve this, the catalytic diversity needs to be analysed alongside the C_3 cycle enzyme sequences, to identify potential catalytic switches for improving photosynthesis and productivity. As an example of the potential of this approach, a recent study has identified a single nucleotide substitution in the SBPase gene in rice that results in a negative effect on plant growth and grain yield. This work indicates a Glycine residue at position 364, at the C-terminal of SBPase, could play an important role in SBPase function not only in rice as this amino acid is conserved across a wide range of species (Li et al., 2020).

Metabolite profiling allows and enables the analysis of interspecies or environmental effects on a pathway. Metabolite profiling of C_3 intermediates has not been applied extensively and has been restricted to single species (see Stitt et al., 2010, for examples of C_3, and Stitt and Heldt, 1985, and Leegood and Von Caemmerer, 1988, for examples of C_4). One reason for this is the technical limitations of these approaches. Until about 15 years ago, it was impossible to measure many C_3 intermediates by standard enzymatic assays. These can now be quantified routinely using HPLC MS/MS (Arrivault et al., 2009, 2019; Borghi et al., 2019; Cruz et al., 2008; Treves et al., 2022). Most of these studies use a combination between enzymatic assays and HPLC MS/MS. An added difficulty is the way to label safely and consistently plant leaves using P or CO_2 isotopes before quenching (see Stitt et al., 2021). Although it is not needed to analyse steady-state C_3 intermediates profile, labelling is necessary when focussing on fluxes of intermediates going through the C_3 and downstream pathways. Another drawback is the very short half-life of C_3 intermediates (<1 s and many closer to 0.1 s; Arrivault et al., 2009). This means that even slight fluctuations in irradiance or CO_2 levels during the quenching process lead to changes in metabolite levels. This prompted the development of quick quenching methods (see Stitt et al., 2021 for examples).

Recently, comparisons of metabolite profiles from C_3 and C_4 and even algae were achieved using these strategies. Interestingly, it revealed specificity and diversity in C_3 cycle operations between C_3, C_4 and algae, as expected, but more unexpectedly diversity was also evident between C_3 species. Variation between different C_3 species was driven by many metabolites (including RuBP, 3PGA, triose phosphate, FBP, F6P, S7P, Ru5P+Xu5P and 2PG) pointing to different operating modes of the C_3 cycle (Arrivault et al., 2019). They have shared features between C_3 species; for example, restriction on flux at FBPase and SBPase at low irradiance between Arabidopsis and rice and how these two C_3 species prioritise different reactions when submitted to different irradiance (Arrivault et al., 2019; Borghi et al., 2019; Stitt et al., 2021). The implication of these findings is that strategies to improve photosynthesis are likely to vary depending on the crop, and it highlights the need for systemic analyses of target species and probably cultivars within the same species. A recent study of the highly performant alga of *Chlorella ohadii* (Treves et al., 2020) and a comparison with higher plants and other algae in different light regimes (Treves et al., 2022) has highlighted distinct flux patterns between algae and higher plants. They noticed faster labelling of SBP, pentose phosphate and RuBP in *Chlorella ohadii* and higher flux from the C_3 through lower glycolysis to PEP expanding to all downstream pathways leading to a faster growth. Investigating what mechanisms support faster protein and lipid synthesis would provide useful information to improve crop photosynthesis and yield. It is however unlikely that metabolic profiling could be used to screen a wide array of different species or cultivars as it does not easily lend itself to high throughput, automated approaches. This type of study is likely to be most beneficial to target individual species in different environments or to provide data that can be built into models.

A more promising approach to provide inputs for modelling crop performance could be done via flux balance analysis (FBA). FBA uses information about accumulation rates of measured metabolites, combined with stoichiometric coefficients of unmeasured metabolic reactions and known transcriptional processes/regulation, in order to estimate steady-state metabolic fluxes through a network to explorer the impact of metabolic changes on a specific objective trait/function (e.g. biomass). For example, researchers have investigated fluxes of central metabolism contributing to the development of tomato fruits (*Solanum lycopersicum*) using constraint-based FBA. This study has unveiled details such as the dynamics of carbon partitioning that supports fruit development and the energetic adaptations prior to fruit ripening that leads to dissipation of ATP (consistently with the concept of 'climacteric crisis') (Colombié et al., 2015). Traditional FBA approaches rely on the assumption that the system investigated is in a steady state - in other words, a system in which the state of metabolite pools and relevant traits does not change in time. This can be a pitfall when investigating highly dynamic crop systems which are exposed to constant genetic selection and environmental fluctuations. For these reasons, FBA-like approaches that can provide information on changes in metabolite abundances over time have been the focus of recent development. An extensive review of different modelling techniques available can be found in Clark et al. (2020).

Numerous models incorporating the C_3 cycle have been established, and they allow the systematic alterations of parameters and to draw general conclusions about processes regulations (see Section 6.2). The strength of these mathematical models is that they can offer a tool to study a range of scenarios and numerous simulations that would not be possible to do experimentally. Recently a new modular photosynthesis model based on the interactions and interdependencies of the photosynthetic electron transport chain (PETC) and the C_3 cycle, with a focus on investigating the supply-demand control of photosynthesis, was developed (Matuszyńska et al., 2019). A fresh update to this model also includes ROS formation and metabolism highlighting the significant impact of SBPase on the regulation of ROS scavenging mechanisms, the C_3 cycle and the PETC (Saadat et al., 2021). These models provide a powerful framework for targeted engineering of enzymes for photosynthesis improvement.

Despite the fact that the C_3 cycle is found in every basic plant textbook, our understanding of this pathway at the genetic, biochemical and physiological levels or how these parameters vary between species, including our most important crop plants, is still incomplete. However, with the advent of new approaches, we are presented with the opportunity to explore this crucial pathway and this new knowledge and contribute to the generation of higher yielding crops.

8 Conclusion

All photosynthetic organisms use the C_3 (Calvin-Benson) cycle to fix atmospheric CO_2 into organic compounds needed for the development and growth of the plant. The reactions of this pathway in higher plants takes place in the chloroplast stroma of green tissues, predominantly in the leaves.

Given that the C_3 cycle is highly conserved, it is reasonable to predict that lessons learned from model plants will be transferable to crops. We provide an update on the current understanding of the biochemistry, regulation and limitations of the C_3 cycle, and future questions to be addressed to enable improvement of the cycle to boost crop yields.

9 Where to look for further information

The following articles provide a good overview of the subject:

- Raines, C. A. (2003). The Calvin cycle revisited. *Photosynthesis Research* 75, 1-10.
- Gurrieri, L., Fermani, S., Zaffagnini, M., Sparla, F. and Trost, P. (2021). Calvin-Benson cycle regulation is getting complex. *Trends in Plant Science, 26*, 898–912.

10 References

Afamefule, C. and Raines, C. A. 2020. Insights into the regulation of the expression pattern of Calvin-Benson-Bassham cycle enzymes in C3 and C4 grasses. *Frontiers in Plant Science* 11, 570436.

Akana, J., Fedorov, A. A., Fedorov, E., Novak, W. R. P., Babbitt, P. C., Almo, S. C. and Gerlt, J. A. 2006. d-ribulose 5-phosphate 3-epimerase: functional and structural relationships to members of the Ribulose-Phosphate Binding (β/α)8-barrel superfamily. *Biochemistry* 45(8), 2493-2503.

Arrivault, S., Alexandre Moraes, T., Obata, T., Medeiros, D. B., Fernie, A. R., Boulouis, A., Ludwig, M., Lunn, J. E., Borghi, G. L., Schlereth, A., Guenther, M. and Stitt, M. 2019. Metabolite profiles reveal interspecific variation in operation of the Calvin-Benson cycle in both C4 and C3 plants. *Journal of Experimental Botany* 70(6), 1843-1858.

Arrivault, S., Guenther, M., Ivakov, A., Feil, R., Vosloh, D., Van Dongen, J. T., Sulpice, R. and Stitt, M. 2009. Use of reverse-phase liquid chromatography, linked to tandem mass spectrometry, to profile the Calvin cycle and other metabolic intermediates in Arabidopsis rosettes at different carbon dioxide concentrations. *The Plant Journal: For Cell and Molecular Biology* 59(5), 826-839.

Bar-On, Y. M. and Milo, R. 2019. The global mass and average rate of RuBisCO. *Proceedings of the National Academy of Sciences of the United States of America* 116(10), 4738-4743.

Bathellier, C., Tcherkez, G., Lorimer, G. H. and Farquhar, G. D. 2018. RuBisCO is not really so bad. *Plant, Cell and Environment* 41(4), 705-716.

Batista-Silva, W., Da Fonseca-Pereira, P., Martins, A. O., Zsögön, A., Nunes-Nesi, A. and Araújo, W. L. 2020. Engineering improved photosynthesis in the era of synthetic biology. *Plant Communications* 1(2), 100032.

Borghi, G. L., Moraes, T. A., Günther, M., Feil, R., Mengin, V., Lunn, J. E., Stitt, M. and Arrivault, S. 2019. Relationship between irradiance and levels of Calvin-Benson cycle and other intermediates in the model eudicot Arabidopsis and the model monocot rice. *Journal of Experimental Botany* 70(20), 5809-5825.

Buchanan, B. B. 1980. Role of light in the regulation of chloroplast enzymes. *Annual Review of Plant Physiology* 31(1), 341-374.

Buchanan, B. B. 1991. Regulation of CO2 assimilation in oxygenic photosynthesis: the ferredoxin/thioredoxin system: perspective on its discovery, present status, and future development. *Archives of Biochemistry and Biophysics* 288(1), 1-9.

Burgess, S. J., Reyna-Llorens, I., Stevenson, S. R., Singh, P., Jaeger, K. and Hibberd, J. M. 2019. Genome-wide transcription factor binding in leaves from C3 and C4 grasses. *The Plant Cell* 31(10), 2297-2314.

Carmo-Silva, A. E. and Salvucci, M. E. 2013. The regulatory properties of RuBisCO Activase differ among species and affect photosynthetic induction during light transitions. *Plant Physiology* 161(4), 1645-1655.

Carrera, D. Á., George, G. M., Fischer-Stettler, M., Galbier, F., Eicke, S., Truernit, E., Streb, S. and Zeeman, S. C. 2021. Distinct plastid fructose bisphosphate aldolases function in photosynthetic and non-photosynthetic metabolism in Arabidopsis. *Journal of Experimental Botany* 72(10), 3739-3755.

Castro-Torres, E., Jiménez-Sandoval, P., Romero-Romero, S., Fuentes-Pascacio, A., López-Castillo, L. M., Díaz-Quezada, C., Fernández-Velasco, D. A., Torres-Larios, A. and Brieba, L. G. 2019. Structural basis for the modulation of plant cytosolic triosephosphate isomerase activity by mimicry of redox-based modifications. *The Plant Journal: For Cell and Molecular Biology* 99(5), 950-964.

Chibani, K., Couturier, J., Selles, B., Jacquot, J. P. and Rouhier, N. 2010. The chloroplastic thiol reducing systems: dual functions in the regulation of carbohydrate metabolism and regeneration of antioxidant enzymes, emphasis on the poplar redoxin equipment. *Photosynthesis Research* 104(1), 75-99.

Chibani, K., Pucker, B., Dietz, K. J. and Cavanagh, A. 2021. Genome-wide analysis and transcriptional regulation of the typical and atypical thioredoxins in Arabidopsis thaliana. *FEBS Letters* 595(21), 2715–2730.

Clark, T. J., Guo, L., Morgan, J. and Schwender, J. 2020. Modeling plant metabolism: from network reconstruction to mechanistic models. *Annual Review of Plant Biology* 71, 303–326.

Colombié, S., Nazaret, C., Bénard, C., Biais, B., Mengin, V., Solé, M., Fouillen, L., Dieuaide-Noubhani, M., Mazat, J. P., Beauvoit, B. and Gibon, Y. 2015. Modelling central metabolic fluxes by constraint-based optimization reveals metabolic reprogramming of developing Solanum Lycopersicum (tomato) fruit. *The Plant Journal: For Cell and Molecular Biology* 81(1), 24–39.

Cruz, J. A., Emery, C., Wüst, M., Kramer, D. M. and Lange, B. M. 2008. Metabolite profiling of Calvin cycle intermediates by Hplc-Ms using mixed-mode stationary phases. *The Plant Journal: For Cell and Molecular Biology* 55(6), 1047–1060.

Davidi, D., Shamshoum, M., Guo, Z., Bar-On, Y. M., Prywes, N., Oz, A., Jablonska, J., Flamholz, A., Wernick, D. G., Antonovsky, N., De Pins, B., Shachar, L., Hochhauser, D., Peleg, Y., Albeck, S., Sharon, I., Mueller-Cajar, O. and Milo, R. 2020. Highly active rubiscos discovered by systematic interrogation of natural sequence diversity. *The EMBO Journal* 39(18), e104081.

Degen, G. E., Orr, D. J. and Carmo-Silva, E. 2021. Heat-induced changes in the abundance of wheat RuBisCO Activase isoforms. *New Phytologist* 229(3), 1298–1311.

Degen, G. E., Worrall, D. and Carmo-Silva, E. 2020. An isoleucine residue acts as a thermal and regulatory switch in wheat RuBisCO Activase. *The Plant Journal: For Cell and Molecular Biology* 103(2), 742–751.

Driever, S. M., Simkin, A. J., Alotaibi, S., Fisk, S. J., Madgwick, P. J., Sparks, C. A., Jones, H. D., Lawson, T., Parry, M. A. J. and Raines, C. A. 2017. Increased Sbpase activity improves photosynthesis and grain yield in wheat grown in greenhouse conditions. *Philosophical Transactions of the Royal Society of London. Series B, Biological Sciences* 372(1730), 20160384.

Fell, D. A. 1997. *Understanding the Control of Metabolism*. London: Portland Press.

Fermani, S., Ripamonti, A., Sabatino, P., Zanotti, G., Scagliarini, S., Sparla, F., Trost, P. and Pupillo, P. 2001. Crystal structure of the non-regulatory A4 isoform of spinach chloroplast glyceraldehyde-3-phosphate dehydrogenase complexed with NADP. *Journal of Molecular Biology* 314(3), 527-542.

Flis, A., Mengin, V., Ivakov, A. A., Mugford, S. T., Hubberten, H. M., Encke, B., Krohn, N., Höhne, M., Feil, R., Hoefgen, R., Lunn, J. E., Millar, A. J., Smith, A. M., Sulpice, R. and Stitt, M. 2019. Multiple circadian clock outputs regulate diel turnover of carbon and nitrogen reserves. *Plant, Cell and Environment* 42(2), 549–573.

Galmés, J., Hermida-Carrera, C., Laanisto, L. and Niinemets, Ü. 2016. A compendium of temperature responses of RuBisCO kinetic traits: variability among and within photosynthetic groups and impacts on photosynthesis modeling. *Journal of Experimental Botany* 67(17), 5067–5091.

Gerhardt, S., Echt, S., Busch, M., Freigang, J. R., Auerbach, G. N., Bader, G., Martin, W. F., Bacher, A., Huber, R. and Fischer, M. 2003. Structure and properties of an engineered transketolase from maize. *Plant Physiology* 132(4), 1941–1949.

Graf, A., Schlereth, A., Stitt, M. and Smith, A. M. 2010. Circadian control of carbohydrate availability for growth in *Arabidopsis* plants at night. *Proceedings of the National Academy of Sciences* 107(20), 9458–9463.

Gurrieri, L., Del Giudice, A., Demitri, N., Falini, G., Pavel, N. V., Zaffagnini, M., Polentarutti, M., Crozet, P., Marchand, C. H., Henri, J., Trost, P., Lemaire, S. D., Sparla, F. and Fermani, S. 2019. Arabidopsis and *Chlamydomonas* phosphoribulokinase crystal structures complete the redox structural proteome of the Calvin-Benson cycle. *Proceedings of the National Academy of Sciences of the United States of America* 116(16), 8048-8053.

Gurrieri, L., Fermani, S., Zaffagnini, M., Sparla, F. and Trost, P. 2021. Calvin-Benson cycle regulation is getting complex. *Trends in Plant Science* 26(9), 898-912.

Gütle, D. D., Roret, T., Hecker, A., Reski, R. and Jacquot, J. P. 2017. Dithiol disulphide exchange in redox regulation of chloroplast enzymes in response to evolutionary and structural constraints. *Plant Science: An International Journal of Experimental Plant Biology* 255, 1-11.

Gütle, D. D., Roret, T., Müller, S. J., Couturier, J., Lemaire, S. D., Hecker, A., Dhalleine, T., Buchanan, B. B., Reski, R., Einsle, O. and Jacquot, J. P. 2016. Chloroplast Fbpase and Sbpase are thioredoxin-linked enzymes with similar architecture but different evolutionary histories. *Proceedings of the National Academy of Sciences of the United States of America* 113(24), 6779-6784.

Henkes, S., Sonnewald, U., Badur, R., Flachmann, R. and Stitt, M. 2001. A small decrease of plastid transketolase activity in antisense tobacco transformants has dramatic effects on photosynthesis and phenylpropanoid metabolism. *The Plant Cell* 13(3), 535-551.

Hermida-Carrera, C., Kapralov, M. V. and Galmés, J. 2016. RuBisCO catalytic properties and temperature response in crops. *Plant Physiology* 171(4), 2549-2561.

Howard, T. P., Fryer, M. J., Singh, P., Metodiev, M., Lytovchenko, A., Obata, T., Fernie, A. R., Kruger, N. J., Quick, W. P., Lloyd, J. C. and Raines, C. A. 2011. Antisense suppression of the small chloroplast protein Cp12 in tobacco alters carbon partitioning and severely restricts growth. *Plant Physiology* 157(2), 620-631.

Howard, T. P., Metodiev, M., Lloyd, J. C. and Raines, C. A. 2008. Thioredoxin-mediated reversible dissociation of a stromal multiprotein complex in response to changes in light availability. *Proceedings of the National Academy of Sciences of the United States of America* 105(10), 4056-4061.

Kappachery, S., Baniekal-Hiremath, G., Yu, J. W. and Park, S. W. 2015. Effect of over-and under-expression of glyceraldehyde 3-phosphate dehydrogenase on tolerance of plants to water-deficit stress. *Plant Cell, Tissue and Organ Culture (PCTOC)* 121(1), 97-107.

Khozaei, M., Fisk, S., Lawson, T., Gibon, Y., Sulpice, R., Stitt, M., Lefebvre, S. C. and Raines, C. A. 2015. Overexpression of plastid transketolase in tobacco results in a thiamine auxotrophic phenotype. *The Plant Cell* 27(2), 432-447.

Kim, S. Y., Slattery, R. A. and Ort, D. R. 2021. A role for differential RuBisCO Activase isoform expression in C4 bioenergy grasses at high temperature. *GCB Bioenergy* 13(1), 211-223.

Kubien, D. S., Whitney, S. M., Moore, P. V. and Jesson, L. K. 2008. The biochemistry of RuBisCO in Flaveria. *Journal of Experimental Botany* 59(7), 1767-1777.

Lawlor, D. 2001. *Photosynthesis* (3rd edn.). Oxford: Bios Scientific Publishers Ltd.

Le Moigne, T., Crozet, P., Lemaire, S. D. and Henri, J. 2020. High-resolution crystal structure of chloroplastic ribose-5-phosphate isomerase from Chlamydomonas reinhardtii–an enzyme involved in the photosynthetic Calvin-Benson cycle. *International Journal of Molecular Sciences* 21(20), 7787.

Le Moigne, T., Gurrieri, L., Crozet, P., Marchand, C. H., Zaffagnini, M., Sparla, F., Lemaire, S. D. and Henri, J. 2021. Crystal structure of chloroplastic thioredoxin z defines a type-specific target recognition. *The Plant Journal: For Cell and Molecular Biology* 107(2), 434–447.

Leegood, R. C. 1990. 2-Enzymes of the Calvin cycle. In Lea, P. J. (Ed.) *Methods in Plant Biochemistry*, Academic Press, 15–37.

Leegood, R. C. and Von Caemmerer, S. 1988. The relationship between contents of photosynthetic metabolites and the rate of photosynthetic carbon assimilation in leaves of Amaranthus edulis L. *Planta* 174(2), 253-262.

Lefebvre, S., Lawson, T., Zakhleniuk, O. V., Lloyd, J. C., Raines, C. A. and Fryer, M. 2005. Increased Sedoheptulose-1,7-Bisphosphatase activity in transgenic tobacco plants stimulates photosynthesis and growth from an early stage in development. *Plant Physiology* 138(1), 451–460.

Lemaire, S. D., Michelet, L., Zaffagnini, M., Massot, V. and Issakidis-Bourguet, E. 2007. Thioredoxins in chloroplasts. *Current Genetics* 51(6), 343–365.

Li, C., Li, N., Huang, R., Chen, C., Guo, J., Yang, X., Zhang, X., Sun, C., Deng, X. and Wang, P. 2020. A single nucleotide substitution at the 3′-end of Sbpase gene involved in Calvin cycle severely affects plant growth and grain yield in rice. *BMC Plant Biology* 20(1), 345.

Li, K., Qiu, H., Zhou, M., Lin, Y., Guo, Z. and Lu, S. 2018. Chloroplast Protein 12 expression alters growth and chilling tolerance in tropical forage Stylosanthes guianensis (Aublet) Sw. *Frontiers in Plant Science* 9, 1319.

Li, R., Qiu, Z., Wang, X., Gong, P., Xu, Q., Yu, Q. B. and Guan, Y. 2019. Pooled Crispr/Cas9 reveals redundant roles of plastidial phosphoglycerate kinases in carbon fixation and metabolism. *The Plant Journal: For Cell and Molecular Biology* 98(6), 1078–1089.

Long, S. P., Zhu, X. G., Naidu, S. L. and Ort, D. R. 2006. Can improvement in photosynthesis increase crop yields? *Plant, Cell and Environment* 29(3), 315-330.

López-Calcagno, P., Omar Abuzaid, A., Lawson, T. and Anne Raines, C. 2017. Arabidopsis Cp12 mutants have reduced levels of phosphoribulokinase and impaired function of the Calvin–Benson cycle. *Journal of Experimental Botany* 68(9), 2285–2298.

López-Calcagno, P. E., Brown, K. L., Simkin, A. J., Fisk, S. J., Vialet-Chabrand, S., Lawson, T. and Raines, C. A. 2020. Stimulating photosynthetic processes increases productivity and water-use efficiency in the field. *Nature Plants* 6(8), 1054–1063.

López-Castillo, L. M., Jiménez-Sandoval, P., Baruch-Torres, N., Trasviña-Arenas, C. H., Díaz-Quezada, C., Lara-González, S., Winkler, R. and Brieba, L. G. 2016. Structural basis for redox regulation of cytoplasmic and chloroplastic triosephosphate isomerases from Arabidopsis thaliana. *Frontiers in Plant Science* 7, 1817.

Marri, L., Sparla, F., Pupillo, P. and Trost, P. 2004. Co-ordinated gene expression of photosynthetic glyceraldehyde-3-phosphate dehydrogenase, phosphoribulokinase, and Cp12 in Arabidopsis thaliana. *Journal of Experimental Botany* 56(409), 73–80.

Marri, L., Zaffagnini, M., Collin, V., Issakidis-Bourguet, E., Lemaire, S. D., Pupillo, P., Sparla, F., Miginiac-Maslow, M., Trost, P. 2009. Prompt and easy activation by specific thioredoxins of Calvin cycle enzymes of Arabidopsis thaliana in the GAPDH/CP12/PRK supramolecular complex, *Mol. Plant*, 2, 259–269.

Massange-Sánchez, J. A., Casados-Vázquez, L. E., Juarez-Colunga, S., Sawers, R. J. H. and Tiessen, A. 2020. The phosphoglycerate kinase (Pgk) gene family of maize (Zea mays var. B73). *Plants* 9(12), 1639.

Matuszyńska, A., Saadat, N. P. and Ebenhöh, O. 2019. Balancing energy supply during photosynthesis - a theoretical perspective. *Physiologia Plantarum* 166(1), 392-402.
Mengin, V., Pyl, E. T., Alexandre Moraes, T., Sulpice, R., Krohn, N., Encke, B. and Stitt, M. 2017. Photosynthate partitioning to starch in Arabidopsis thaliana is insensitive to light intensity but sensitive to photoperiod due to a restriction on growth in the light in short photoperiods. *Plant, Cell and Environment* 40(11), 2608-2627.
Michelet, L., Zaffagnini, M., Morisse, S., Sparla, F., Pérez-Pérez, M. E., Francia, F., Danon, A., Marchand, C. H., Fermani, S., Trost, P. and Lemaire, S. D. 2013. Redox regulation of the Calvin-Benson cycle: something old, something new. *Frontiers in Plant Science* 4, 470.
Morisse, S., Michelet, L., Bedhomme, M., Marchand, C. H., Calvaresi, M., Trost, P., Fermani, S., Zaffagnini, M. and Lemaire, S. D. 2014. Thioredoxin-dependent redox regulation of chloroplastic phosphoglycerate kinase from Chlamydomonas reinhardtii. *Journal of Biological Chemistry* 289(43), 30012-30024.
Naranjo, B., Diaz-Espejo, A., Lindahl, M. and Cejudo, F. J. 2016. Type-f thioredoxins have a role in the short-term activation of carbon metabolism and their loss affects growth under short-day conditions in Arabidopsis thaliana. *Journal of Experimental Botany* 67(6), 1951-1964.
Nikkanen, L., Toivola, J. and Rintamäki, E. 2016. Crosstalk between chloroplast thioredoxin systems in regulation of photosynthesis. *Plant, Cell and Environment* 39(8), 1691-1705.
Okegawa, Y. and Motohashi, K. 2015. Chloroplastic thioredoxin m functions as a major regulator of Calvin cycle enzymes during photosynthesis in vivo. *The Plant Journal: For Cell and Molecular Biology* 84(5), 900-913.
Orr, D. J., Alcântara, A., Kapralov, M. V., Andralojc, P. J., Carmo-Silva, E. and Parry, M. A. J. 2016. Surveying RuBisCO diversity and temperature response to improve crop photosynthetic efficiency. *Plant Physiology* 172(2), 707-717.
Perdomo, J. A., Cavanagh, A. P., Kubien, D. S. and Galmés, J. 2015. Temperature dependence of in vitro RuBisCO kinetics in species of Flaveria with different photosynthetic mechanisms. *Photosynthesis Research* 124(1), 67-75.
Petersen, J., Brinkmann, H. and Cerff, R. 2003. Origin, evolution, and metabolic role of a novel glycolytic Gapdh enzyme recruited by land plant plastids. *Journal of Molecular Evolution* 57(1), 16-26.
Pilkington, S. M., Encke, B., Krohn, N., Höhne, M., Stitt, M. and Pyl, E. T. 2015. Relationship between starch degradation and carbon demand for maintenance and growth in Arabidopsis thaliana in different irradiance and temperature regimes. *Plant, Cell and Environment* 38(1), 157-171.
Poolman, M. G., Ölçer, H., Lloyd, J. C., Raines, C. A. and Fell, D. A. 2001. Computer modelling and experimental evidence for two steady states in the photosynthetic Calvin cycle. *European Journal of Biochemistry* 268(10), 2810-2816.
Portis, A. R. 2003. RuBisCO Activase - RuBisCO's catalytic chaperone. *Photosynthesis Research* 75(1), 11-27.
Raines, C. A. 2003. The Calvin cycle revisited. *Photosynthesis Research* 75(1), 1-10.
Raines, C. A. 2022. Improving plant productivity by re-tuning the regeneration of the RuBP in the Calvin-Benson-Bassham cycle, New Phytologist 2022, 236,350-356.
Rocha, A. G., Mehlmer, N., Stael, S., Mair, A., Parvin, N., Chigri, F., Teige, M. and Vothknecht, U. C. 2014. Phosphorylation of Arabidopsis transketolase at Ser428 provides a potential paradigm for the metabolic control of chloroplast carbon metabolism. *Biochemical Journal* 458(2), 313-322.

Rodermel, S. R., Abbott, M. S. and Bogorad, L. 1988. Nuclear-organelle interactions: nuclear antisense gene inhibits ribulose bisphosphate carboxylase enzyme levels in transformed tobacco plants. *Cell* 55(4), 673-681.

Rosa-Téllez, S., Anoman, A. D., Flores-Tornero, M., Toujani, W., Alseek, S., Fernie, A. R., Nebauer, S. G., Muñoz-Bertomeu, J., Segura, J. and Ros, R. 2017. Phosphoglycerate kinases are co-regulated to adjust metabolism and to optimize growth. *Plant Physiology* 176(2), 1182-1198.

Saadat, N. P., Nies, T., Van Aalst, M., Hank, B., Demirtas, B., Ebenhöh, O. and Matuszyńska, A. 2021. Computational analysis of alternative photosynthetic electron flows linked With oxidative stress. *Frontiers in Plant Science* 12, 750580.

Scafaro, A. P., Bautsoens, N., Den Boer, B., Van Rie, J. and Gallé, A. 2019. A conserved sequence from heat-adapted species improves RuBisCO Activase thermostability in Wheat1. *Plant Physiology* 181(1), 43-54.

Schürmann, P. and Buchanan, B. B. 2008. The ferredoxin/thioredoxin system of oxygenic photosynthesis. *Antioxidants and Redox Signaling* 10(7), 1235-1274.

Schürmann, P. and Jacquot, J. P. 2000. Plant thioredoxin systems revisted. *Annual Review of Plant Physiology and Plant Molecular Biology* 51, 371-400.

Scialdone, A., Mugford, S. T., Feike, D., Skeffington, A., Borrill, P., Graf, A., Smith, A. M. and Howard, M. 2013. Arabidopsis plants perform arithmetic division to prevent starvation at night. *eLife* 2, e00669.

Sharkey, T. D. 2019. Discovery of the canonical Calvin-Benson cycle. *Photosynthesis Research* 140(2), 235-252.

Sharwood, R. E., Ghannoum, O., Kapralov, M. V., Gunn, L. H. and Whitney, S. M. 2016. Temperature responses of RuBisCO from Paniceae grasses provide opportunities for improving C3 photosynthesis. *Nature Plants* 2, 16186.

Simkin, A. J., López-Calcagno, P. E. and Raines, C. A. 2019. Feeding the world: improving photosynthetic efficiency for sustainable crop production. *Journal of Experimental Botany* 70(4), 1119-1140.

Smith, A. M. and Stitt, M. 2007. Coordination of carbon supply and plant growth. *Plant, Cell and Environment* 30(9), 1126-1149.

Sparla, F., Pupillo, P. and Trost, P. 2002. The C-terminal extension of glyceraldehyde-3-phosphate dehydrogenase Subunit B acts as an autoinhibitory domain regulated by thioredoxins and nicotinamide adenine dinucleotide. *Journal of Biological Chemistry* 277(47), 44946-44952.

Stitt, M. and Heldt, H. W. 1985. Generation and maintenance of concentration gradients between the mesophyll and bundle sheath in maize leaves. *Biochimica et Biophysica Acta (Bba) - Bioenergetics* 808(3), 400-414.

Stitt, M., Luca Borghi, G. and Arrivault, S. 2021. Targeted metabolite profiling as a top-down approach to uncover interspecies diversity and identify key conserved operational features in the Calvin-Benson cycle. *Journal of Experimental Botany* 72(17), 5961-5986.

Stitt, M., Lunn, J. and Usadel, B. 2010. Arabidopsis and primary photosynthetic metabolism - more than the icing on the cake. *The Plant Journal: For Cell and Molecular Biology* 61(6), 1067-1091.

Stitt, M. and Schulze, D. 1994. Does RuBisCO control the rate of photosynthesis and plant growth? An exercise in molecular ecophysiology. *Plant, Cell and Environment* 17(5), 465-487.

Sulpice, R., Flis, A., Ivakov, A. A., Apelt, F., Krohn, N., Encke, B., Abel, C., Feil, R., Lunn, J. E. and Stitt, M. 2014. Arabidopsis coordinates the diurnal regulation of carbon allocation and growth across a Wide Range of photoperiods. *Molecular Plant* 7(1), 137–155.

Szabó, J., Varga, A., Flachner, B., Konarev, P. V., Svergun, D. I., Závodszky, P. and Vas, M. 2008. Communication between the nucleotide site and the main molecular hinge of 3-phosphoglycerate kinase. *Biochemistry* 47(26), 6735–6744.

Thormählen, I., Zupok, A., Rescher, J., Leger, J., Weissenberger, S., Groysman, J., Orwat, A., Chatel-Innocenti, G., Issakidis-Bourguet, E., Armbruster, U. and Geigenberger, P. 2017. Thioredoxins play a crucial role in dynamic acclimation of photosynthesis in fluctuating light. *Molecular Plant* 10(1), 168–182.

Treves, H., Küken, A., Arrivault, S., Ishihara, H., Hoppe, I., Erban, A., Höhne, M., Moraes, T. A., Kopka, J., Szymanski, J., Nikoloski, Z. and Stitt, M. 2022. Carbon flux through photosynthesis and central carbon metabolism show distinct patterns between algae, C3 and C4 plants. *Nature Plants* 8(1), 78–91.

Treves, H., Siemiatkowska, B., Luzarowska, U., Murik, O., Fernandez-Pozo, N., Moraes, T. A., Erban, A., Armbruster, U., Brotman, Y., Kopka, J., Rensing, S. A., Szymanski, J. and Stitt, M. 2020. Multi-omics reveals mechanisms of total resistance to extreme illumination of a desert alga. *Nature Plants* 6(8), 1031–1043.

Tsukamoto, Y., Fukushima, Y., Hara, S. and Hisabori, T. 2013. Redox control of the activity of phosphoglycerate kinase in Synechocystis sp. PCC6803. *Plant and Cell Physiology* 54(4), 484–491.

von Caemmerer, S. 2000. *Biochemical Models of Leaf Photosynthesis*. Collingwood, Australia: Csiro.

von Caemmerer, S. and Farquhar, G. D. 1981. Some relationships between the biochemistry of photosynthesis and the gas exchange of leaves. *Planta* 153(4), 376–387.

Wang, L., Xie, J., Du, Q., Song, F., Xiao, L., Quan, M. and Zhang, D. 2019. Transcription factors involved in the regulatory networks governing the Calvin-Benson-Bassham cycle. *Tree Physiology* 39(7), 1159–1172.

Whitney, S. M., Sharwood, R. E., Orr, D., White, S. J., Alonso, H. and Galmés, J. 2011. Isoleucine 309 acts as a C_4 catalytic switch that increases ribulose-1,5-bisphosphate carboxylase/oxygenase (RuBisCO) carboxylation rate in Flaveria. *Proceedings of the National Academy of Sciences* 108(35), 14688–14693.

Wullschleger, S. D. 1993. Biochemical limitations to carbon assimilation in C3 plants—A retrospective analysis of the A/Ci curves from 109 species. *Journal of Experimental Botany* 44(5), 907–920.

Xu, J., Bräutigam, A., Weber, A. P. M. and Zhu, X. G. 2016. Systems analysis of cis-regulatory motifs in C4 photosynthesis genes using maize and rice leaf transcriptomic data during a process of de-etiolation. *Journal of Experimental Botany* 67(17), 5105–5117.

Yokochi, Y., Fukushi, Y., Wakabayashi, K.-I., Yoshida, K. and Hisabori, T. 2021. Oxidative regulation of chloroplast enzymes by thioredoxin and thioredoxin-like proteins in *Arabidopsis thaliana*. *Proceedings of the National Academy of Sciences* 118, e2114952118.

Yoshida, K., Hara, A., Sugiura, K., Fukaya, Y. and Hisabori, T. 2018. Thioredoxin-like2/2-Cys peroxiredoxin redox cascade supports oxidative thiol modulation in chloroplasts. *Proceedings of the National Academy of Sciences of the United States of America* 115(35), E8296–E8304.

Zhang, N. and Portis, A. R. 1999. Mechanism of light regulation of RuBisCO: a specific role for the larger RuBisCO Activase isoform involving reductive activation by thioredoxin-f. *Proceedings of the National Academy of Sciences of the United States of America* 96(16), 9438–9443.

Zhao, H., Tang, Q., Chang, T., Xiao, Y. and Zhu, X.-G. 2020. Why an increase in activity of an enzyme in the Calvin–Benson cycle does not always lead to an increased photosynthetic CO2 uptake rate?–a theoretical analysis. *In Silico Plants* 3(1): 1–13.

Zhu, X. G., De Sturler, E. and Long, S. P. 2007. Optimizing the distribution of resources between enzymes of carbon metabolism can dramatically increase photosynthetic rate: a numerical simulation using an evolutionary algorithm. *Plant Physiology* 145(2), 513–526.

Zhu, X. G., Govindjee, G., Baker, N. R., Desturler, E., Ort, D. O. and Long, S. P. 2005. Chlorophyll a fluorescence induction kinetics in leaves predicted from a model describing each discrete step of excitation energy and electron transfer associated with photosystem II. *Planta* 223(1), 114–133.

Zhu, X. G., Wang, Y., Ort, D. R. and Long, S. P. 2013. e-photosynthesis: a comprehensive dynamic mechanistic model of C3 photosynthesis: from light capture to sucrose synthesis. *Plant, Cell and Environment* 36(9), 1711–1727.

Chapter 2

Understanding the genetics of C_3 photosynthesis in crop plants

P. Carvalho, G. Elias da Silva and N. J. M. Saibo, Instituto de Tecnologia Química e Biológica António Xavier da Universidade Nova de Lisboa (ITQB NOVA), Portugal

1 Introduction

Photosynthesis has the potential to increase crop yield, and designing strategies to improve photosynthesis requires a good understanding of its regulation at different levels, including gene expression. Photosynthesis involves two phases: the light reactions, which produce O_2, ATP, and NADPH; and the dark reactions, Calvin–Benson–Bassham (CBB) cycle, which reduce CO_2 to carbohydrate and consume the ATP and NADPH produced in the light reactions (Buchanan et al., 2015). To carry out photosynthesis, plants require a large cohort of genes, which encode proteins to capture light energy, store energy in sugars, and build the subcellular structures needed to capture this energy. The function of many of these genes is known, but the regulatory networks that coordinate their expression are poorly understood.

Plant genomes encode approximately 3000 proteins that are needed to carry out photosynthesis (Wang et al., 2017). These *c.* 3000 proteins include

http://dx.doi.org/10.19103/AS.2022.0119.03

all proteins needed to build, replicate, maintain and operate chloroplasts, including the cytosolic enzymes, as well as enzymes of the mitochondrion and peroxisome. Most of these proteins are encoded by the nuclear genome, and the respective cohort of genes is commonly known as photosynthesis-associated nuclear genes (PhANG). Although the vast majority of the photosynthesis-associated proteins are present in the chloroplast, only *c.* 100 are encoded in the chloroplast genome, by genes known as photosynthesis-associated plastid genes (PhAPGs). This partitioning of genes between nuclear and chloroplast genomes requires a tight orchestration of nuclear and chloroplast gene expression (Kleine and Leister, 2016; Woodson and Chory, 2008). PhANG and PhAPGs are quite well-identified in the model plant *Arabidopsis thaliana*, and many of them have been functionally characterized (Berry et al., 2013; Leister and Schneider, 2003). However, in crops very little has been reported regarding both the identification of those genes and their functional characterization.

The expression of the photosynthesis-associated genes is tightly regulated by environmental cues (e.g. light, nitrogen, abiotic and biotic stresses) as well as by internal signals (e.g. redox species, development, circadian clock, end products of photosynthesis). Retrograde (from chloroplast to nucleus) and anterograde (from nucleus to chloroplast) signaling underpin the highly coordinated gene expression in both nuclear and chloroplast genomes. The retrograde signaling uses chloroplast-generated signals, e.g. ROS or tetrapyrroles, to regulate nuclear gene expression, while anterograde signaling utilizes nuclear-encoded transcriptional and post-transcriptional regulators, e.g. sigma factors or RNA-binding proteins, to regulate the expression of genes encoded by the chloroplast genome (Berry et al., 2013). Designing photosynthesis improvement strategies requires a detailed understanding of how the photosynthetic-associated genes are regulated. It is important to understand how their expression pattern is regulated in response to specific stimulus and the gene regulatory networks involved. However, all this information is very scarce, particularly for crop plants.

In addition to better understand photosynthesis-associated gene regulation, it is important to know how the expression of these genes can be modulated. The manipulation of nuclear-encoded genes is quite well established for many different crop plants, but changing chloroplast-encoded gene expression using transplastomics is still challenging, particularly in cereals. Although efficient transplastomic methods have been optimized for crop species such as tomato (*Solanum lycopersicum*), potato (*Solanum tuberosum*) and lettuce (*Lactuca sativa*), the proper selection and regeneration of protocols is still the biggest bottleneck for cereals, the world's most important staple crops (Bock, 2014, 2015).

2 Photosynthesis-associated genes

The photosynthetic reactions take place in the chloroplast, an organelle that arose through endosymbiosis from cyanobacteria (Ku et al., 2015). This organelle possesses its own genome, which encodes photosynthesis-related genes and the machinery required for its gene expression. Nevertheless, given that over the course of evolution most genes from the prokaryotic ancestors of chloroplasts were transferred into the nucleus, most chloroplast proteins are encoded in the nucleus (Table 1). Thus, genes encoding proteins involved in the photosynthetic process fall into two main categories: chloroplast-encoded genes (also known as PhAPG) and nuclear-encoded genes (also known as PhANG). Chloroplast-encoded genes are transcribed, translated and regulated within this organelle, and their names begin with a lowercase letter, e.g. *psa*A, *psb*A or *rcb*L (see Table 1). Nuclear-encoded genes are transcribed in the nucleus and their transcripts are translated by ribosomes in the cytoplasm. The resultant proteins have a chloroplast transit peptide that will guide them to the chloroplast. The names of the nuclear-encoded genes begin with an uppercase letter, e.g. *Lhca*, *Psb*S or *Rbc*S (see Table 1).

3 Regulation of photosynthesis-associated genes by different signals

The photosynthesis-associated genes are regulated at transcriptional, post-transcriptional, translational, and post-translational levels. All these levels play important roles in photosynthesis regulation, but this chapter will mainly focus on the transcriptional regulation of photosynthesis. Different signals are known to modulate the expression of the photosynthesis-associated genes, and the most relevant ones will be discussed, trying to focus on what is known for crop plants.

3.1 Light regulation

Light is strictly required for chloroplast development in all angiosperms and directly regulates the expression of hundreds of chloroplast-related genes (Cackett et al., 2022), being the primary environmental determinant of photosynthetic gene expression. In addition, light modulates the levels of several hormones, including brassinosteriods, cytokinins, auxins and gibberellins, which themselves control chloroplast development, particularly during the early stages of plant development (Müller and Munné-Bosch, 2021). Light is therefore extremely important to modulate gene expression that leads to an increased chloroplast abundance and putative higher photosynthetic efficiency of crops.

Little is known about the light regulation of photosynthetic gene expression in crop plants. Nevertheless, several chloroplast genes from barley, such as *rbc*L, *psb*A, and *psb*D-*psb*C (Christopher et al., 1992; Klein and Mullet, 1990; Thum et al., 2001), among others, have been reported as being upregulated by light.

Table 1 Representative photosynthetic genes, their associated proteins, where they are encoded, and their function

Gene	Protein	Encoded	Function
*Lhca*1(*CabI*)-5	Antenna proteins of PSI	Nucleus	Light harvesting
*Lhcb*1(*CabII*)-7	Antenna proteins of PSII	Nucleus	
*psa*A-C, I, J	PSI components	Chloroplast	Electron transport
*Psa*D-H, K, L, N, O		Nucleus	
*psb*A-N, T, Z, 27, 28	PSII components	Chloroplast	
*Psb*O-S, W-Y		Nucleus	
*pet*A, B, D, G, L	Subunits of cytochrome b6f complex	Chloroplast	
*Pet*C, M		Nucleus	
*Pet*E	Plastocyanin	Nucleus	
*Pet*F	Plastid ferredoxin	Nucleus	
*Pet*H	Ferredoxin-NADPH-oxidoreductase	Nucleus	
*atp*F, H, I	CF0 ATPase subunits	Chloroplast	ATP synthesis
*Atp*G		Nucleus	
*atp*A, B, E	CF1 ATPase subunits	Chloroplast	
*Atp*C, D		Nucleus	
*rbc*L	Rubisco Large subunit	Chloroplast	CBB cycle
*Rbc*S	Rubisco Small subunit	Nucleus	
Rca	Rubisco activase	Nucleus	
PGK	3-Phosphoglycerate kinase	Nucleus	
GAPA	Glyceraldehyde-3-phosphate dehydrogenase (subunit A)	Nucleus	
GAPB	Glyceraldehyde-3-phosphate dehydrogenase (subunit A)	Nucleus	
TPI	Triose-phosphate isomerase	Nucleus	
FBA	Fructose-1,6-biosphosphate aldolase	Nucleus	
cpFBP	Chloroplastic fructose-1,6-bisphosphatase	Nucleus	
TKL	Transketolase	Nucleus	
SBP	Sedoheptulose-1,7-bisphosphatase	Nucleus	
RPI	Ribose-5-phosphate isomerase	Nucleus	
RPE	Ribulose-5-phosphate 3-epimerase	Nucleus	
PRK	Phosphoribulokinase	Nucleus	

The light-induced transcription of these genes is regulated by sigma factors and a number of other regulatory factors. Several sigma factors encoding genes are known to be upregulated by light (Chi et al., 2015; Onda et al., 2008; Tan and Troxler, 1999). In addition, chloroplast DNA-binding factor 1 (CDF1) in maize and *rbc*L promoter-binding protein (RBLP) in tobacco are examples of other regulators that bind to the *rbc*L promoter to induce its transcription in response to light (Kim et al., 2002; Lam et al., 1988). The nuclear-encoded de-etiolated 1 (DET1) was shown to negatively regulate light-induced plastid transcription. Its mechanism of action is not clear, but it was shown to hinder the light-activation of *psb*D transcription in *Arabidopsis* (Christopher and Hoffer, 1998) and, in rice, the mutation of its orthologue (*OsDET1*) increased chlorophyll content (Huang et al., 2013).

The light regulation of PhANGs is even less understood than the light regulation of PhAPGs. In *Arabidopsis*, HY5 is known to directly regulate the transcription of photosynthesis-related genes during seedling photomorphogenic development (Lee et al., 2007), and more recently it was shown to act antagonistically with MYC2 mediating the blue light-inducible expression of the genes *CAB1* and *RbcS-1A* (Chakraborty et al., 2019). The expression of the maize genes *Pep1*, *Pdk1* and *PsaD1* was shown to be differentially regulated by different light wavelengths (Liu et al., 2017), but the regulators of this response are unknown. The rice transcription factor (TF) phytochrome interacting factor-like 1 (OsPIL1) is a light-signaling mediator that regulates photosynthesis. It upregulates the expression of two chlorophyll biosynthesis genes, *OsPORB* and *OsCAO1*, both directly binding to their promoter and indirectly through the TFs GOLDEN2-like1 (OsGLK1) and OsGLK2 (Sakuraba et al., 2017). Through these TFs, OsPIL1 also regulates genes encoding light-harvesting complex proteins, probably promoting their transcription.

3.2 Redox state

Redox regulation is central to the operation of photosynthesis. However, the roles of different oxidants and antioxidants in the regulation of photosynthetic gene expression remain poorly understood. Many studies have investigated changes in gene expression patterns in response to various redox signals such as hydrogen peroxide, singlet oxygen or redox signals from the photosynthetic electron transport chain (Pfannschmidt et al., 2009). Most of these studies indicate that photosynthetic genes are responsive to such signals (Fey et al., 2005; Piippo et al., 2006). In addition, it was shown that genes encoding components of both photosystem II (PSII) repair cycle components and thylakoid NAD(P)H (nicotinamide adenine dinucleotide (phosphate)) dehydrogenases (NDH) were downregulated in mutants with low glutathione and/or ascorbate (Queval and Foyer, 2012). Regulation of genes encoding specific components of the photosynthetic electron transport chains by hydrogen peroxide, ascorbate and glutathione may serve to balance non-cyclic and cyclic electron flow pathways in relation to oxidant production and reductant availability (Foyer et al., 2012; Queval and Foyer, 2012).

The major problem in fully understanding redox-regulated gene expression is to distinguish between genes that are controlled directly by a redox signal (primary target gene) and genes that are controlled indirectly, e.g. by the effects resulting from expression changes in primary target genes or posttranslational modifications (PTM) of regulatory proteins. It has been suggested that in most cases, the redox regulation of the photosynthesis-associated genes is an indirect regulation. A good example of this is the regulation of the photosynthesis reaction center genes *psb*A (encoding the D1 protein of PSII) and *psa*A/B (encoding the P700 apoprotein of PSI) by the sigma factor Sig1. This regulation depends on the phosphorylation of Sig1, which in turn is under the control of the plastoquinone (PQ) redox state (Shimizu et al., 2010).

Many TFs function through different mechanisms that affect their subcellular localization and or activity. Transcription factors are well-known to be regulated by PTMs, such as phosphorylation, acetylation, ubiquitylation, and SUMOylation, which are known to regulate their function (Skelly et al., 2016). Since it has been shown that the cellular redox status can regulate TF activity through oxidative posttranslational modifications (oxPTMs) (Dietz, 2014), this might be the way that the redox state modulates photosynthetic gene expression.

3.3 Circadian rhythm

The interaction between the circadian clock and photosynthesis regulation is well documented and known to occur in both directions. Photosynthesis is regulated by the circadian clocks (Dodd et al., 2015; Venkat and Muneer, 2022), and the photosynthesis end products contribute to entrain the circadian rhythm (Haydon et al., 2013). A substantial photosynthetic advantage is conferred by the correct matching of the circadian clock period with that of the environmental light–dark cycle (Dodd et al., 2005). However, how photosynthesis is regulated by the circadian clock, particularly at the gene level, is largely unknown.

It has long been reported that the expression of many genes encoding light-harvesting chlorophyll a/b-binding proteins (LHCP) shows diurnal oscillation in different crop plants that were first grown under light–dark cycles and then transferred to continuous light or darkness, indicating control by an endogenous circadian clock (Kellmann et al., 1993; Nagy et al., 1988; Tavladoraki et al., 1989). The expression of *psb*D from barley and wheat was shown to be highly regulated by the circadian clock (Nakahira et al., 1998; Thum et al., 2001). However, the molecular mechanisms underlying this regulation are unknown. The gene expression of the nuclear-encoded sigma factor A (SigA) from wheat was found to be circadian-regulated, and it is therefore a putative regulator of the circadian rhythm shown by *psb*D or other plastid-encoded genes (Morikawa et al., 1999).

Regarding the circadian regulation of the genes encoding enzymes from the CBB cycle, it is known that the transcript levels of both *Rubisco activase* (*Rca*) and the *small subunit of Rubisco* (*Rcb*S) also oscillate (Pilgrim and McClung, 1993; Watillon et al., 1993). It has been suggested that *Rca* is also under circadian regulation through a different mechanism. The mRNA encoding *Rca* is alternatively spliced, thus originating two proteins with different regulatory features depending on light intensity (Zhang et al., 2002). Since the ratio between these two isoforms shows a circadian rhythm (Sanchez et al., 2010), it has been suggested that the clock contributes to the fine-tuning of this efficient system, having a physiological relevance.

3.4 Sink-source regulation

It has long been known that photosynthesis efficiency is downregulated by its end products in a large number of crop plants (Azcón-Bieto, 1983; Goldschmidt and Huber, 1992), a process well known as sink regulation of photosynthesis (Paul and Foyer, 2001). However, very little is known about the molecular mechanisms underlying this regulation, particularly regarding the regulation of photosynthetic genes. The first studies on the regulation of these genes by the end products, such as glucose and sucrose, showed that photo-assimilates could decrease the transcript levels of *Rbc*S, *Lhcb* and *Atp-D* (Cheng et al., 1992; Harter et al., 1993; Krapp et al., 1993). Given that exposure to elevated CO_2 can reduce the photosynthetic capacity of plants due to the accumulation of carbohydrates in the leaf (Ainsworth and Long, 2005; Chen et al., 2005), plants grown under elevated CO_2 have been largely used to explore the carbohydrate control of photosynthetic gene expression as a mechanism of sink regulation of photosynthesis. It was observed that tomato plants under elevated CO_2 have decreased levels of nuclear-encoded transcripts, such as *RcbS*, *Lhcb* and *Rca* (Van Oosten and Besford, 1995). In *Arabidopsis* and wheat, elevated CO_2 decreases both *Rbc*S and *rbc*L transcripts (Cheng et al., 1998; Nie et al., 1995). In wheat, *PGKase* gene expression also shows a decrease in response to elevated CO_2. More recently, a transcriptomic study in maize under elevated CO_2 showed significant downregulation for genes encoding *Psa*D, *Psa*F, *Psa*G, *Psa*K, *Psa*O subunits in PSI, and *Psb*Y, *psb*27, *psb*28, *Pet*E and *Pet*F in PSII (Huang et al., 2019a). It would be interesting to identify the transcriptional gene network associated with this regulation.

3.5 Nitrogen deficiency

Nitrogen is one of the essential macronutrients required for plant growth and development. Therefore, it is often a major limiting factor for plant productivity and crop yield. It is well known that photosynthesis capacity is mainly dependent on the N content of chloroplasts in the leaf (Evans, 1989) and that regulation of nuclear-encoded photosynthetic proteins by nitrogen is mainly at the transcriptional level (Plumley and Schmidt, 1989). It was initially shown in *Chlamydomonas reinhardtii* that genes encoding LHCII and RbcS subunits are downregulated in nitrogen-limited cells (Plumley and Schmidt, 1989).

A number of very interesting gene expression studies have been carried out in wheat to understand what are the molecular mechanisms underlying the plant response to nitrogen deficiency (Curci et al., 2017;Liu et al., 2020a; Sultana et al., 2020). They show that different genes encoding important photosynthetic proteins are downregulated by nitrogen starvation. Though showing slightly different results in terms of genes differentially regulated, these studies show that different genes encoding components of the PSI, PSII, cytochrome b6/ complex, photosynthetic electron transport chain, F-type ATPase and Rubisco subunits are downregulated under nitrogen deficiency. The misregulation observed may impede the proper function of the photosynthetic electron transport chain, photosystem repair and CBB cycle. Sultana et al. (2020) also show that most wheat genes encoding light-harvesting chlorophyll complex proteins are downregulated by nitrogen deficiency.

N deficiency also promotes significant changes in the expression patterns of a large number of tomato genes. Most of the genes downregulated encode proteins from the NDH complex, photosystems, LHCP complexes and chlorophyll biosynthesis (Renau-Morata et al., 2021). The enrichment in genes related to the cyclic electron flow (CEF) under N deprivation conditions may function to readjust the balance between ATP/NADPH for cellular metabolism. A similar response to N deficiency has been reported for *Chlamydomonas reinhardtii* (Saroussi et al., 2016).

Understanding N deficiency-responsive photosynthetic genes, particularly their transcriptional regulation, might provide a base to improve plant N deficiency tolerance as well as photoprotection capacity under stress conditions. Nevertheless, one cannot forget that photosynthesis regulation is dependent on the nitrogen status, as well as on the carbon status, of the plant.

3.6 Other adverse environmental conditions

Many abiotic stresses, such as drought, heat, salinity or high light, are known to induce photoinhibition and reduce photosynthesis efficiency. These stresses have a primary effect on photosynthesis disrupting different photosynthetic components, such as stomatal conductance, electron transport, PSI and PSII, ATP-generating system and CBB cycle. A rapid response of the plants is key for their survival in a fluctuating environment, and plants respond to stress by a coordinate chloroplast and nuclear-encoded gene expression. Trying to reduce photoinhibition and avoid cell death, most genes involved in photosynthesis are downregulated in response to stress. However, depending on the plant, the stress and its severity, photosynthetic gene regulation can vary enormously.

Regarding water deficit, it is well known that most genes involved in photosynthesis are downregulated in response to drought. This has been shown in different C_3 crops, such as barley (Daszkowska-Golec et al., 2019; Ozturk et al., 2002), rice (Degenkolbe et al., 2009; Wang et al., 2011) and purging nut (*Jatropha curcas* L.) (Zhang et al., 2015), among others. Plants exposed to drought show downregulation mainly in genes involved in LHC proteins PSI, PSII and F-type ATPase, but genes encoding RbcS, Rca, phosphoglycerate kinase (PGK), glyceraldehyde-3-phosphate dehydrogenase (GAPA), phosphoribulokinase (PRK), triosephosphate isomerase (TPI) and fructose-1,6-bisphosphatase have also been shown to be downregulated. Besides identifying the photosynthesis-related genes regulated by drought in barley, Daszkowska-Golec et al. (2019) also unveiled 17 possible regulatory interactions between those genes and their TFs. This study provides a very interesting list of candidate genes for future genetic research and improvement of drought tolerance by targeting photosynthesis. In addition to the transcriptional regulation in response to drought, it was reported that positive selection occurring mainly in photosynthetic genes may be involved in drought tolerance improvement (Zhang et al., 2016). This study provides a group of functional markers and candidate genes associated with photosynthesis that might be very useful for future genetic research and improvement of drought tolerance.

Analyzing the transcriptomic response of different C_3 plants to heat stress, we observe that many photosynthetic genes encoding LHCP, PSII, PSI, redox chain and CBB cycle proteins are differentially regulated. In contrast to other stresses, under which most photosynthesis-associated genes are downregulated, it has been reported that high temperature can also induce the expression of many of these genes. For instance, in poplar, among the 56 photosynthesis-related genes differentially regulated under heat stress, 21 were upregulated (Song et al., 2014), and in wheat, under short heat stress, photosynthesis was among the most enriched biological processes in upregulated genes (Arenas-M et al., 2022). However, it is important to notice that different transcript profiles can be observed for plants subjected to short or long heat stress. For instance, 41 rice genes encoding LHC, PSI and PSII and oxygen-evolving enhancer proteins were induced by short heat stress, but repressed by long heat stress (Liu et al., 2020b). The poplar genes encoding PetA, PetM, PetB and

AtpA are also upregulated by short heat stress and downregulated by long heat stress (Song et al., 2014). The gene upregulation observed under short heat stress may be a way to counteract thylakoid membrane leakiness and provide protection against irreversible damage. The electron transport around PSI, which may be improved by the upregulation of *Pet*A, *Pet*M and *Pet*B, is known to provide acclimatory plasticity to the photosynthetic machinery under adverse environmental conditions (Suorsa, 2015). The specific photosynthesis-related genes that are upregulated by heat stress can differ among species, but *Rca* is induced by heat stress in many different species (Arenas-M et al., 2022; Chaudhary et al., 2021; Rangan et al., 2020; Song et al., 2014), and its overexpression alone (Chaudhary et al., 2021) or in combination with Rubisco can improve heat tolerance (Qu et al., 2021).

It has been well documented that photosynthesis is highly susceptible to salt stress and that plant responses to this stress involve transcriptional regulation of the photosynthesis-related genes. However, the information regarding the transcriptional responses of these genes to high salinity is not straightforward to interpret, as different plant species and particularly many different stress conditions (duration of the treatment and concentration of NaCl) have been used. Nevertheless, most reports show that most genes encoding the LHCs as well as components of the PSII, PSI, Cytb6f and ATP synthase are downregulated under salt stress conditions in different plant species (Amirbakhtiar et al., 2021; Duarte-Delgado et al., 2020; Song et al., 2020; Sui et al., 2015; Zhang et al., 2020a). This is very clear in the study performed on watermelon (Song et al., 2020). There are however exceptions. For instance, in leaves of a salt-tolerant barley landrace subjected to salt stress for 8 h, genes encoding components of PSI and PSII, including PSI assembly protein and PSII reaction center proteins required for stability and/or assembly, PSI P700 chlorophyll apoproteins, PSI iron–sulfur center, PSII D2 protein, and PSII 10 kDa polypeptide essential for photochemical activities were upregulated (Ouertani et al., 2021). This transcript profile might be associated with either the tolerant phenotype of the barley landrace investigated or the specificities of the treatment applied. Specific expression patterns have been associated with the level of tolerance of the plant. In sorghum, genes encoding photosystem and electron transport components were less affected by salt stress in a tolerant compared to a salt-sensitive variety (Sui et al., 2015). In wheat, among the 101 photosynthesis-related genes downregulated in a salt-susceptible variety, only 11 are salt-responsive in a tolerant variety (Duarte-Delgado et al., 2020). This suggests that an appropriated regulation of the photosynthesis-related genes may lead to improved salt stress tolerance. The effect of salt stress on CBB cycle encoding genes is very scarce, maybe due to the slight transcriptional changes caused by salt stress that are normally observed for genes encoding proteins such as Rca or Rubisco. Nevertheless, Bayramov and Guliyev (2014) showed that Rca-α/β ratio is upregulated at a post-transcriptional level in *Brachypodium distachyon* plants under salinity stress and, in barley, Rubisco genes were shown to be downregulated in a salt-sensitive but not in salt-tolerant variety (Nefissi Ouertani et al., 2022).

As expected, light stress is well known to modulate photosynthesis efficiency and the physiological plant responses to light are well documented, mainly the responses to high light (Shi et al., 2022; Szymańska et al., 2017). However, knowledge regarding the genetic response of plants to different light intensities is still scarce. It was reported that transcription of *psb*A (encoding the D1 protein of PSII) plays an important role in high-light stress responses (Zhang et al., 2020b). In tomato, *psb*A is known to be regulated by WHIRLY1 (Zhuang et al., 2019), a TF recently shown to be required for the acclimation of barley leaves to high light (Saeid Nia et al., 2022). In *Arabidopsis*, it was shown that most nuclear genes encoding the core LHCPI, LHCPII, PSI, PSII, as well as CBB cycle enzymes, are repressed by high light, indicating an adjustment in the light-harvesting system and reaction centers to capture less light energy for photosynthesis and thus avoid photoinhibition (Huang et al., 2019b). The same study shows that among the few genes upregulated by high light, there is *PsbS*, which encodes a protein recognized in higher plants as an important component in dissipating excess light energy via regulation of non-photochemical quenching. In addition to high light, low light is known to reduce photosynthesis efficiency and growth in different plants, such as rice (Venkateswarlu, 1977). However, there are plants, such as wheat and barley, which acclimate to low-irradiance conditions to keep or even increase yield (Arenas-Corraliza et al., 2019). Barley showed a reduction of the carotenoids/chlorophyll ratio under low irradiance compared to full-light conditions and adjusted the chlorophyll a/b ratio, PSII quantum efficiency, electron transport rate and non-photochemical quenching to shade conditions. It has been suggested that low-light tolerance can be associated with an accelerated expression of genes encoding chlorophyll biosynthesis, LHC and PSII, and PSI proteins (Jing et al., 2022). Upregulation of the SBPase encoding gene has also been associated with rice varieties more tolerant to low light (Kumar et al., 2020; Sekhar et al., 2019). Nevertheless, a better understanding of the connection between photosynthesis-related gene expression and light responses is crucial for breeding (low or high) light stress-tolerant crops better able to cope with the rapidly changing environment.

In addition to abiotic stress, biotic damage is known to reduce photosynthesis capacity (Barón et al., 2011) and it was shown that biotic stress globally downregulates photosynthesis-related genes (Bilgin et al., 2010). When analyzing 22 different forms of biotic stress and 8 different plant species, it was determined that 84% of the genes encoding chloroplast-targeted proteins were downregulated by the biotic damage (Bilgin et al., 2010).

Genes coding for proteins in LHC, PSI, PSII, ATP synthase and CBB cycle were downregulated by biotic damage. In contrast to the light reactions-associated genes, only slightly more than half of the genes coding for CBB cycle enzymes were downregulated following biotic stress. This may result from the fact that some CBB cycle enzymes have a dual role in CBB cycle and other metabolic pathways. Genes encoding enzymes unique to the CBB cycle, such as Rubisco, Rca, PRK, CP12 and SBPase, are downregulated following biotic damage (Bilgin et al., 2010). Noteworthy, the majority of the photosynthesis-related genes downregulated by biotic stress shared common regulatory elements, such as G-box (CACGTG), T-box (ACTTTG) and SORLIP (GCCAC) motifs (Bilgin et al., 2010). This suggests a universal and highly coordinated transcriptional response to biotic stress to modulate photosynthesis. The question is whether this is an adaptative response and how it is regulated. This response allows plants to invest resources in defense without debilitating near-term photosynthetic losses (Kangasjärvi et al., 2014), but it has also been shown that chloroplast-targeted effectors can reprogram transcription of photosynthesis-related genes to destabilize PSII and thus decrease ROS production (Zabala et al., 2015). A better understanding of the role of the photosynthesis-related genes regulated by biotic damage in crop plants will facilitate the development of their improvement.

4 Photosynthetic gene regulators

Understanding the transcriptional regulators of photosynthesis-associated gene expression has been proposed to be essential to increase plant photosynthetic efficiency (Batista-Silva et al., 2020). To date, most knowledge regarding the key regulators of the photosynthesis-associated genes is restricted to the model plant *A. thaliana*. In this section, we aim to focus mainly on their homologues in rice and tomato, economically important models for other C_3 crops, but other crops may also be discussed.

4.1 Major regulators of the photosynthetic nuclear-encoded genes

Little is known about PhANG regulators, but some TFs have already been described for this function (Rossi et al., 2015; Wang et al., 2017). In this section, we aim to discuss the major regulators of photosynthesis-associated genes for *Arabidopsis* and rice, particularly the ones identified in Wang et al. (2017), and include as well their tomato orthologues. As shown in Table 2, the number of orthologues for these regulators is different among species and some are missing for tomato.

Table 2 Major regulators of photosynthesis in *Arabidopsis*, rice and tomato

Major regulators of photosynthesis				
Arabidopsis thaliana		*Oryza sativa*		*Solanum lycopersium*
Gene ID	Protein	Gene ID	Protein	Gene ID
AT5G11260	HY5	LOC_Os06g39960	OsbZIP48	Solyc08g061130
AT3G17609	HYH	LOC_Os02g10860	OsbZIP18	
		LOC_Os01g07880	OsbZIP01/OsRE1	
AT2G20570	GLK1	LOC_Os06g24070	OsGLK1	Solyc07g053630
AT5G44190	GLK2	LOC_Os01g13740	OsGLK2	Solyc10g008160
AT4G26150 AT5G56860	CGA1	LOC_Os02g12790	OsCGA1	Solyc07g038160
	GNC	LOC_Os06g37450	OsGATA16	
AT1G01720	ATAF1	LOC_Os05g32830	not expressed	No homologue
		LOC_Os01g66120	OsNAC6/ SNAC2	
AT5G39610	ORE1	LOC_Os12g41680	OsNAC60	Solyc02g088180
		LOC_Os02g36880	OMTN1	
		LOC_Os04g38720	OsNAC2 / OsORE1	
AT4G23750	CRF2	LOC_Os01g46870	OsERF54	Solyc08g081960
AT3G13960	GRF5	LOC_Os02g53690	OsGRF1/ Rhd1	No homologue
		LOC_Os06g02560	OsGRF5	

Arabidopsis and rice genes encoding these regulators were described by Wang et al. (2017) and tomato orthologues were identified using Shoot.bio (Emms and Kelly, 2022) interface. Two *Arabidopsis* genes showed no orthologous in the tomato genome.

4.1.1 HY5/HYH

Elongated hypocotyl 5 (HY5) was first described in *Arabidopsis* as a key regulator of cell proliferation, cell elongation and chloroplast development (Oyama et al., 1997). The *hy5* mutant presented defects in photomorphogenesis, with longer hypocotyls under light, and lower chloroplast number (Oyama et al., 1997). HY5 has also been characterized as a direct regulator of genes encoding LHC proteins, such as *Lhcb1.3* and *Lhca4* (Catalá et al., 2011; Toledo-Ortiz et al., 2014), and other important regulators of chloroplast development and function, namely *RbcS-1A*, *GLK2* and *genomes uncoupled 1* (*GUN1*) (Chattopadhyay et al., 1998; Kobayashi et al., 2012; Toledo-Ortiz et al., 2014). The function of HY5 was shown to be partially overlapping with a close homologue, HY5-homologue (HYH) (Holm et al., 2002). Contrary to HY5, which is under phytochrome regulation, HYH appears to function downstream of cryptochromes.

Regarding crop plants, three HY5 homologues (OsbZIP18, OsbZIP49 and OsbZIP01/OsRE1) were identified and characterized in rice (Sun et al., 2022). Similar to HY5, OsZIP48 seems to be a positive regulator of chlorophyll biosynthesis as its overexpression leads to an increased chlorophyll content (Burman et al., 2018; Sun et al., 2022). OsbZIP18 was found to positively regulate enzymes involved in the branched-chain amino acids (Sun et al., 2022) and OsRE1 was shown to promote rice heading date by regulation of the *Early heading date 1* (*Ehd1*) gene (Chai et al., 2021); however, the association of these regulators with photosynthesis-associated genes is yet to be investigated. In tomato, only one HY5 homologue (SlHY5) was identified (Table 2). SlHY5 has a similar role to HY5 by repressing hypocotyl growth under light (Qiu et al., 2019) and plays an important role in anthocyanin biosynthesis and fruit ripening (Qiu et al., 2019; Wang et al., 2021b); nevertheless, no function on photosynthesis regulation has been reported.

4.1.2 GLK1/GLK2

The first characterized member of the *GLK* gene family was *GOLDEN2* (*G2*) in maize (Langdale and Kidner, 1994). *G2* mutant plants showed a pale green phenotype and have smaller chloroplast than wild-type plants. The function of G2 was shown to be important for chloroplast development in C_3-like and bundle sheath chloroplasts, with its homologue ZmGLK1 being proposed to regulate the chloroplast development in mesophyll cells. In *Arabidopsis*, two orthologues have been identified, *GLK1* and *GLK2*. Phylogenetic analysis shows that the duplication event occurred after monocots and dicots divergence, indicating that different mechanisms of action for the paralogue pair might be observed in *Arabidopsis* and rice, for instance (Fitter et al., 2002). Single mutants for each *GLK* gene showed no phenotype in *Arabidopsis* in most photosynthetic tissues, but double mutants revealed a pale green phenotype, which indicates some level of functional redundancy between the paralogue genes. Detailed analysis revealed that the chloroplasts of the double mutant display reduced thylakoid grana and downregulation of several photosynthesis-associated genes (Fitter et al., 2002; Waters et al., 2009).

In rice, it was first shown that overexpression of either *OsGLK1* or *OsGLK2* induces chloroplast differentiation in embryogenic calli, which display developed chloroplasts, high chlorophyll levels and are capable of autotrophic growth (Nakamura et al., 2009). The rice lines overexpressing *OsGLK1* showed ectopic chloroplast development in the vascular bundle and bundle sheath cells and a large number of PhANGs are upregulated (Nakamura et al., 2009). It has also been shown that the expression of the maize *GLK1* and *GLK2* can improve the photosynthesis and yield of transgenic rice plants (Li et al., 2020; Yeh et al., 2022).

In tomato, the most studied GLK homologue is SlGLK2, which is part of the *uniform ripening* allele that has been selected in most commercially available tomato varieties. Though two *GLK* are present in tomato, only *SlGLK2* is expressed in the fruit, where it is responsible for the control of chlorophyll levels (Nguyen et al., 2014; Powell et al., 2012). It was shown that overexpression of *SlGLK2* increases the photosynthetic rate in the fruit, contributing to increased sugar content (Nguyen et al., 2014; Powell et al., 2012). Although the function of SlGLK1 has not been fully characterized, its encoding gene has been described as a direct target of SlARF6A, alongside *RbcS*, *CAB1* and *CAB2*. Misexpression of *SlARF6A* was shown to alter chloroplast number, on both fruits and leaves, as well as altered photosynthetic rate (Yuan et al., 2019). GLKs have also been described in other crops. In barley, HvGLK2 mutants show impaired chloroplast development in the spike, which leads to a severe reduction in spike photosynthesis and seed yield (Taketa et al., 2021).

In conclusion, the role of GLKs in regulating chloroplast development and function has been highly conserved throughout plant species and is an interesting target for crop improvement, being able to raise photosynthetic efficiency in both leaves and fruits.

4.1.3 CGA1/GNC

In *Arabidopsis*, two different GATA transcription factors, GATA nitrate-inducible carbon-metabolism-involved (GNC) and cytokinin-responsive GATA1 (CGA1), were shown to regulate chloroplast development. The double mutant *gnc cga1* shows a chlorophyll-deficient phenotype and smaller chloroplasts in the hypocotyl. It was shown that each of these genes is sufficient to induce chloroplast biogenesis. Ectopic overexpression of either *GNC* or *CGA1* induces the biogenesis of chloroplasts in the hypocotyl cortex and root pericycle root cells (Chiang et al., 2012). It has been shown that these two transcription factors and the *GLK* pair, described earlier, have redundant functions regarding the regulation of chloroplast development. The double mutant for each pair, as well as the quadruple mutant, shows severe impairment in chloroplast development, and the quadruple mutant shows the most severe impact but still assembles chloroplasts (Zubo et al., 2018).

In rice, two orthologues for the *CGA1* and *GNC* pair have been identified (*OsCGA1* and *OsGATA16*), but only OsCGA1 was shown to regulate chloroplast development (Hudson et al., 2013). Misexpression of *OsCGA1* is able to alter chloroplast development, either increasing chloroplast number when it is overexpressed or reducing when it is silenced (Hudson et al., 2013). Interestingly, when *OsCGA1* expression is driven to the rice bundle sheath cells, it induces chloroplast development in these cells (Lee et al., 2021). The other rice CGA1/GNC homologue, OsGATA16, has been mainly associated with abiotic stress tolerance traits (Le et al., 2021; Zhang et al., 2021), but its role regulating chlorophyll content was also observed (Zhao et al., 2019).

In tomato, only one CGA1/GNC homologue was identified, but no information could be found regarding its role regulating photosynthesis-associated genes.

4.1.4 *ATAF1*

A. thaliana activating factor 1 (ATAF1) was first described as being involved in stress tolerance response in *Arabidopsis* (Jensen et al., 2008; Lu et al., 2006). It was shown that ATAF1 can alter the levels of ABA in plants and it is regulated by different stress signals, including H_2O_2 (Allu et al., 2014; Balazadeh et al., 2010). However, it was also shown that ATAF1 is able to regulate directly *GLK1* and *ORESARA 1 (ORE1)* gene expression, which are known to regulate chloroplast maintenance and senescence. The role of ATAF1 was described as a bridge between stress signals and chloroplast maintenance, acting upstream of chloroplast senescence regulators. *ATAF1* expression is regulated by both ABA and H_2O_2 levels, and when activated under stress, it acts to repress *GLK1* expression and activate *ORE1*, thus shifting the physiological balance to senescence (Garapati et al., 2015).

The rice genome contains two *ATAF1* orthologues (Table 2), but only one seems to be transcribed (*OsNAC6/SNAC2*). Similar to what has been shown in *Arabidopsis*, OsNAC6 was shown to be involved in the response to different abiotic stresses (Nakashima et al., 2007; Ohnishi et al., 2005; Takasaki et al., 2010) and its overexpression leads to increased cold and salt tolerance (Hu et al., 2008; Rachmat et al., 2014). However, in contrast to *Arabidopsis*, in rice no association with chloroplast maintenance has been reported. In tomato, no *ATAF1* orthologue could be identified.

4.1.5 ORE1

ORE1 was described as being involved in the regulation of senescence, with *ore1 Arabidopsis* mutants showing a delayed senescence phenotype (Woo et al., 2004). It was found that ORE1 is able to interact with both GLK1 and GLK2, antagonizing their function and shifting the process from chloroplast maintenance, promoted by GLKs, to chloroplast senescence, promoted by ORE1 (Rauf et al., 2013). ORE1 was also shown to directly bind and activate chlorophyll catabolic genes as well as ethylene biosynthesis genes (Qiu et al., 2015). Together with EIN3, it forms a forward-feedback loop to regulate chlorophyll levels in an ethylene-dependent fashion.

The three ORE1 homologues found in rice (OsNAC60, OMTN1, OsORE1/OsNAC2) have been reported to be targeted by miRNA164, which is involved in abiotic and biotic stress responses (Fang et al., 2014; Jia et al., 2020; Wang et al., 2018, 2021a; Zhong et al., 2020). OsNAC2 has been shown to be involved in abiotic stress responses and different developmental processes (Chen et al., 2015; Jiang et al., 2019; Mao et al., 2007, 2020). Regarding the regulation of photosynthesis-related traits, it was found that OsNAC2 can act similarly to ORE1 by upregulating genes involved in chlorophyll catabolism and inducing leaf senescence (Mao et al., 2017). In tomato, a single ORE1 homologue was identified and its role in the regulation of senescence has also been studied. SlORE1 was shown to activate senescence and can physically interact with SlGLKs, and, similar to rice, it is also regulated by miRNA164. Interestingly, knockdown plants for SlORE1 were shown to have higher carbon assimilation and improved fruit quality (Lira et al., 2017).

The phenotypes reported in both rice and tomato show that modulation of ORE1 homologues may be interesting for photosynthesis improvement, since in both plants they seem to be able to regulate leaf chlorophyll content. Further studies on how these genes can be manipulated may bring further knowledge for crop improvement.

4.1.6 CRF2

Cytokinin-responsive factors (CRF) have been shown to be involved in leaf senescence, but it was found that there is a large redundancy between them. Only multiple *CRF* mutants show delayed leaf senescence, with quadruple *crf* mutants being embryo lethal (Raines et al., 2016). CRF2 has been shown to be involved in regulating chloroplast division by the activation of *plastid division 2* (PDV2) promoting chloroplast division (Okazaki et al., 2009). It was recently shown that CRF2 is also involved in the salt stress response by regulating chlorophyll levels and photosynthesis-associated genes (Keshishian et al., 2022). In both tomato and rice, a single homologue for CRF2 can be identified (Table 2), but a link between their function and the regulation of photosynthesis-associated genes has not yet been investigated.

4.1.7 GRF5

Growth-regulating factors (GRF) were first described as being regulators of leaf size and morphology (Kim et al., 2003). GRF5 was shown to regulate leaf size, working in coordination with *angustifolia* 3 (*AN3*), by promoting cell size (Horiguchi et al., 2005). Later, it was reported that GRF5 could also regulate chloroplast number by promoting chloroplast division, which translated into altered photosynthesis efficiency in *GRF5* overexpressing plants (Vercruyssen et al., 2015). In addition, GRF5 was shown to alter leaf senescence, increasing leaf lifetime when overexpressed (Vercruyssen et al., 2015). Rice genome harbors two *GRF5* orthologues *(OsGRF1/rdh1* and *OsGRF5*). OsGRF1 has been reported to regulate heading date and grain size (Luo et al., 2005) as well as leaf morphology and shoot growth (Lu et al., 2020; Yang et al., 2021), but no direct link between OsGRF1 and photosynthesis has been reported to date. OsGRF5 has been associated with drought tolerance (Yu et al., 2012), but its possible role in photosynthesis regulation in rice is yet to be identified. In tomato, no GRF5 orthologue was found.

4.2 Nuclear-encoded sigma factors

The prokaryotic origin of chloroplast makes it to retain some characteristics of prokaryotic transcription regulation, namely its regulation by sigma factors (SIG). The existence of nuclear-encoded SIGs allows the nuclear control of chloroplast gene expression. *Arabidopsis* genome encodes six SIGs and most of them have been characterized (Chi et al., 2015). It has been shown that SIGs can have both the redundant and specific role regulating plastid gene expression. Both rice and tomato genome contain orthologues for most of the *Arabidopsis* SIG encoding genes (Table 3). In tomato, there are two orthologues for the *Arabidopsis SIG1* gene and rice presents two homologues for *Arabidopsis SIG2* gene. In this section we summarize the role of each *Arabidopsis* SIG regulating photosynthesis-associated genes and compare it to what is known in both tomato and rice. Despite our efforts, no information was found regarding the role of tomato SIGs regulating photosynthesis-associated genes.

Table 3 *Arabidopsis* SIGs and their orthologues in both rice and tomato

Nuclear-encoded sigma factors				
A. thaliana		*Oryza sativa*		*Solanum lycopersium*
Gene ID	Protein	Gene ID	Protein	Gene ID
AT1G64860	SIG1	LOC_Os08g06630	OsSIG1	Solyc06g072710
				Solyc03g097320
AT1G08540	SIG2	LOC_Os11g26160	SIG2A	Solyc01g081490
		LOC_Os03g16430	SIG2B	
AT3G53920	SIG3	LOC_Os05g51150	-	Solyc08g065970
AT5G13730	SGI4	na	-	Solyc01g087690
AT5G24120	SIG5	LOC_Os05g50930	OsSIG5	Solyc03g007370
AT2G36990	SIG6	LOC_Os08g14450	OsSIG6	Solyc09g008040

Orthologues of *Arabidopsis* SIG in rice are the ones described by Wang et al. (2017). Tomato homologues were identified using Shoot.bio (Emms and Kelly, 2022).

4.2.1 *SIG1*

In *Arabidopsis*, SIG1 was found to directly bind the promotor of *rbcL* and *psbA* operons (Hakimi et al., 2000; Homann and Link, 2003). The activity of SIG1 has been shown to be regulated by phosphorylation, dependent on the redox state of the plastoquinone pool. The changes in phosphorylation of SIG1 were shown to alter the transcription levels of *psbA* and *psaA* as well as the *psaA/psbA* transcript ratio, being proposed as having a role in regulating the stoichiometry of photosystems in response to different light intensities (Shimizu et al., 2010). Chromatin immunoprecipitation (ChIP) allowed the identification of additional targets of SIG1, showing its role regulating other photosynthesis-associated genes (Hanaoka et al., 2012).

In rice, a knockout mutant for *OsSIG1* showed reduced chlorophyll content and altered levels of PSI core proteins, with a striking reduction of *psaB* transcript levels, leading to a reduced electron transport rate between PSII and PSI (Tozawa et al., 2007). The effects of *OsSIG1* knockout were more pronounced when plants were grown under low-light and low-temperature conditions (Du et al., 2021). Gene expression analysis showed that *OsSIG1* is only expressed later during leaf development, when compared with *OsSIG2A* and *OsSIG2B*, indicating that different SIG proteins may have different roles during leaf development (Kasai et al., 2004).

4.2.2 *SIG2*

Downregulation or knockout of *SIG2* in *Arabidopsis* leads to a chlorophyll deficiency, phenotype that was proposed to be linked to the repression of several tRNAs, and downregulation of *psbA*, *psaJ* and *psbD* (Kanamaru et al., 2001; Privat et al., 2003). Interestingly, the pale green phenotype observed was restricted to the cotyledons, with full green leaves being present in older plants. By the analysis of gene and protein expression of SIG1, SIG2 and SIG3, it was observed that *SIG2* transcripts were present at a high level in the cotyledons but downregulated later during development. It was proposed that other SIGs, namely SIG3, could act in later developmental stages to regulate chloroplast development (Kanamaru et al., 2001; Privat et al., 2003). Interestingly, rice *SIG2* orthologues (*OsSIG2A* and *OsSIG2B*) were also found to be expressed earlier than *OsSIG1* (Kasai et al., 2004), indicating that the same stepwise function of SIG proteins may be conserved among different species. Moreover, it was shown that OsSIG2A is essential for chloroplast development under low temperatures (23°C), with *OsSIG2A* silencing lines showing a severe reduction in chlorophyll content (Yu et al., 2019).

4.2.3 SIG3 and SIG4

In *Arabidopsis*, both *sig3* and *sig4* mutants showed no clear phenotype regarding chloroplast content and/or photosynthesis efficiency. The only differences observed were in terms of transcript levels for *psbN*, *atpH* and *psbB* in *sig3* mutant, and for *NADH dehydrogenase F* (*ndhF*) in *sig4* mutant (Favory et al., 2005; Zghidi et al., 2007). Although SIG3 and SIG4 were shown to regulate several genes, they do not appear to have a crucial role in chloroplast development/function. In rice, no information regarding the function of OsSIG3 was found and there appears to be no homologue for SIG4.

4.2.4 SIG5

In *Arabidopsis*, SIG5 has been associated with the blue light-responsive regulation of *psbD-psbC* operon, which encodes D2 reaction center protein as well as the chlorophyll-binding antenna protein C43 of PSII. Although it is known that *SIG5* expression is induced by blue light, as well as by multiple stress signals, the signal transduction mechanisms remain unknown (Nagashima et al., 2004; Tsunoyama et al., 2002, 2004). The induction of *SIG5* under multiple stresses was shown to be important for the recovery of damaged PSII reactions center, being delayed in *sig5* mutant plants (Nagashima et al., 2004).

Apart from the blue light regulation, *SIG5* was also shown to be under the influence of the circadian clock. It was shown that SIG5 can regulate the circadian oscillation of plastid genes, possibly being involved in the integration of both light and circadian information into the plastid genome, in both *Arabidopsis* and *Marchantia polymorpha* (Ichikawa et al., 2008; Noordally et al., 2013). In rice, *OsSIG5* has been described as a gene that is expressed in a tissue-specific and light-dependent manner (Kubota et al., 2007). Given that SIG5 circadian regulation has been observed in both higher plants and mosses, it is expected that its function has been conserved during evolution. Nevertheless, whether SIG5 from other crops has similar roles needs to be investigated.

4.2.5 SIG6

The *Arabidopsis sig6* mutant shows a chlorophyll-deficient phenotype that appears restricted to the cotyledon stage (Ishizaki et al., 2005; Loschelder et al., 2006). In the early developmental stages, plastid-encoded RNA polymerase (PEP)-dependent transcripts were found to be downregulated in *sig6* mutant but recover to wild-type levels in the seedling stage (Ishizaki et al., 2005; Loschelder et al., 2006). Some plastid operons (*atpB/E, ndhC/psbG/ndhJ*) were found to be downregulated until later stages of development, namely in the rossette stage, although no phenotype was found (Schweer et al., 2006). This indicates that SIG6 plays an important role during the early stages of plant development, namely in the cotyledon, but its role is compensated by other factor(s) in later developmental stages.

Chloroplast and nucleus gene regulation can work in both directions. Although most of the results for sigma factors involve anterograde signaling (nucleus to chloroplast), sigma factors have also been shown to play a role in retrograde signaling (chloroplast to nucleus). The involvement in retrograde signaling of SIG6 was hypothesized from the reduction of PhANGs differential expressed observed in the *sig6* mutant. Detailed analysis showed that the role of SIG6 in retrograde signaling is dependent on GUN1, given that *sig6* mutant phenotype is rescued by the knockout of *GUN1* (Woodson et al., 2013). OsSIG6, similar to OsSIG5, was also shown to be regulated by both light- and tissue-specific expression (Kubota et al., 2007), but its functional characterization as the regulator of PhAPG is lacking.

5 Use of transcriptional regulators to regulate photosynthesis in C_3 crops in the field

One of the few examples of a major photosynthesis regulator in rice is the *higher yield rice* (*HYR*) AP2/ERF transcription factor. The analysis of plants overexpressing HYR showed its capacity to increase rice photosynthesis efficiency, which was more pronounced in higher CO_2 concentrations and high-light intensities (Ambavaram et al., 2014). Moreover, HYR expression led to an increase in chlorophyll content and starch grain size in the chloroplasts, namely in the parenchyma cell of the flag leaf that serves as important carbohydrate storage for the grain. HYR was shown to increase plant productivity under optimal and stress conditions, with HYR overexpressing plants being more tolerant to drought stress than WT plants. It was further shown that HYR binds to the promoters of genes either involved in photosynthesis, namely genes encoding PSII, LHC2 and ferredoxin 3 proteins, or encoding TFs known to regulate central carbon metabolism genes, namely ARF1 and GASR2. HYR was shown to work as an activator of gene expression for all these genes, although it works as a repressor of the TF WRKY72, which regulates root growth and development (Ambavaram et al., 2014).

Negative regulator of photosynthesis 1 (NRP1), a bHLH TF, was identified as being negatively correlated with photosynthesis-associated genes (Chen et al., 2021). The knockout of *NRP1* led to an upregulation of these genes, namely PS components and LHCP complexes, as well as the CBB cycle. This upregulation of photosynthesis genes was accompanied by an increase in photosynthetic rate under field conditions in the *nrp1* mutant lines and significantly decreased in *NRP1* overexpressing lines. The changes in photosynthesis observed in the lines misexpressing *NRP1* lead to changes in plant biomass. In conclusion, NRP1 seems to function as a negative regulator of photosynthetic genes, leading to a reduction of photosynthesis rate when it is present. The mechanism by which it acts on these genes remains unknown, analysis of its bHLH domain showed that NRP1 has a novel bHLH domain, which is not yet described in terms of DNA-binding preferences, but it seems to work as a transcriptional activator (Chen et al., 2021). Further understanding of NRP1 targets could clarify its function and serve to uncover novel targets for the improvement of photosynthesis in rice.

6 How can the expression of the photosynthesis-associated genes be modulated?

In plants, photosynthesis-associated genes are encoded by two different genomes, nuclear and chloroplast, which offer challenges but also possibilities when it comes to the transformation of crop plants. Nuclear genome transformation has been extensively used in different plants to characterize gene function and/or to improve photosynthetic efficiency (Driever et al., 2017; Ermakova et al., 2019; Feng et al., 2007; Kromdijk et al., 2016; Qu et al., 2021). However, for photosynthesis studies and engineering, chloroplast transformation offers several advantages compared to nuclear transformation. First, in angiosperms, chloroplasts are maternally inherited, limiting the outcrossing of modified plants. Second, plastid-encoded proteins accumulate at much higher levels than those encoded in the nucleus. Third, gene editing and gene insertion are precise and less susceptible to silencing (Kumar and Ling, 2021; Olejniczak et al., 2016).

Given the previously-mentioned advantages, why is chloroplast transformation not extensively used in crops? This is mainly due to the several technical and plant-specific constraints that limit it. Biolistic transformation is well known to be the prime method of chloroplast transformation, but the selection of transplastomic events and the regeneration represent a major obstacle to expanding it to different species (Ruf et al., 2019). Polyethylene glycol-mediated transformation of protoplasts can also be used, but it presents the same constraints in terms of selection and regeneration and is more laborious (Bock, 2014). The most common selection marker in transplastomics is the *aadA* gene from *Escherichia* coli that codes for an aminoglycoside 3′-adenylyl transferase, which grants resistance to spectinomycin, an inhibitor of chloroplast protein synthesis by chloroplasts' 70S ribosomes (Maliga and Bock, 2011). Although the *aadA* transgene and spectinomycin combination, as well as the regeneration, proved efficient in some crop species (e.g. tomato and potato), the proper selection and regeneration protocols for chloroplast-transformed cells is still the biggest bottleneck for most staple crops, including cereals (Bock, 2014).

Chloroplasts retained part of the prokaryotic genome/gene structure and expression from the ancestral endosymbiont. Most plastid genes are organized in operons, clusters of several genes co-expressed from the same mRNA. Although the organization of the genome resembles the prokaryotic model, noted by the small intergenic region, plastid gene expression gained other features such as RNA splicing of the polycistronic mRNAs (Barkan, 2011). A good example is the splicing of the *psb*B gene cluster mRNA that co-expresses five genes involved in PSII and cyt b6/f complex (Westhoff and Herrmann, 1988). Another example is the splicing of the *ndhH-D* operon, which leads to a higher accumulation of the monocistronic *psaC* mRNA (encoding the PsaC protein of photosystem I) in comparison to the six *ndh* genes (Del Campo et al., 2002). These examples suggest that chloroplast gene expression is predominately regulated at the post-transcriptional level.

The possibility of stacking genes in operons for modifying existing or implementing novel metabolic pathways (Bock, 2013, 2022) offers possibilities in photosynthesis engineering. Multiple genes are often needed to remodel photosynthesis-related pathways or novel alternative pathways, for instance the glycolate metabolism photorespiration bypass (South et al., 2019) or the improved non-photochemical quenching (NPQ) relaxation (Kromdijk et al., 2016). Gene stacking in chloroplast genome may solve problems associated with gene expression level, silencing, transgene insertion, and outcrossing of the modified plants. Homologous recombination between two or more vectors and the plastid genome is possible with a single transformation (Herz et al., 2005), and it might be necessary when stacking many genes and or when altering native photosynthesis-associated genes that are encoded in different operons.

Since some photosynthesis-associated nuclear-encoded genes may not be expressed at adequate (high) levels, transplastomics may also be a promising strategy to express these genes in the chloroplast. A good example of this is the work reported by Martin-Avila et al. (2020), who were able to redesign the plastid *rbcL* locus to combine with the nuclear-encoded *RbcS*. This strategy allows the expression of the two subunits of the Rubisco enzyme in the chloroplast, leading to a higher Rubisco production.

7 Conclusion and future trends in research

Understanding how photosynthesis-associated genes are regulated is essential when we aim at improving photosynthesis. This chapter summarizes what is known regarding the transcriptional regulation of these genes by different internal and external signals, particularly in crop plants. Particularly during the last decade, the massification of the transcriptomic studies has allowed identifying a number of photosynthesis-associated genes that are regulated at transcriptional levels by different signals (e.g. light, redox state, a/biotic stress, etc.) in *Arabidopsis* and also in a few crop plants. In order to better understand how photosynthesis is regulated and thus design genetic strategies to improve its efficiency, under optimal or suboptimal environmental conditions, more transcriptomics studies need to be performed for different crops and more focused on photosynthetic-associated genes. Nevertheless, there are already many transcriptomic data sets publicly available that can be further explored focusing on photosynthesis regulation. It is also essential to characterize the function of the photosynthesis differential regulated genes in crop plants. CRISPR/Cas9 technology has proved to be a very efficient technique to knockout genes in different crops such as wheat (Li et al., 2021), rice (Chen et al., 2021), barley (Lawrenson and Harwood, 2019) or tomato (Reem and Van Eck, 2019) and is an excellent tool for the functional characterization of those genes. In addition, CRISPR/Cas9-mediated gene targeting (Nishizawa-Yokoi et al., 2020), which allows the precise modification of a target sequence, including the introduction of point mutations, knock-in of a reporter gene and/or swapping of a functional domain, may also be used to either characterize the function or implement different photosynthesis improvement strategies.

A number of main regulators for the photosynthesis-associated nuclear-encoded genes have been reported in *Arabidopsis* and rice (Wang, et al., 2017); however, very little is known regarding their function. In this chapter, we have discussed what is known about the role of these regulators in photosynthesis performance in crops, particularly in rice and tomato. Nevertheless, to better understand their function in regulating photosynthesis, it is essential to investigate what are the target genes of these regulators (using, for instance, a ChiP-seq approach). We have also discussed the role of sigma factors regulating chloroplast-encoded genes. Since little is known regarding their function, it would again be extremely important to characterize their function in crops as well as to identify their target genes.

The literature tells us that very little is known about the transcriptional regulation of PhANGs and PhAPGs. Therefore, it is urgent to know more about the transcript profile of these genes, under optimal and adverse environmental conditions, and also unveil their transcriptional regulators. A deeper understanding of gene regulation will definitely allow a better design of the strategies to improve photosynthesis in crop plants.

Most strategies used so far to improve plant photosynthesis are based on nuclear genome modification. Chloroplast transformation is a very promising technique and has several advantages over nuclear genome modification to improve photosynthesis; however, the limitations to implement it in cereals make it less suitable for this purpose. It is therefore of utmost importance to investigate and overcome these limitations in order to make transplastomics possible and routinely used in different crops.

8 Acknowledgements

This work was supported by EU Horizon-2020 through the project PhotoBoost (GA No. 862127) and by Fundação para a Ciência e a Tecnologia (FCT) through the project GREEN-IT Bioresources for Sustainability R&D (Unit UIDB/04551/2020 and UIDP/04551/2020) and the fellowships PD/BD/128403/2017 (PC) and PD/BD/138096/2018 (GES).

9 Where to look for further information

The following articles provide a good overview of the subject:

- Berry, J. O., Yerramsetty, P., Zielinski, A. M. and Mure, C. M. (2013), Photosynthetic gene expression in higher plants, *Photosynth Res.*, 117(1), 91–120. DOI: 10.1007/s11120-013-9880-8.
- Wang, P., Hendron, R. W. and Kelly, S. (2017), Transcriptional control of photosynthetic capacity: conservation and divergence from Arabidopsis to rice. *New Phytol.*, 216(1), 32–45. DOI: 10.1111/nph.14682.

And the book: Shevela, D., Olof Björn, L. and Govindjee (2019), *Photosynthesis Solar Energy for Life*, World Scientific Publishing Company. ISBN-10: 9813223103 ISBN-13: 978-9813223103.

10 References

Ainsworth, E. A. and Long, S. P. (2005), What have we learned from 15 years of free-air CO2 enrichment (FACE)? A meta-analytic review of the responses of photosynthesis, canopy properties and plant production to rising CO2, *New Phytol.*, 165(2), 351-371. DOI: 10.1111/j.1469-8137.2004.01224.x.

Allu, A. D., Soja, A. M., Wu, A., Szymanski, J. and Balazadeh, S. (2014), Salt stress and senescence: identification of cross-talk regulatory components, *J. Exp. Bot.*, 65(14), 3993–4008. DOI: 10.1093/jxb/eru173.

Ambavaram, M. M. R., Basu, S., Krishnan, A., Ramegowda, V., Batlang, U., Rahman, L., Baisakh, N. and Pereira, A. (2014), Coordinated regulation of photosynthesis in rice increases yield and tolerance to environmental stress, *Nat. Commun.*, 5(1), 5302. DOI: 10.1038/ncomms6302.

Amirbakhtiar, N., Ismaili, A., Ghaffari, M. R., Mirdar Mansuri, R., Sanjari, S. and Shobbar, Z. S. (2021), Transcriptome analysis of bread wheat leaves in response to salt stress, *PLoS ONE*, 16(7), e0254189. DOI: 10.1371/journal.pone.0254189.

Arenas, M. A., Castillo, F. M., Godoy, D., Arenas-M, A., Castillo, F. M., Godoy, D., Canales, J. and Calderini, D. F. (2022), Transcriptomic and physiological response of durum wheat grain to short-term heat stress during early grain filling, *Plants*, 11(1), 59. DOI: 10.3390/plants11010059.

Arenas-Corraliza, M. G., Rolo, V., López-Díaz, M. L. and Moreno, G. (2019), Wheat and barley can increase grain yield in shade through acclimation of physiological and morphological traits in Mediterranean conditions, *Sci. Rep.*, 9(1), 9547. DOI: 10.1038/s41598-019-46027-9.

Azcón-Bieto, J. (1983), Inhibition of photosynthesis by carbohydrates in wheat leaves, *Plant Physiol.*, 73(3), 681–686. DOI: 10.1104/pp.73.3.681.

Balazadeh, S., Wu, A. and Mueller-Roeber, B. (2010), Salt-triggered expression of the ANAC092-dependent senescence regulon in *Arabidopsis thaliana*, *Plant Signal. Behav.*, 5(6), 733–735. DOI: 10.4161/psb.5.6.11694.

Barkan, A. (2011), Expression of plastid genes: organelle-specific elaborations on a prokaryotic scaffold, *Plant Physiol.*, 155(4), 1520–1532. DOI: 10.1104/pp.110.171231.

Barón, M., Flexas, J. and DeLucia, E. H. (2011), Photosynthetic responses to biotic stress, in *Terrestrial Photosynthesis in a Changing Environment a Molecular, Physiological and Ecological Approach*. DOI: 10.1017/CBO9781139051477.026.

Batista-Silva, W., da Fonseca-Pereira, P., Martins, A. O., Zsögön, A., Nunes-Nesi, A. and Araújo, W. L. (2020), Engineering improved photosynthesis in the era of synthetic biology, *Plant Commun.*, 1(2), 100032. DOI: 10.1016/j.xplc.2020.100032.

Bayramov, S. and Guliyev, N. (2014), Changes in RuBisCO Activase gene expression and polypeptide content in *Brachypodium distachyon*, *Plant Physiol. Biochem.*, 81: 61–66. DOI: 10.1016/j.plaphy.2014.01.013.

Berry, J. O., Yerramsetty, P., Zielinski, A. M. and Mure, C. M. (2013), Photosynthetic gene expression in higher plants, *Photosyn. Res.*, 117(1–3), 91–120. DOI: 10.1007/s11120-013-9880-8.

Bilgin, D. D., Zavala, J. A., Zhu, J., Clough, S. J., Ort, D. R. and DeLucia, E. H. (2010), Biotic stress globally downregulates photosynthesis genes, *Plant Cell Environ.*, 33(10), 1597–1613. DOI: 10.1111/j.1365-3040.2010.02167.x.

Bock, R. (2013), Strategies for metabolic pathway engineering with multiple transgenes, *Plant Mol. Biol.*, 83(1–2), 21–31. DOI: 10.1007/s11103-013-0045-0.

Bock, R. (2014), Genetic engineering of the chloroplast: novel tools and new applications, *Curr. Opin. Biotechnol.*, 26, 7–13. DOI: 10.1016/j.copbio.2013.06.004.

Bock, R. (2015), Engineering plastid genomes: methods, tools, and applications in basic research and biotechnology, *Annu. Rev. Plant Biol.*, 66(1), 211–241. DOI: 10.1146/annurev-arplant-050213-040212.

Bock, R. (2022), Transplastomic approaches for metabolic engineering, *Curr. Opin. Plant Biol.*, 66, 102185. DOI: 10.1016/j.pbi.2022.102185.

Buchanan, B. B., Gruissem, W. and Jones, R. L. (Eds.) (2015), *Biochemistry & Molecular Biology of Plants* (2nd edn.), WILEY Blackwell.

Burman, N., Bhatnagar, A. and Khurana, J. P. (2018), OsbZIP48, a HY5 transcription factor ortholog, exerts pleiotropic effects in light-regulated development, *Plant Physiol.*, 176(2), 1262–1285. DOI: 10.1104/pp.17.00478.

Cackett, L., Luginbuehl, L. H., Schreier, T. B., Lopez-Juez, E. and Hibberd, J. M. (2022), Chloroplast development in green plant tissues: the interplay between light, hormone, and transcriptional regulation, *New Phytol.*, 233(5), 2000–2016. DOI: 10.1111/nph.17839.

Catalá, R., Medina, J. and Salinas, J. (2011), Integration of low temperature and light signaling during cold acclimation response in Arabidopsis, *Proc. Natl. Acad. Sci. U. S. A.*, 108(39), 16475–16480. DOI: 10.1073/pnas.1107161108.

Chai, J., Zhu, S., Li, C., Wang, C., Cai, M., Zheng, X., Zhou, L., Zhang, H., Sheng, P., Wu, M., Jin, X., Cheng, Z., Zhang, X., Lei, C., Ren, Y., Lin, Q., Zhou, S., Guo, X., Wang, J., Zhao, Z. and Wan, J. (2021), OsRE1 interacts with OsRIP1 to regulate rice heading date by finely modulating *Ehd1* expression, *Plant Biotechnol. J.*, 19(2), 300–310. DOI: 10.1111/pbi.13462.

Chakraborty, M., Gangappa, S. N., Maurya, J. P., Sethi, V., Srivastava, A. K., Singh, A., Dutta, S., Ojha, M., Gupta, N., Sengupta, M., Ram, H. and Chattopadhyay, S. (2019), Functional interrelation of MYC2 and HY5 plays an important role in Arabidopsis seedling development, *Plant J.*, 99(6), 1080–1097. DOI: 10.1111/tpj.14381.

Chattopadhyay, S., Ang, L. H., Puente, P., Deng, X. W. and Wei, N. (1998), Arabidopsis bZIP protein HY5 directly interacts with light-responsive promoters in mediating light control of gene expression, *Plant Cell*, 10(5), 673–683. DOI: 10.1105/tpc.10.5.673.

Chaudhary, C., Sharma, N. and Khurana, P. (2021), Decoding the wheat awn transcriptome and overexpressing *TaRca1β* in rice for heat stress tolerance, *Plant Mol. Biol.*, 105(1–2), 133–146. DOI: 10.1007/s11103-020-01073-0.

Chen, X., Lu, S., Wang, Y., Zhang, X., Lv, B., Luo, L., Xi, D., Shen, J., Ma, H. and Ming, F. (2015), *OsNAC2* encoding a NAC transcription factor that affects plant height through mediating the gibberellic acid pathway in rice, *Plant Cell Biotechnol. Mol. Biol.*, 82(2), 302–314. DOI: 10.1111/tpj.12819.

Chen, G. Y., Yong, Z. H., Liao, Y., Zhang, D. Y., Chen, Y., Zhang, H. B., Chen, J., Zhu, J. G. and Xu, D. Q. (2005), Photosynthetic acclimation in rice leaves to free-air CO2 enrichment related to both ribulose-1,5-bisphosphate carboxylation limitation and ribulose-1,5-bisphosphate regeneration limitation, *Plant Cell Physiol.*, 46(7), 1036–1045. DOI: 10.1093/pcp/pci113.

Chen, F., Zheng, G., Qu, M., Wang, Y., Lyu, M. A. and Zhu, X. G. (2021), Knocking out *Negative Regulator of Photosynthesis 1* increases rice leaf photosynthesis and biomass production in the field, *J. Exp. Bot.*, 72(5), 1836–1849. DOI: 10.1093/jxb/eraa566.

Cheng, C. L., Acedo, G. N., Cristinsin, M. and Conkling, M. A. (1992), Sucrose mimics the light induction of Arabidopsis nitrate reductase gene transcription, *Proc. Natl. Acad. Sci. U. S. A.*, 89(5), 1861–1864. DOI: 10.1073/pnas.89.5.1861.

Cheng, S. H., Moore, B. D. and Seemann, J. R. (1998), Effects of short- and long-term elevated CO2 on the expression of *ribulose-1,5-bisphosphate carboxylase/oxygenase* genes and carbohydrate accumulation in leaves of *Arabidopsis thaliana* (L.) Heynh, *Plant Physiol.*, 116(2), 715–723. DOI: 10.1104/pp.116.2.715.

Chi, W., He, B., Mao, J., Jiang, J. and Zhang, L. (2015), Plastid sigma factors: their individual functions and regulation in transcription, *BBA*, 1847(9), 770–778. DOI: 10.1016/j.bbabio.2015.01.001.

Chiang, Y. H., Zubo, Y. O., Tapken, W., Kim, H. J., Lavanway, A. M., Howard, L., Pilon, M., Kieber, J. J. and Schaller, G. E. (2012), Functional characterization of the GATA transcription factors GNC and CGA1 reveals their key role in chloroplast development, growth, and division in Arabidopsis, *Plant Physiol.*, 160(1), 332–348. DOI: 10.1104/pp.112.198705.

Christopher, D. A. and Hoffer, P. H. (1998), DET1 represses a chloroplast blue light-responsive promoter in a developmental and tissue-specific manner in *Arabidopsis thaliana*, *Plant J.*, 14(1), 1–11. DOI: 10.1046/j.1365-313X.1998.00078.x.

Christopher, D. A., Kim, M. and Mullet, J. E. (1992), A novel light-regulated promoter is conserved in cereal and dicot chloroplasts, *Plant Cell*, 4(7), 785–798. DOI: 10.1105/tpc.4.7.785.

Curci, P. L., Aiese Cigliano, R., Zuluaga, D. L., Janni, M., Sanseverino, W. and Sonnante, G. (2017), Transcriptomic response of durum wheat to nitrogen starvation, *Sci. Rep.*, 7(1), 1176. DOI: 10.1038/s41598-017-01377-0.

Daszkowska-Golec, A., Collin, A., Sitko, K., Janiak, A., Kalaji, H. M. and Szarejko, I. (2019), Genetic and physiological dissection of photosynthesis in barley exposed to drought stress, *Int. J. Mol. Sci.*, 20(24), 6341. DOI: 10.3390/ijms20246341.

Degenkolbe, T., Do, P. T., Zuther, E., Repsilber, D., Walther, D., Hincha, D. K. and Köhl, K. I. (2009), Expression profiling of rice cultivars differing in their tolerance to long-term drought stress, *Plant Mol. Biol.*, 69(1-2), 133-153. DOI: 10.1007/s11103-008-9412-7.

Del Campo, E. M., Sabater, B. and Martín, M. (2002), Post-transcriptional control of chloroplast gene expression: accumulation of stable *psaC* mRNA is due to downstream RNA cleavages in the *ndhD* gene, *J. Biol. Chem.*, 277(39), 36457-36464. DOI: 10.1074/jbc.M204500200.

Dietz, K. J. (2014), Redox regulation of transcription factors in plant stress acclimation and development, *Antioxid. Redox Signal.*, 21(9), 1356-1372. DOI: 10.1089/ars.2013.5672.

Dodd, A. N., Belbin, F. E., Frank, A. and Webb, A. A. (2015), Interactions between circadian clocks and photosynthesis for the temporal and spatial coordination of metabolism, *Front. Plant Sci.*, 6: 245. DOI: 10.3389/fpls.2015.00245.

Dodd, A. N., Salathia, N., Hall, A., Kévei, E., Tóth, R., Nagy, F., Hibberd, J. M., Millar, A. J. and Webb, A. A. (2005), Plant circadian clocks increase photosynthesis, growth, survival, and competitive advantage, *Science*, 309(5734), 630-633. DOI: 10.1126/science.1115581.

Driever, S. M., Simkin, A. J., Alotaibi, S., Fisk, S. J., Madgwick, P. J., Sparks, C. A., Jones, H. D., Lawson, T., Parry, M. A. J. and Raines, C. A. (2017), Increased SBPase activity improves photosynthesis and grain yield in wheat grown in greenhouse conditions, *Philos. Trans. R. Soc. Lond., B, Biol. Sci.*, 372(1730), 20160384. DOI: 10.1098/rstb.2016.0384.

Du, Y., Mo, W., Ma, T., Tang, W., Tian, L. and Lin, R. (2021), A pentatricopeptide repeat protein DUA1 interacts with sigma factor 1 to regulate chloroplast gene expression in rice, *Photosyn. Res.*, 147(2), 131-143. DOI: 10.1007/s11120-020-00793-0.

Duarte-Delgado, D., Dadshani, S., Schoof, H., Oyiga, B. C., Schneider, M., Mathew, B., Léon, J. and Ballvora, A. (2020), Transcriptome profiling at osmotic and ionic phases of salt stress response in bread wheat uncovers trait-specific candidate genes, *BMC Plant Biol.*, 20(1), 428. DOI: 10.1186/s12870-020-02616-9.

Emms, D. M. and Kelly, S. (2022), SHOOT: phylogenetic gene search and ortholog inference, *Genome Biol.*, 23(1), 85. DOI: 10.1186/s13059-022-02652-8.

Ermakova, M., Lopez-Calcagno, P. E., Raines, C. A., Furbank, R. T. and von Caemmerer, S. (2019), Overexpression of the Rieske FeS protein of the cytochrome b 6 f complex increases C4 photosynthesis in *Setaria viridis*, *Commun. Biol.*, 2(1), 314. DOI: 10.1038/s42003-019-0561-9.

Evans, J. R. (1989), Photosynthesis and nitrogen relationships in leaves of C3 plants, *Oecologia*, 78(1), 9-19. DOI: 10.1007/BF00377192.

Fang, Y., Xie, K. and Xiong, L. (2014), Conserved miR164-targeted NAC genes negatively regulate drought resistance in rice, *J. Exp. Bot.*, 65(8), 2119-2135. DOI: 10.1093/jxb/eru072.

Favory, J. J., Kobayshi, M., Tanaka, K., Peltier, G., Kreis, M., Valay, J. G. and Lerbs-Mache, S. (2005), Specific function of a plastid sigma factor for *ndhF* gene transcription, *Nucleic Acids Res.*, 33(18), 5991-5999. DOI: 10.1093/nar/gki908.

Feng, L., Han, Y., Liu, G., An, B., Yang, J., Yang, G., Li, Y. and Zhu, Y. (2007), Overexpression of sedoheptulose-1,7-bisphosphatase enhances photosynthesis and growth under salt stress in transgenic rice plants, *Funct. Plant Biol.*, 34(9), 822–834. DOI: 10.1071/FP07074.

Fey, V., Wagner, R., Bräutigam, K., Wirtz, M., Hell, R., Dietzmann, A., Leister, D., Oelmüller, R. and Pfannschmidt, T. (2005), Retrograde plastid redox signals in the expression of nuclear genes for chloroplast proteins of *Arabidopsis thaliana*, *J. Biol. Chem.*, 280(7), 5318-5328. DOI: 10.1074/jbc.M406358200.

Fitter, D. W., Martin, D. J., Copley, M. J., Scotland, R. W. and Langdale, J. A. (2002), *GLK* gene pairs regulate chloroplast development in diverse plant species, *Plant Cell Biotechnol. Mol. Biol.*, 31(6), 713–727. DOI: 10.1046/j.1365-313X.2002.01390.x.

Foyer, C. H., Neukermans, J., Queval, G., Noctor, G. and Harbinson, J. (2012), Photosynthetic control of electron transport and the regulation of gene expression, *J. Exp. Bot.*, 63(4), 1637–1661. DOI: 10.1093/jxb/ers013.

Garapati, P., Xue, G. P., Munné-Bosch, S. and Balazadeh, S. (2015), Transcription factor ATAF1 in Arabidopsis promotes senescence by direct regulation of key chloroplast maintenance and senescence transcriptional cascades, *Plant Physiol.*, 168(3), 1122-1139. DOI: 10.1104/pp.15.00567.

Goldschmidt, E. E. and Huber, S. C. (1992), Regulation of photosynthesis by end-product accumulation in leaves of plants storing starch, sucrose, and hexose sugars, *Plant Physiol.*, 99(4), 1443–1448. DOI: 10.1104/pp.99.4.1443.

Hakimi, M. A., Privat, I., Valay, J. G. and Lerbs-Mache, S. (2000), Evolutionary conservation of C-terminal domains of primary sigma70-type transcription factors between plants and bacteria, *J. Biol. Chem.*, 275(13), 9215-9221. DOI: 10.1074/jbc.275.13.9215.

Hanaoka, M., Kato, M., Anma, M. and Tanaka, K. (2012), SIG1, a sigma factor for the chloroplast RNA polymerase, differently associates with multiple DNA regions in the chloroplast chromosomes *in vivo*, *Int. J. Mol. Sci.*, 13(10), 12182–12194. DOI: 10.3390/ijms131012182.

Harter, K., Talke-Messerer, C., Barz, W. and Schafer, E. (1993), Light- and sucrose-dependent gene expression in photomixotrophic cell suspension cultures and protoplasts of rape (*Brassica napus* L.), *Plant J.*, 4(3), 507–516. DOI: 10.1046/j.1365-313X.1993.04030507.x.

Haydon, M. J., Mielczarek, O., Robertson, F. C., Hubbard, K. E. and Webb, A. A. (2013), Photosynthetic entrainment of the *Arabidopsis thaliana* circadian clock, *Nature*, 502(7473), 689-692. DOI: 10.1038/nature12603.

Herz, S., Füßl, M., Steiger, S. and Koop, H. U. (2005), Development of novel types of plastid transformation vectors and evaluation of factors controlling expression, *Transgenic Res.*, 14(6), 969-982. DOI: 10.1007/s11248-005-2542-7.

Holm, M., Ma, L. G., Qu, L. J. and Deng, X. W. (2002), Two interacting bZIP proteins are direct targets of COP1-mediated control of light-dependent gene expression in Arabidopsis, *Genes Dev.*, 16(10), 1247–1259. DOI: 10.1101/gad.969702.

Homann, A. and Link, G. (2003), DNA-binding and transcription characteristics of three cloned sigma factors from mustard (*Sinapis alba* L.) suggest overlapping and distinct roles in plastid gene expression, *Eur. J. Biochem.*, 270(6), 1288–1300. DOI: 10.1046/j.1432-1033.2003.03494.x.

Horiguchi, G., Kim, G. T. and Tsukaya, H. (2005), The transcription factor AtGRF5 and the transcription coactivator AN3 regulate cell proliferation in leaf primordia of *Arabidopsis thaliana*, *Plant Cell Biotechnol. Mol. Biol.*, 43(1), 68-78. DOI: 10.1111/j.1365-313X.2005.02429.x.

Hu, H., You, J., Fang, Y., Zhu, X., Qi, Z. and Xiong, L. (2008), Characterization of transcription factor gene SNAC2 conferring cold and salt tolerance in rice, *Plant Mol. Biol.*, 67(1-2), 169-181. DOI: 10.1007/s11103-008-9309-5.

Huang, Y., Fang, R., Li, Y., Liu, X., Wang, G., Yin, K., Jin, J. and Herbert, S. J. (2019a), Warming and elevated CO_2 alter the transcriptomic response of maize (*Zea mays* L.) at the silking stage, *Sci. Rep.*, 9(1), 1. DOI: 10.1038/s41598-019-54325-5.

Huang, J., Zhao, X. and Chory, J. (2019b), The Arabidopsis transcriptome responds specifically and dynamically to high light stress, *Cell Rep.*, 29(12), 4186–4199.e3. DOI: 10.1016/j.celrep.2019.11.051.

Huang, J., Qin, F., Zang, G., Kang, Z., Zou, H., Hu, F., Yue, C., Li, X. and Wang, G. (2013), Mutation of OsDET1 increases chlorophyll content in rice, *Plant Sci.*, 210: 241–249. DOI: 10.1016/j.plantsci.2013.06.003.

Hudson, D., Guevara, D. R., Hand, A. J., Xu, Z., Hao, L., Chen, X., Zhu, T., Bi, Y. M. and Rothstein, S. J. (2013), Rice cytokinin Gata Transcription Factor1 regulates chloroplast development and plant architecture, *Plant Physiol.*, 162(1), 132–144. DOI: 10.1104/pp.113.217265.

Ichikawa, K., Shimizu, A., Okada, R., Satbhai, S. B. and Aoki, S. (2008), The plastid sigma factor SIG5 is involved in the diurnal regulation of the chloroplast gene *psbD* in the moss *Physcomitrella patens*, *FEBS Lett.*, 582(3), 405–409. DOI: 10.1016/j.febslet.2007.12.034.

Ishizaki, Y., Tsunoyama, Y., Hatano, K., Ando, K., Kato, K., Shinmyo, A., Kobori, M., Takeba, G., Nakahira, Y. and Shiina, T. (2005), A nuclear-encoded sigma factor, Arabidopsis SIG6, recognizes sigma-70 type chloroplast promoters and regulates early chloroplast development in cotyledons, *Plant Cell Biotechnol. Mol. Biol.*, 42(2), 133–144. DOI: 10.1111/j.1365-313X.2005.02362.x.

Jensen, M. K., Hagedorn, P. H., de Torres-Zabala, M., Grant, M. R., Rung, J. H., Collinge, D. B. and Lyngkjaer, M. F. (2008), Transcriptional regulation by an NAC (NAM-ATAF1,2-CUC2) transcription factor attenuates ABA signalling for efficient basal defence towards *Blumeria graminis* f. sp. *hordei* in Arabidopsis, *Plant Cell Biotechnol. Mol. Biol.*, 56(6), 867–880. DOI: 10.1111/j.1365-313X.2008.03646.x.

Jia, Y., Li, C., Li, Q., Liu, P., Liu, D., Liu, Z., Wang, Y., Jiang, G. and Zhai, W. (2020), Characteristic dissection of *Xanthomonas oryzae* pv. *oryzae* responsive microRNAs in rice, *Int. J. Mol. Sci.*, 21(3), 785. DOI: 10.3390/ijms21030785.

Jiang, D., Zhou, L., Chen, W., Ye, N., Xia, J. and Zhuang, C. (2019), Overexpression of a microRNA-targeted NAC transcription factor improves drought and salt tolerance in Rice via ABA-mediated pathways, *Rice*, 12(1), 76. DOI: 10.1186/s12284-019-0334-6.

Jing, A., Xiaoli, W. and Honghao, H. (2022). Transcriptome analysis reveals the accelerated expression of genes related to photosynthesis and chlorophyll biosynthesis contribution to shadetolerant in *Phoebe bournei*. *BMC Plant Biology* 22(2681): 268. DOI: https://doi.org/10.1186/s12870-022-03657-y.

Kanamaru, K., Nagashima, A., Fujiwara, M., Shimada, H., Shirano, Y., Nakabayashi, K., Shibata, D., Tanaka, K. and Takahashi, H. (2001), An Arabidopsis sigma factor (SIG2)-dependent expression of plastid-encoded tRNAs in chloroplasts, *Plant Cell Physiol.*, 42(10), 1034–1043. DOI: 10.1093/pcp/pce155.

Kangasjärvi, S., Tikkanen, M., Durian, G. and Aro, E. M. (2014), Photosynthetic light reactions - an adjustable hub in basic production and plant immunity signaling, *Plant Physiol. Biochem.*, 81: 128–134. DOI: 10.1016/j.plaphy.2013.12.004.

Kasai, K., Kawagishi-Kobayashi, M., Teraishu, M., Ito, Y., Ochi, K., Wakasa, K. and Tozawa, Y. (2004), Differential expression of three plastidial sigma factors, *OsSIG1*, *OsSIG2A*, and *OsSIG2B*, during leaf development in rice, *Biosci. Biotechnol. Biochem.*, 68(4), 973-977. DOI: 10.1271/bbb.68.973.

Kellmann, J. W., Merforth, N., Wiese, M., Pichersky, E. and Piechulla, B. (1993), Concerted circadian oscillations in transcript levels of nineteen *Lha/b* (*cab*) genes in *Lycopersicon esculentum* (tomato), *Mol. Gen. Genet.*, 237(3), 439-448. DOI: 10.1007/BF00279449.

Keshishian, E. A., Cliver, B. R., McLaughlin, W. F., Hallmark, H. T., Plačková, L., Goertzen, L. R., Novák, O., Cobine, P. A., Leisner, C. P. and Rashotte, A. M. (2022), Cytokinin Response Factor2 is involved in modulating the salt stress response, *Plant Cell Biotechnol. Mol. Biol.*, 110(4), 1097-1110. DOI: 10.1111/tpj.15726.

Kim, J. H., Choi, D. and Kende, H. (2003), The AtGRF family of putative transcription factors is involved in leaf and cotyledon growth in Arabidopsis, *Plant Cell Biotechnol. Mol. Biol.*, 36(1), 94-104. DOI: 10.1046/j.1365-313X.2003.01862.x.

Kim, J. W., Park, J. K., Kim, B. H., Lee, J. S. and Sim, W. S. (2002), Molecular analysis of the accumulation of the transcripts of the large subunit gene of ribulose-1,5-bisphosphate carboxylase/oxygenase by light, *Mol. Cells*, 14(2), 281-287.

Kleine, T. and Leister, D. (2016), Retrograde signalling: organelles go networking, *BBA*, 1857(8), 1313-1325. DOI: 10.1016/j.bbabio.2016.03.017.

Klein, R. R. and Mullet, J. E. (1990), Light-induced transcription of chloroplast genes. psbA transcription is differentially enhanced in illuminated barley, *J. Biol. Chem.*, 265(4), 1895-1902. DOI: 10.1016/s0021-9258(19)39915-6.

Kobayashi, K., Baba, S., Obayashi, T., Sato, M., Toyooka, K., Keränen, M., Aro, E. M., Fukaki, H., Ohta, H., Sugimoto, K. and Masuda, T. (2012), Regulation of root greening by light and auxin/cytokinin signaling in Arabidopsis, *Plant Cell*, 24(3), 1081-1095. DOI: 10.1105/tpc.111.092254.

Krapp, A., Hofmann, B., Schäfer, C. and Stitt, M. (1993), Regulation of the expression of *RbcS* and other photosynthetic genes by carbohydrates: a mechanism for the "sink regulation" of photosynthesis? *Plant J.*, 3(6), 817-828. DOI: 10.1111/j.1365-313X.1993.00817.x.

Kromdijk, J., Głowacka, K., Leonelli, L., Gabilly, S. T., Iwai, M., Niyogi, K. K. and Long, S. P. (2016), Improving photosynthesis and crop productivity by accelerating recovery from photoprotection, *Science*, 354(6314), 857-861. DOI: 10.1126/science.aai8878.

Ku, C., Nelson-Sathi, S., Roettger, M., Sousa, F. L., Lockhart, P. J., Bryant, D., Hazkani-Covo, E., McInerney, J. O., Landan, G. and Martin, W. F. (2015), Endosymbiotic origin and differential loss of eukaryotic genes, *Nature*, 524(7566), 427-432. DOI: 10.1038/nature14963.

Kubota, Y., Miyao, A., Hirochika, H., Tozawa, Y., Yasuda, H., Tsunoyama, Y., Niwa, Y., Imamura, S., Shirai, M. and Asayama, M. (2007), Two novel nuclear genes, *OsSIG5* and *OsSI*, encoding potential plastid sigma factors of RNA polymerase in rice: tissue-specific and light-responsive gene expression, *Plant Cell Physiol.*, 48(1), 186-192. DOI: 10.1093/pcp/pcl050.

Kumar, A. U. and Ling, A. P. K. (2021), Gene introduction approaches in chloroplast transformation and its applications, *J. Genet. Eng. Biotechnol.*, 19(1), 1-10. DOI: 10.1186/s43141-021-00255-7.

Kumar, A., Panda, D., Mohanty, S., Biswal, M., Dey, P., Dash, M., Sah, R. P., Kumar, S., Baig, M. J. and Behera, L. (2020), Role of sedoheptulose-1,7 bisphosphatase in low light tolerance of rice (*Oryza sativa* L.), *Physiol. Mol. Biol. Plants*, 26(12), 2465–2485. DOI: 10.1007/s12298-020-00905-z.

Lam, E., Hanley-Bowdoin, L. and Chua, N. H. (1988), Characterization of a chloroplast sequence-specific DNA binding factor, *J. Biol. Chem.*, 263(17), 8288–8293. DOI: 10.1016/s0021-9258(18)68475-3.

Langdale, J. A. and Kidner, C. A. (1994), Bundle sheath defective, a mutation that disrupts cellular differentiation in maize leaves, *Development*, 120(3), 673–681. DOI: 10.1242/dev.120.3.673.

Lawrenson, T. and Harwood, W. A. (2019), Creating targeted gene knockouts in barley using CRISPR/Cas9, *Methods Mol. Biol.*, 1900, 217–232. DOI: 10.1007/978-1-4939-8944-7_14.

Le, T. D., Gathignol, F., Vu, H. T., Nguyen, K. L., Tran, L. H., Vu, H. T. T., Dinh, T. X., Lazennec, F., Pham, X. H., Véry, A. A., Gantet, P. and Hoang, G. T. (2021), Genome-wide association mapping of salinity tolerance at the seedling stage in a panel of Vietnamese landraces reveals new valuable QTLs for salinity stress tolerance breeding in rice, *Plants*, 10(6), 1088. DOI: 10.3390/plants10061088.

Lee, J., He, K., Stolc, V., Lee, H., Figueroa, P., Gao, Y., Tongprasit, W., Zhao, H., Lee, I. and Deng, X. W. (2007), Analysis of transcription factor HY5 genomic binding sites revealed its hierarchical role in light regulation of development, *Plant Cell*, 19(3), 731–749. DOI: 10.1105/tpc.106.047688.

Lee, D. Y., Hua, L., Khoshravesh, R., Giuliani, R., Kumar, I., Cousins, A., Sage, T. L., Hibberd, J. M. and Brutnell, T. P. (2021), Engineering chloroplast development in rice through cell-specific control of endogenous genetic circuits, *Plant Biotechnol. J.*, 19(11), 2291–2303. DOI: 10.1111/pbi.13660.

Leister, D. and Schneider, A. (2003), From genes to photosynthesis in *Arabidopsis thaliana*, *Int. Rev. Cytol.*, 228, 31–83. DOI: 10.1016/S0074-7696(03)28002-5.

Li, J., Li, Y. and Ma, L. (2021), Recent advances in CRISPR/Cas9 and applications for wheat functional genomics and breeding, *aBIOTECH*, 2(4), 375–385. DOI: 10.1007/s42994-021-00042-5.

Li, X., Wang, P., Li, J., Wei, S., Yan, Y., Yang, J., Zhao, M., Langdale, J. A. and Zhou, W. (2020), Maize *GOLDEN2-LIKE* genes enhance biomass and grain yields in rice by improving photosynthesis and reducing photoinhibition, *Commun. Biol.*, 3(1), 151. DOI: 10.1038/s42003-020-0887-3.

Lira, B. S., Gramegna, G., Trench, B. A., Alves, F. R. R., Silva, E. M., Silva, G. F. F., Thirumalaikumar, V. P., Lupi, A. C. D., Demarco, D., Purgatto, E., Nogueira, F. T. S., Balazadeh, S., Freschi, L. and Rossi, M. (2017), Manipulation of a senescence-associated gene improves fleshy fruit yield, *Plant Physiol.*, 175(1), 77–91. DOI: 10.1104/pp.17.00452.

Liu, X., Yin, C., Xiang, L., Jiang, W., Xu, S. and Mao, Z. (2020a), Transcription strategies related to photosynthesis and nitrogen metabolism of wheat in response to nitrogen deficiency, *BMC Plant Biol.*, 20(1), 448. DOI: 10.1186/s12870-020-02662-3.

Liu, G., Zha, Z., Cai, H., Qin, D., Jia, H., Liu, C., Qiu, D., Zhang, Z., Wan, Z., Yang, Y., Wan, B., You, A. and Jiao, C. (2020b), Dynamic transcriptome analysis of anther response to heat stress during anthesis in thermotolerant rice (*Oryza sativa* L.), *Int. J. Mol. Sci.*, 21(3), 1155. DOI: 10.3390/ijms21031155.

Liu, T. D., Zhang, X. W., Xu, Y., Liu, S. Q. and Chen, X. W. (2017), Light quality modifies the expression of photosynthetic genes in maize seedlings, *Photosynthetica*, 55(2), 360–367. DOI: 10.1007/s11099-016-0227-5.

Loschelder, H., Schweer, J., Link, B. and Link, G. (2006), Dual temporal role of plastid sigma factor 6 in Arabidopsis development, *Plant Physiol.*, 142(2), 642-650. DOI: 10.1104/pp.106.085878.

Lu, P. L., Chen, N. Z., An, R., Su, Z., Qi, B. S., Ren, F., Chen, J. and Wang, X. C. (2006), A novel drought-inducible gene, ATAF1, encodes a NAC family protein that negatively regulates the expression of stress-responsive genes in Arabidopsis, *Plant Mol. Biol.*, 63(2), 289–305. DOI: 10.1007/s11103-006-9089-8.

Lu, Y., Meng, Y., Zeng, J., Luo, Y., Feng, Z., Bian, L. and Gao, S. (2020), Coordination between Growth-Regulating Factor1 and GRF-interacting factor1 plays a key role in regulating leaf growth in rice, *BMC Plant Biol.*, 20(1), 200. DOI: 10.1186/s12870-020-02417-0.

Luo, A.-D., Liu, L., Tang, Z.-S., Bai, X., Cao, S. and Chu, C. (2005), Down-regulation of *OsGRF1* Gene in rice *rhd1* mutant results in reduced heading date, *J. Integr. Plant Biol.*, 47(6), 745–752. DOI: 10.1111/j.1744-7909.2005.00071.x.

Maliga, P. and Bock, R. (2011), Plastid biotechnology: food, fuel, and medicine for the 21st century, *Plant Physiol.*, 155(4), 1501–1510. DOI: 10.1104/pp.110.170969.

Mao, C., Ding, W., Wu, Y., Yu, J., He, X., Shou, H. and Wu, P. (2007), Overexpression of a NAC-domain protein promotes shoot branching in rice, *New Phytol.*, 176(2), 288–298. DOI: 10.1111/j.1469-8137.2007.02177.x.

Mao, C., He, J., Liu, L., Deng, Q., Yao, X., Liu, C., Qiao, Y., Li, P. and Ming, F. (2020), OsNAC2 integrates auxin and cytokinin pathways to modulate rice root development, *Plant Biotechnol. J.*, 18(2), 429–442. DOI: 10.1111/pbi.13209.

Mao, C., Lu, S., Lv, B., Zhang, B., Shen, J., He, J., Luo, L., Xi, D., Chen, X. and Ming, F. (2017), A rice NAC transcription factor promotes leaf senescence via ABA biosynthesis, *Plant Physiol.*, 174(3), 1747–1763. DOI: 10.1104/pp.17.00542.

Martin-Avila, E., Lim, Y. L., Birch, R., Dirk, L. M. A., Buck, S., Rhodes, T., Sharwood, R. E., Kapralov, M. V. and Whitney, S. M. (2020), Modifying plant photosynthesis and growth via simultaneous chloroplast transformation of RuBisCO large and small subunits, *Plant Cell*, 32(9), 2898–2916. DOI: 10.1105/tpc.20.00288.

Morikawa, K., Ito, S., Tsunoyama, Y., Nakahira, Y., Shiina, T. and Toyoshima, Y. (1999), Circadian-regulated expression of a nuclear-encoded plastid σ factor gene (*sigA*) in wheat seedlings, *FEBS Lett.*, 451(3), 275–278. DOI: 10.1016/S0014-5793(99)00593-1.

Müller, M. and Munné-Bosch, S. (2021), Hormonal impact on photosynthesis and photoprotection in plants, *Plant Physiol.*, 185(4), 1500–1522. DOI: 10.1093/plphys/kiaa119.

Nagashima, A., Hanaoka, M., Shikanai, T., Fujiwara, M., Kanamaru, K., Takahashi, H. and Tanaka, K. (2004), The multiple-stress responsive plastid sigma factor, SIG5, directs activation of the *psbD* blue light-responsive promoter (BLRP) in *Arabidopsis thaliana*, *Plant Cell Physiol.*, 45(4), 357–368. DOI: 10.1093/pcp/pch050.

Nagy, F., Kay, S. A. and Chua, N.-H. (1988), A circadian clock regulates transcription of the wheat *Cab-1* gene, *Genes Develop.*, 2(4), 376–382. DOI: 10.1101/gad.2.4.376.

Nakahira, Y., Baba, K., Yoneda, A., Shiina, T. and Toyoshima, Y. (1998), Circadian-regulated transcription of the *psbD* light-responsive promoter in wheat chloroplasts, *Plant Physiol.*, 118(3), 1079–1088. DOI: 10.1104/pp.118.3.1079.

Nakamura, H., Muramatsu, M., Hakata, M., Ueno, O., Nagamura, Y., Hirochika, H., Takano, M. and Ichikawa, H. (2009), Ectopic overexpression of the transcription factor OsGLK1 induces chloroplast development in non-green rice cells, *Plant Cell Physiol.*, 50(11), 1933-1949. DOI: 10.1093/pcp/pcp138.

Nakashima, K., Tran, L. S., van Nguyen, D., Fujita, M., Maruyama, K., Todaka, D., Ito, Y., Hayashi, N., Shinozaki, K. and Yamaguchi-Shinozaki, K. (2007), Functional analysis of a NAC-type transcription factor OsNAC6 involved in abiotic and biotic stress-responsive gene expression in rice, *Plant Cell Biotechnol. Mol. Biol.*, 51(4), 617–630. DOI: 10.1111/j.1365-313X.2007.03168.x.

Nefissi Ouertani, R., Arasappan, D., Ruhlman, T. A., Ben Chikha, M., Abid, G., Mejri, S., Ghorbel, A. and Jansen, R. K. (2022), Effects of salt stress on transcriptional and physiological responses in barley leaves with contrasting salt tolerance, *Int. J. Mol. Sci.*, 23(9). DOI: 10.3390/ijms23095006.

Nguyen, C. V., Vrebalov, J. T., Gapper, N. E., Zheng, Y., Zhong, S., Fei, Z. and Giovannoni, J. J. (2014), Tomato *GOLDEN2-LIKE* transcription factors reveal molecular gradients that function during fruit development and ripening, *Plant Cell*, 26(2), 585–601. DOI: 10.1105/tpc.113.118794.

Nie, G., Hendrix, D. L., Webber, A. N., Kimball, B. A. and Long, S. P. (1995), Increased accumulation of carbohydrates and decreased photosynthetic gene transcript levels in wheat grown at an elevated CO_2 concentration in the field, *Plant Physiol.*, 108(3), 975–983. DOI: 10.1104/pp.108.3.975.

Nishizawa-Yokoi, A., Mikami, M. and Toki, S. (2020), A universal system of CRISPR/Cas9-mediated gene targeting using all-in-one vector in plants, *Front. Genome Editing*, 2: 604289. DOI: 10.3389/fgeed.2020.604289.

Noordally, Z. B., Ishii, K., Atkins, K. A., Wetherill, S. J., Kusakina, J., Walton, E. J., Kato, M., Azuma, M., Tanaka, K., Hanaoka, M. and Dodd, A. N. (2013), Circadian control of chloroplast transcription by a nuclear-encoded timing signal, *Science*, 339(6125), 1316–1319. DOI: 10.1126/science.1230397.

Ohnishi, T., Sugahara, S., Yamada, T., Kikuchi, K., Yoshiba, Y., Hirano, H. Y. and Tsutsumi, N. (2005), OsNAC6, a member of the NAC gene family, is induced by various stresses in rice, *Genes Genet. Syst.*, 80(2), 135–139. DOI: 10.1266/ggs.80.135.

Okazaki, K., Kabeya, Y., Suzuki, K., Mori, T., Ichikawa, T., Matsui, M., Nakanishi, H. and Miyagishima, S. Y. (2009), The Plastid Division1 and 2 components of the chloroplast division machinery determine the rate of chloroplast division in land plant cell differentiation, *Plant Cell*, 21(6), 1769–1780. DOI: 10.1105/tpc.109.067785.

Olejniczak, S. A., Łojewska, E., Kowalczyk, T. and Sakowicz, T. (2016), Chloroplasts: state of research and practical applications of plastome sequencing, *Planta*, 244(3), 517–527. DOI: 10.1007/s00425-016-2551-1.

Onda, Y., Yagi, Y., Saito, Y., Takenaka, N. and Toyoshima, Y. (2008), Light induction of Arabidopsis SIG1 and SIG5 transcripts in mature leaves: differential roles of cryptochrome 1 and cryptochrome 2 and dual function of SIG5 in the recognition of plastid promoters, *Plant Cell Biotechnol. Mol. Biol.*, 55(6), 968–978. DOI: 10.1111/j.1365-313X.2008.03567.x.

Ouertani, R. N., Arasappan, D., Abid, G., Ben Chikha, M., Jardak, R., Mahmoudi, H., Mejri, S., Ghorbel, A., Ruhlman, T. A. and Jansen, R. K. (2021), Transcriptomic analysis of salt-stress-responsive genes in barley roots and leaves, *Int. J. Mol. Sci.*, 22(15), 8155. DOI: 10.3390/ijms22158155.

Oyama, T., Shimura, Y. and Okada, K. (1997), The Arabidopsis *HY5* gene encodes a bZIP protein that regulates stimulus-induced development of root and hypocotyl, *Genes Develop.*, 11(22), 2983–2995. DOI: 10.1101/gad.11.22.2983.

Ozturk, Z. N., Talamé, V., Deyholos, M., Michalowski, C. B., Galbraith, D. W., Gozukirmizi, N., Tuberosa, R. and Bohnert, H. J. (2002), Monitoring large-scale changes in transcript abundance in drought- and salt-stressed barley, *Plant Mol. Biol.*, 48(5-6), 551-573. DOI: 10.1023/A:1014875215580.

Paul, M. J. and Foyer, C. H. (2001), Sink regulation of photosynthesis, *J. Exp. Bot.*, 52(360), 1383-1400. DOI: 10.1093/jexbot/52.360.1383.

Pfannschmidt, T., Bräutigam, K., Wagner, R., Dietzel, L., Schröter, Y., Steiner, S. and Nykytenko, A. (2009), Potential regulation of gene expression in photosynthetic cells by redox and energy state: approaches towards better understanding, *Ann. Bot.*, 103(4), 599-607. DOI: 10.1093/aob/mcn081.

Piippo, M., Allahverdiyeva, Y., Paakkarinen, V., Suoranta, U. M., Battchikova, N. and Aro, E. M. (2006), Chloroplast-mediated regulation of nuclear genes in *Arabidopsis thaliana* in the absence of light stress, *Physiol. Genom.*, 25(1), 142-152. DOI: 10.1152/physiolgenomics.00256.2005.

Pilgrim, M. L. and McClung, C. R. (1993), Differential involvement of the circadian clock in the expression of genes required for ribulose-1,5-bisphosphate carboxylase/oxygenase synthesis, assembly, and activation in *Arabidopsis thaliana*, *Plant Physiol.*, 103(2), 553-564. DOI: 10.1104/pp.103.2.553.

Plumley, F. G. and Schmidt, G. W. (1989), Nitrogen-dependent regulation of photosynthetic gene expression, *Proc. Natl. Acad. Sci. U. S. A.*, 86(8), 2678-2682. DOI: 10.1073/pnas.86.8.2678.

Powell, A. L. T., Nguyen, C. V. v., Hill, T., Cheng, K. L., Figueroa-Balderas, R., Aktas, H., Ashrafi, H., Pons, C., Fernández-Muñoz, R., Vicente, A., Lopez-Baltazar, J., Barry, C. S., Liu, Y., Chetelat, R., Granell, A., Van Deynze, A., Giovannoni, J. J. and Bennett, A. B. (2012), *Uniform ripening* encodes a *golden 2-like* transcription factor regulating tomato fruit chloroplast development, *Science*, 336(6089), 1711-1715. DOI: 10.1126/science.1222218.

Privat, I., Hakimi, M. A., Buhot, L., Favory, J. J. and Mache-Lerbs, S. (2003), Characterization of Arabidopsis plastid sigma-like transcription factors SIG1, SIG2 and SIG3, *Plant Mol. Biol.*, 51(3), 385-399. DOI: 10.1023/A:1022095017355.

Qiu, K., Li, Z., Yang, Z., Chen, J., Wu, S., Zhu, X., Gao, S., Gao, J., Ren, G., Kuai, B. and Zhou, X. (2015), EIN3 and ORE1 accelerate degreening during ethylene-mediated leaf senescence by directly activating chlorophyll catabolic genes in Arabidopsis, *PLoS Genet.*, 11(7), e1005399. DOI: 10.1371/journal.pgen.1005399.

Qiu, Z., Wang, H., Li, D., Yu, B., Hui, Q., Yan, S., Huang, Z., Cui, X. and Cao, B. (2019), Identification of candidate HY5-dependent and -independent regulators of anthocyanin biosynthesis in tomato, *Plant Cell Physiol.*, 60(3), 643-656. DOI: 10.1093/pcp/pcy236.

Qu, Y., Sakoda, K., Fukayama, H., Kondo, E., Suzuki, Y., Makino, A., Terashima, I. and Yamori, W. (2021), Overexpression of both RuBisCO and RuBisCO Activase rescues rice photosynthesis and biomass under heat stress, *Plant Cell Environ.*, 44(7), 2308-2320. DOI: 10.1111/pce.14051.

Queval, G. and Foyer, C. H. (2012), Redox regulation of photosynthetic gene expression, *Philos. Trans. R Soc. Lond. B Biol. Sci.*, 367(1608), 3475-3485. DOI: 10.1098/rstb.2012.0068.

Rachmat, A., Nugroho, S., Sukma, D., Aswidinnoor, H. and Sudarsono, S. (2014), Overexpression of OsNAC6 transcription factor from Indonesia rice cultivar enhances drought and salt tolerance, *Emir. J. Food Agric.*, 26(6), 519. DOI: 10.9755/ejfa.v26i6.17672.

Raines, T., Shanks, C., Cheng, C. Y., McPherson, D., Argueso, C. T., Kim, H. J., Franco-Zorrilla, J. M., López-Vidriero, I., Solano, R., Vaňková, R., Schaller, G. E. and Kieber, J. J. (2016), The cytokinin response factors modulate root and shoot growth and promote leaf senescence in Arabidopsis, *Plant Cell Biotechnol. Mol. Biol.*, 85(1), 134–147. DOI: 10.1111/tpj.13097.

Rangan, P., Furtado, A. and Henry, R. (2020), Transcriptome profiling of wheat genotypes under heat stress during grain-filling, *J. Cereal Sci.*, 91: 102895. DOI: 10.1016/j.jcs.2019.102895.

Rauf, M., Arif, M., Dortay, H., Matallana-Ramírez, L. P., Waters, M. T., Gil Nam, H., Lim, P. O., Mueller-Roeber, B. and Balazadeh, S. (2013), ORE1 balances leaf senescence against maintenance by antagonizing G2-like-mediated transcription, *EMBO Rep.*, 14(4), 382–388. DOI: 10.1038/embor.2013.24.

Reem, N. T. and Van Eck, J. (2019), Application of CRISPR/Cas9-mediated gene editing in tomato, *Methods Mol. Biol.*, 1917: 171–182. DOI: 10.1007/978-1-4939-8991-1_13.

Renau-Morata, B., Molina, R. V., Minguet, E. G., Cebolla-Cornejo, J., Carrillo, L., Martí, R., García-Carpintero, V., Jiménez-Benavente, E., Yang, L., Cañizares, J., Canales, J., Medina, J. and Nebauer, S. G. (2021), Integrative transcriptomic and metabolomic analysis at organ scale reveals gene modules involved in the responses to suboptimal nitrogen supply in tomato, *Agronomy*, 11(7), 1320. DOI: 10.3390/agronomy11071320.

Rossi, M., Bermudez, L. and Carrari, F. (2015), Crop yield: challenges from a metabolic perspective, *Curr. Opin. Plant Biol.*, 25: 79–89. DOI: 10.1016/j.pbi.2015.05.004.

Ruf, S., Forner, J., Hasse, C., Kroop, X., Seeger, S., Schollbach, L., Schadach, A. and Bock, R. (2019), High-efficiency generation of fertile transplastomic Arabidopsis plants, *Nat. Plants*, 5(3), 282–289. DOI: 10.1038/s41477-019-0359-2.

Saeid Nia, M., Repnik, U., Krupinska, K. and Bilger, W. (2022), The plastid-nucleus localized DNA-binding protein WHIRLY1 is required for acclimation of barley leaves to high light, *Planta*, 255(4), 84. DOI: 10.1007/s00425-022-03854-x.

Sakuraba, Y., Kim, E. Y., Han, S. H., Piao, W., An, G., Todaka, D., Yamaguchi-Shinozaki, K. and Paek, N. C. (2017), Rice Phytochrome-Interacting Factor-Like1 (OsPIL1) is involved in the promotion of chlorophyll biosynthesis through feed-forward regulatory loops, *J. Exp. Bot.*, 68(15), 4103–4114. DOI: 10.1093/jxb/erx231.

Sanchez, S. E., Petrillo, E., Beckwith, E. J., Zhang, X., Rugnone, M. L., Hernando, C. E., Cuevas, J. C., Godoy Herz, M. A., Depetris-Chauvin, A., Simpson, C. G., Brown, J. W., Cerdán, P. D., Borevitz, J. O., Mas, P., Ceriani, M. F., Kornblihtt, A. R. and Yanovsky, M. J. (2010), A methyl transferase links the circadian clock to the regulation of alternative splicing, *Nature*, 468(7320), 112–116. DOI: 10.1038/nature09470.

Saroussi, S. I., Wittkopp, T. M. and Grossman, A. R. (2016), The Type II NADPH dehydrogenase facilitates cyclic electron flow, energy-dependent quenching, and chlororespiratory metabolism during acclimation of *Chlamydomonas reinhardtii* to nitrogen deprivation, *Plant Physiol.*, 170(4), 1975–1988. DOI: 10.1104/pp.15.02014.

Schweer, J., Loschelder, H. and Link, G. (2006), A promoter switch that can rescue a plant sigma factor mutant, *FEBS Lett.*, 580(28–29), 6617–6622. DOI: 10.1016/j.febslet.2006.11.010.

Sekhar, S., Panda, D., Kumar, J., Mohanty, N., Biswal, M., Baig, M. J., Kumar, A., Umakanta, N., Samantaray, S., Pradhan, S. K., Shaw, B. P., Swain, P. and Behera, L. (2019), Comparative transcriptome profiling of low light tolerant and sensitive rice varieties induced by low light stress at active tillering stage, *Sci. Rep.*, 9(1), 5753. DOI: 10.1038/s41598-019-42170-5.

Shi, Y., Ke, X., Yang, X., Liu, Y. and Hou, X. (2022), Plants response to light stress, *J. Genet. Genomics*, 49(8), 735–747. DOI: 10.1016/j.jgg.2022.04.017.

Shimizu, M., Kato, H., Ogawa, T., Kurachi, A., Nakagawa, Y. and Kobayashi, H. (2010), Sigma factor phosphorylation in the photosynthetic control of photosystem stoichiometry, *Proc. Natl. Acad. Sci. U. S. A.*, 107(23), 10760–10764. DOI: 10.1073/pnas.0911692107.

Skelly, M. J., Frungillo, L. and Spoel, S. H. (2016), Transcriptional regulation by complex interplay between post-translational modifications, *Curr. Opin. Plant Biol.*, 33: 126–132. DOI: 10.1016/j.pbi.2016.07.004.

Song, Y., Chen, Q., Ci, D., Shao, X. and Zhang, D. (2014), Effects of high temperature on photosynthesis and related gene expression in poplar, *BMC Plant Biol.*, 14(1), 111. DOI: 10.1186/1471-2229-14-111.

Song, Q., Joshi, M. and Joshi, V. (2020), Transcriptomic analysis of short-term salt stress response in watermelon seedlings, *Int. J. Mol. Sci.*, 21(17), 6036. DOI: 10.3390/ijms21176036.

South, P. F., Cavanagh, A. P., Liu, H. W. and Ort, D. R. (2019), Synthetic glycolate metabolism pathways stimulate crop growth and productivity in the field, *Science*, 363(6422), eaat9077. DOI: 10.1126/science.aat9077.

Sui, N., Yang, Z., Liu, M. and Wang, B. (2015), Identification and transcriptomic profiling of genes involved in increasing sugar content during salt stress in sweet sorghum leaves, *BMC Genomics*, 16(1), 534. DOI: 10.1186/s12864-015-1760-5.

Sultana, N., Islam, S., Juhasz, A., Yang, R., She, M., Alhabbar, Z., Zhang, J. and Ma, W. (2020), Transcriptomic study for identification of major nitrogen stress responsive genes in Australian bread wheat cultivars, *Front. Genet.*, 11: 583785. DOI: 10.3389/fgene.2020.583785.

Sun, Y., Wang, B., Ren, J., Zhou, Y., Han, Y., Niu, S., Zhang, Y., Shi, Y., Zhou, J., Yang, C., Ma, X., Liu, X., Luo, Y., Jin, C. and Luo, J. (2022), OsbZIP18, a positive regulator of serotonin biosynthesis, negatively controls the UV-B tolerance in rice, *Int. J. Mol. Sci.*, 23(6), 3215. DOI: 10.3390/ijms23063215.

Suorsa, M. (2015), Cyclic electron flow provides acclimatory plasticity for the photosynthetic machinery under various environmental conditions and developmental stages, *Front. Plant Sci.*, 6: 800. DOI: 10.3389/fpls.2015.00800.

Szymańska, R., Ślesak, I., Orzechowska, A. and Kruk, J. (2017), Physiological and biochemical responses to high light and temperature stress in plants, *Environ. Exp. Bot.*, 139: 165–177. DOI: 10.1016/j.envexpbot.2017.05.002.

Tan, S. and Troxler, R. F. (1999), Characterization of two chloroplast RNA polymerase sigma factors from *Zea mays*: photoregulation and differential expression, *Proc. Natl. Acad. Sci. U. S. A.*, 96(9), 5316–5321. DOI: 10.1073/pnas.96.9.5316.

Takasaki, H., Maruyama, K., Kidokoro, S., Ito, Y., Fujita, Y., Shinozaki, K., Yamaguchi-Shinozaki, K. and Nakashima, K. (2010), The abiotic stress-responsive NAC-type transcription factor OsNAC5 regulates stress-inducible genes and stress tolerance in rice, *Mol. Genet. Genomics*, 284(3), 173–183. DOI: 10.1007/s00438-010-0557-0.

Taketa, S., Hattori, M., Takami, T., Himi, E. and Sakamoto, W. (2021), Mutations in a *Golden2-Like* gene cause reduced seed weight in barley *albino lemma 1* mutants, *Plant Cell Physiol.*, 62(3), 447–457. DOI: 10.1093/pcp/pcab001.

Tavladoraki, P., Kloppstech, K. and Argyroudi-Akoyunoglou, J. (1989), Circadian rhythm in the expression of the mRNA coding for the apoprotein of the light-harvesting complex of photosystem II: phytochrome control and persistent far red reversibility, *Plant Physiol.*, 90(2), 665–672. DOI: 10.1104/pp.90.2.665.

Thum, K. E., Kim, M., Morishige, D. T., Eibl, C., Koop, H. U. and Mullet, J. E. (2001), Analysis of barley chloroplast *psbD* light-responsive promoter elements in transplastomic tobacco, *Plant Mol. Biol.*, 47(3), 353–366. DOI: 10.1023/A:1011616400264.

Toledo-Ortiz, G., Johansson, H., Lee, K. P., Bou-Torrent, J., Stewart, K., Steel, G., Rodríguez-Concepción, M. and Halliday, K. J. (2014), The HY5-PIF regulatory module coordinates light and temperature control of photosynthetic gene transcription, *PLoS Genet.*, 10(6), e1004416. DOI: 10.1371/journal.pgen.1004416.

Tozawa, Y., Teraishi, M., Sasaki, T., Sonoike, K., Nishiyama, Y., Itaya, M., Miyao, A. and Hirochika, H. (2007), The plastid sigma factor SIG1 maintains photosystem I activity via regulated expression of the *psaA* operon in rice chloroplasts, *Plant Cell Biotechnol. Mol. Biol.*, 52(1), 124–132. DOI: 10.1111/j.1365-313X.2007.03216.x.

Tsunoyama, Y., Ishizaki, Y., Morikawa, K., Kobori, M., Nakahira, Y., Takeba, G., Toyoshima, Y. and Shiina, T. (2004), Blue light-induced transcription of plastid-encoded *psbD* gene is mediated by a nuclear-encoded transcription initiation factor, AtSig5, *Proc. Natl. Acad. Sci. U. S. A.*, 101(9), 3304–3309. DOI: 10.1073/pnas.0308362101.

Tsunoyama, Y., Morikawa, K., Shiina, T. and Toyoshima, Y. (2002), Blue light specific and differential expression of a plastid σ factor, Sig5 in *Arabidopsis thaliana*, *FEBS Lett.*, 516(1–3), 225–228. DOI: 10.1016/S0014-5793(02)02538-3.

Van Oosten, J.-J. and Besford, R. T. (1995), Some relationships between the gas exchange, biochemistry and molecular biology of photosynthesis during leaf development of tomato plants after transfer to different carbon dioxide concentrations, *Plant Cell Environ.*, 18(11), 1253–1266. DOI: 10.1111/j.1365-3040.1995.tb00185.x.

Venkat, A. and Muneer, S. (2022), Role of circadian rhythms in major plant metabolic and signaling pathways, *Front. Plant Sci.*, 13: 836244–836244. DOI: 10.3389/fpls.2022.836244.

Venkateswarlu, B. (1977), Influence of low light intensity on growth and productivity of rice, *Oryza sativa*, L, *Plant Soil*, 46(3), 713–719. DOI: 10.1007/BF00015937.

Vercruyssen, L., Tognetti, V. B., Gonzalez, N., Van Dingenen, J., De Milde, L., Bielach, A., De Rycke, R., Van Breusegem, F. and Inzé, D. (2015), Growth Regulating Factor5 stimulates *Arabidopsis* chloroplast division, photosynthesis, and leaf longevity, *Plant Physiol.*, 167(3), 817–832. DOI: 10.1104/pp.114.256180.

Wang, P., Hendron, R. W. and Kelly, S. (2017), Transcriptional control of photosynthetic capacity: conservation and divergence from Arabidopsis to rice, *New Phytol.*, 216(1), 32–45. DOI: 10.1111/nph.14682.

Wang, D., Pan, Y., Zhao, X., Zhu, L., Fu, B. and Li, Z. (2011), Genome-wide temporal-spatial gene expression profiling of drought responsiveness in rice, *BMC Genomics*, 12(1), 149. DOI: 10.1186/1471-2164-12-149.

Wang, S., Sun, S., Guo, R., Liao, W. and Shou, H. (2021a), Transcriptomic profiling of Fe-responsive lncRNAs and their regulatory mechanism in rice, *Genes*, 12(4), 567. DOI: 10.3390/genes12040567.

Wang, W., Wang, P., Li, X., Wang, Y., Tian, S. and Qin, G. (2021b), The transcription factor SlHY5 regulates the ripening of tomato fruit at both the transcriptional and translational levels, *Hortic. Res.*, 8(1), 83. DOI: 10.1038/s41438-021-00523-0.

Wang, Z., Xia, Y., Lin, S., Wang, Y., Guo, B., Song, X., Ding, S., Zheng, L., Feng, R., Chen, S., Bao, Y., Sheng, C., Zhang, X., Wu, J., Niu, D., Jin, H. and Zhao, H. (2018), Osa-miR164a targets *OsNAC60* and negatively regulates rice immunity against the blast fungus

Magnaporthe oryzae, *Plant Cell Biotechnol. Mol. Biol.*, 95(4), 584–597. DOI: 10.1111/tpj.13972.

Waters, M. T., Wang, P., Korkaric, M., Capper, R. G., Saunders, N. J. and Langdale, J. A. (2009), GLK transcription factors coordinate expression of the photosynthetic apparatus in Arabidopsis, *Plant Cell*, 21(4), 1109–1128. DOI: 10.1105/tpc.108.065250.

Watillon, B., Kettmann, R., Boxus, P. and Burny, A. (1993), Developmental and circadian pattern of RuBisCO Activase mRNA accumulation in apple plants, *Plant Mol. Biol.*, 23(3), 501–509. DOI: 10.1007/BF00019298.

Westhoff, P. and Herrmann, R. G. (1988), Complex RNA maturation in chloroplasts: the *psbB* operon from spinach, *Eur. J. Biochem.*, 171(3), 551–564. DOI: 10.1111/j.1432-1033.1988.tb13824.x.

Woo, H. R., Kim, J. H., Nam, H. G. and Lim, P. O. (2004), The delayed leaf senescence mutants of Arabidopsis, *ore1*, *ore3*, and *ore9* are tolerant to oxidative stress, *Plant Cell Physiol.*, 45(7), 923–932. DOI: 10.1093/pcp/pch110.

Woodson, J. D. and Chory, J. (2008), Coordination of gene expression between organellar and nuclear genomes, *Nat. Rev. Genet.*, 9(5), 383–395. DOI: 10.1038/nrg2348.

Woodson, J. D., Perez-Ruiz, J. M., Schmitz, R. J., Ecker, J. R. and Chory, J. (2013), Sigma factor-mediated plastid retrograde signals control nuclear gene expression, *Plant Cell Biotechnol. Mol. Biol.*, 73(1), 1–13. DOI: 10.1111/tpj.12011.

Yang, X., Zhao, X., Dai, Z., Ma, F., Miao, X. and Shi, Z. (2021), OsmiR396/growth regulating factor modulate rice grain size through direct regulation of embryo-specific miR408, *Plant Physiol.*, 186(1), 519–533. DOI: 10.1093/plphys/kiab084.

Yeh, S. Y., Lin, H. H., Chang, Y. M., Chang, Y. L., Chang, C. K., Huang, Y. C., Ho, Y. W., Lin, C. Y., Zheng, J. Z., Jane, W. N., Ng, C. Y., Lu, M. Y., Lai, I. L., To, K. Y., Li, W. H. and Ku, M. S. B. (2022), Maize Golden2-like transcription factors boost rice chloroplast development, photosynthesis, and grain yield, *Plant Physiol.*, 188(1), 442–459. DOI: 10.1093/plphys/kiab511.

Yu, S., Liao, F., Wang, F., Wen, W., Li, J., Mei, H. and Luo, L. (2012), Identification of rice transcription factors associated with drought tolerance using the Ecotilling method, *PLoS ONE*, 7(2), e30765. DOI: 10.1371/journal.pone.0030765.

Yu, Y., Zhou, Z., Pu, H., Wang, B., Zhang, Y., Yang, B., Zhao, T. and Xu, D. (2019), OsSIG2A is required for chloroplast development in rice (*Oryza sativa* L.) at low temperature by regulating plastid genes expression, *Funct. Plant Biol.*, 46(8), 766–776. DOI: 10.1071/FP18254.

Yuan, Y., Xu, X., Gong, Z., Tang, Y., Wu, M., Yan, F., Zhang, X., Zhang, Q., Yang, F., Hu, X., Yang, Q., Luo, Y., Mei, L., Zhang, W., Jiang, C. Z., Lu, W., Li, Z. and Deng, W. (2019), Auxin Response Factor 6A regulates photosynthesis, sugar accumulation, and fruit development in tomato, *Hortic. Res.*, 6(1), 85. DOI: 10.1038/s41438-019-0167-x.

Zabala, M. de T., Littlejohn, G., Jayaraman, S., de Torres-Zabala, M., Littlejohn, G., Jayaraman, S., Studholme, D., Bailey, T., Lawson, T., Tillich, M., Licht, D., Bölter, B., Delfino, L., Truman, W., Mansfield, J., Smirnoff, N. and Grant, M. (2015), Chloroplasts play a central role in plant defence and are targeted by pathogen effectors, *Nat. Plants*, 1(6), 1–10. DOI: 10.1038/NPLANTS.2015.74.

Zghidi, W., Merendino, L., Cottet, A., Mache, R. and Lerbs-Mache, S. (2007), Nucleus-encoded plastid sigma factor SIG3 transcribes specifically the *psbN* gene in plastids, *Nucleic Acids Res.*, 35(2), 455–464. DOI: 10.1093/nar/gkl1067.

Zhang, N., Kallis, R. P., Ewy, R. G. and Portis, A. R. (2002), Light modulation of RuBisCO in Arabidopsis requires a capacity for redox regulation of the larger RuBisCO

Activase isoform, *Proc. Natl. Acad. Sci. U. S. A.*, 99(5), 3330–3334. DOI: 10.1073/pnas.042529999.

Zhang, Z. F., Li, Y. Y. and Xiao, B. Z. (2016), Comparative transcriptome analysis highlights the crucial roles of photosynthetic system in drought stress adaptation in upland rice, *Sci. Rep.*, 6(1), 19349. DOI: 10.1038/srep19349.

Zhang, H., Wu, T., Li, Z., Huang, K., Kim, N. E., Ma, Z., Kwon, S. W., Jiang, W. and Du, X. (2021), OsGATA16, a GATA transcription factor, confers cold tolerance by repressing OsWRKY45–1 at the seedling stage in rice, *Rice*, 14(1), 42. DOI: 10.1186/s12284-021-00485-w.

Zhang, X., Yao, Y., Li, X., Zhang, L. and Fan, S. (2020a), Transcriptomic analysis identifies novel genes and pathways for salt stress responses in *Suaeda salsa* leaves, *Sci. Rep.*, 10(1), 4236. DOI: 10.1038/s41598-020-61204-x.

Zhang, Y., Zhang, A., Li, X. and Lu, C. (2020b), The role of chloroplast gene expression in plant responses to environmental stress, *Int. J. Mol. Sci.*, 21(17), 6082. DOI: 10.3390/ijms21176082.

Zhang, C., Zhang, L., Zhang, S., Zhu, S., Wu, P., Chen, Y., Li, M., Jiang, H. and Wu, G. (2015), Global analysis of gene expression profiles in physic nut (*Jatropha curcas* L.) seedlings exposed to drought stress, *BMC Plant Biol.*, 15(1), 17. DOI: 10.1186/s12870-014-0397-x.

Zhao, Y., Qiang, C., Wang, X., Chen, Y., Deng, J., Jiang, C., Sun, X., Chen, H., Li, J., Piao, W., Zhu, X., Zhang, Z., Zhang, H., Li, Z. and Li, J. (2019), New alleles for chlorophyll content and stay-green traits revealed by a genome wide association study in rice (*Oryza sativa*), *Sci. Rep.*, 9(1), 2541. DOI: 10.1038/s41598-019-39280-5.

Zhong, M., Huang, F., Luo, R., Lv, Y., Ali, U., Sheng, Z., Tang, S., Wei, X. and Hu, P. (2020), The effect of cadmium on the microRNAome, degradome and transcriptome of rice seedlings, *Plant Growth Regul.*, 90(1), 15–27. DOI: 10.1007/s10725-019-00547-6.

Zhuang, K., Kong, F., Zhang, S., Meng, C., Yang, M., Liu, Z., Wang, Y., Ma, N. and Meng, Q. (2019), Whirly1 enhances tolerance to chilling stress in tomato via protection of photosystem II and regulation of starch degradation, *New Phytol.*, 221(4), 1998–2012. DOI: 10.1111/nph.15532.

Zubo, Y. O., Blakley, I. C., Franco-Zorrilla, J. M., Yamburenko, M. V., Solano, R., Kieber, J. J., Loraine, A. E. and Schaller, G. E. (2018), Coordination of chloroplast development through the action of the GNC and GLK transcription factor families, *Plant Physiol.*, 178(1), 130–147. DOI: 10.1104/pp.18.00414.

Part 2

Improving photosynthesis: light harvesting

Chapter 3

Interactions between photosynthesis and the circadian system

Marina Viana Queiroz, Universidade de São Paulo, Brazil; and Martin William Battle and Matthew Alan Jones, University of Glasgow, UK;

1 Introduction

Evolution has driven plants to optimise their metabolism so that they can adapt to environmental challenges. This is crucial since day length and light intensity vary during the year, as do maximum/minimum temperatures and the availability of water. Each of these abiotic factors influences photosynthetic efficiency, since plants must balance photosynthetic potential against water retention and the damage induced by completing photosynthesis in suboptimal conditions. Plants have developed several biochemical and physiological strategies to mitigate these challenges including stomatal closure, chloroplast and leaf movements, changes in respiration rate, and photoinhibition (Schulze et al., 2019). However, when should plants utilise these strategies? How can plants distinguish transient changes in illumination (e.g. from sunflecks or cloud cover) from longer-term changes induced by the daily rising and setting of the sun? One solution is provided by the circadian system, a molecular mechanism that provides an endogenous timing reference. This biological clock enables anticipation of environmental signals such as dawn and daily temperature fluctuations as well as providing a yardstick to determine day length. These features enable plants to respond more

http://dx.doi.org/10.19103/AS.2022.0119.06

appropriately to environmental change as well as modulating biochemical and physiological behaviours to optimise growth in varied conditions. In this review, we will discuss how the circadian system regulates photosynthetic metabolism and how these metabolic signals can feed back into the circadian clock itself.

2 The circadian system: a global regulator of metabolism

Daily rhythms of light and temperature have driven the evolution of the circadian system. Each individual cell is capable of entraining its own circadian system, with additional linking between neighbouring cells and tissues to enable organismal responses (Sorkin and Nusinow, 2021). These transcriptional and translation feedback loops are promoted by external signals that coordinate endogenous gene expression and metabolism and maintain their rhythmicity under light and temperature changes (although the circadian system will persist under constant conditions). The core of the molecular circadian clock consists of interlocking positive and negative feedback loops that control gene

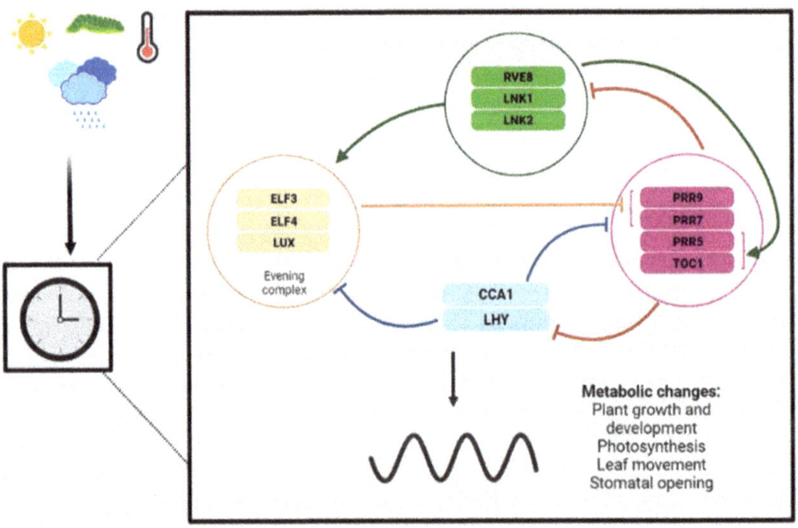

Figure 1 The circadian clock is regulated by several genes acting within transcription-translation feedback loops to generate rhythms that subsequently guide physiological and metabolic processes. REVEILLE8 (RVE8), NIGHT LIGHT-INDUCIBLE AND CLOCK-REGULATED1 (LNK1) and LNK2 are positive regulators of gene expression, whereas PSEUDORESPONSE REGULATORS (PRRs), CIRCADIAN CLOCK ASSOCIATED1 (CCA1), LATE ELONGATED HYPOCOTYL (LHY) and components of the Evening Complex (EARLY FLOWERING3 [ELF3], ELF4 and LUX ARRHYTHMO) are transcriptional repressors. This biological oscillator is sensitive to light and temperature to enable synchronisation between environmental and endogenous signals. Source: Created with Biorender.com.

expression combined with a smaller number of transcriptional activators (Fig. 1; Sorkin and Nusinow, 2021). Morning-phased clock components including *CIRCADIAN CLOCK ASSOCIATED1* (*CCA1*), *LATE ELONGATED HYPOCOTYL* (*LHY*), and *PSEUDO RESPONSE REGULATOR9* (*PRR9*) are activated by light (Ito et al., 2003; Rugnone et al., 2013; Wang and Tobin, 1998), while PRR9 and PRR7 are important for entrainment to temperature (Mizuno et al., 2014; Salomé and McClung, 2005). GIGANTEA regulates ZEITLUPE (ZTL) protein stability and its subsequent regulation of TOC1 (Kim et al., 2007, 2013). CCA1 and LHY repress the expression of *PRR9*, *PRR7*, *PRR5* and *PRR1* (or *TIMING OF CAB EXPRESSION1 - TOC1*), with *TOC1* having the peak expression close to dusk. In turn, the PRRs repress *CCA1/LHY* expression (Fig. 1; Hsu and Harmer, 2014). Myb-like REVEILLE8 (RVE8) is a transcriptional activator and dynamically interacts with NIGHT LIGHT-INDUCIBLE AND CLOCK-REGULATED1 (LNK1) and LNK2 to activate transcription of evening-phased genes (*EARLYFLOWERING3* [ELF3], *ELF4* and *LUX ARRHYTHMO* [*LUX*]) (Xie et al., 2014). Beyond these transcriptional controls, additional layers of regulation, including post-transcriptional and post-translational modification, also contribute to circadian timing (Harmer et al., 2004). For example, reversible phosphorylation can regulate the pace of circadian rhythms by activating, inactivating or providing a targeting signal that contributes to post-translational degradation (Brenna and Albrecht, 2020).

3 The circadian system and its contribution to the regulation of photosynthesis

Although photosynthesis is ultimately dependent upon light, the complex interplay between photosynthesis and other aspects of plant physiology requires coordination of many different aspects of metabolism. One such coordinator is the circadian system which enables timing information to be integrated into biological processes. It has been apparent for many years that photosynthesis is regulated by the circadian system since carbon assimilation follows a rhythmic pattern in plants transferred to constant light (Dodd et al., 2004, 2015; Hennessey and Field, 1991; Salomé et al., 2002). Both wild type plants and plants with a slower circadian clock (such as *ztl-1* seedlings) display circadian rhythms of CO_2 assimilation that correspond to the timing of the molecular timekeeper (Dodd et al., 2004).

One physiological component that likely contributes to these circadian phenotypes is the regulation of guard cells. Guard cells play a major role in regulating stomatal opening by increasing and decreasing their volume via osmotic changes to regulate stomatal pore size (Misra et al., 2015). More importantly, guard cells are responsive to external stimuli like light, heat, water availability and internal signals such as Reactive Oxygen Species (ROS), hormones, lipids and sugar (Misra et al., 2015). Circadian regulation of guard cell movement and stomatal opening (Hassidim et al., 2017; Resco de Dios et al., 2016a) provides a physiological mechanism by which the clock has a fundamental role in carbon assimilation, contributing to photosynthesis by varying intercellular CO_2 availability and consequently regulating the operating efficiency of PSII ($\frac{F_q'}{F_m'}$, or φPSII; Murchie and Lawson, 2013). Several photosynthetic parameters oscillate over circadian time in C_3 and Crassulacean Acid Metabolism (CAM) species, most likely as a consequence of differential gas exchange (Litthauer et al., 2015; Malpas and Jones, 2016; Rascher et al., 2001). Circadian patterns of gas exchange are also observed at the leaf-to-canopy scale, with a significant portion of daytime gas exchange regulated by the clock (Resco de Dios et al., 2016b).

Circadian regulation of guard cell function also confers additional control that can contribute towards Water Use Efficiency (WUE; Hassidim et al., 2017; Lawson and Blatt, 2014; Simon et al., 2020). A primary role for the circadian system appears to be the pre-emptive opening of stomata prior to dawn, which promotes growth at the expense of WUE (Resco de Dios et al., 2016b; Hassidim et al., 2017). However, continued circadian modulation later in the day conversely serves to benefit WUE since *Arabidopsis* seedlings with arhythmic circadian systems tending to have lower WUE (Simon et al., 2020). Interestingly, disruption of the circadian clock preferentially in guard cells can promote WUE in well-watered conditions, although these plants are consequently less tolerant of drought stress overall (Hassidim et al., 2017; Simon et al., 2020). These complex behaviours may represent contrasting requirements to optimise WUE or photosynthetic parameters (Jones, 2017; Resco de Dios et al., 2020). Indeed, modelling guard cell behaviour to optimise either WUE or assimilation in isolation fails to reproduce observed circadian patterns of stomatal opening (Resco de Dios et al., 2020). Instead, it is likely that the circadian system has several competing roles that promote plant growth in suboptimal environments (Hassidim et al., 2017).

3.1 The circadian system and its contribution to the regulation of photosynthesis in C_4 and Crassulacean Acid Metabolism species

Evolutionary pressures to adapt to high temperatures and low water availability have driven the development of C_4 and CAM photosynthesis. These photosynthetic derivations aim to reduce the negative effects of photorespiration by increasing the ribulose-1,5-bisphosphate carboxylase activity through the concentration of CO_2, thereby improving WUE, and photosynthetic efficiency. Ribulose-1,5-bisphosphate carboxylase-oxygenase (Rubisco) is the main enzyme that drives the balance between photorespiration and CO_2 fixation (Voet et al., 2016). The affinity of Rubisco to CO_2 or O_2 is affected by temperature, with Rubisco oxygenase activity increasing at elevated temperatures, leading to increased photorespiration (Dusenge et al., 2019). The occurrence of photorespiration is significantly higher in C_3 plants, compared to C_4 and CAM metabolism, which leads to a decrease in the photosynthetic efficiency of these plants and consequently reduced growth and development in high temperatures (Schulze et al., 2019).

Although both C_4 and CAM mechanisms limit photorespiration, C_4 photosynthesis benefits from the anatomical separation in the leaves, while in CAM plants photosynthetic processes are divided by time (Black and Osmond, 2003; Hatch et al., 1988). C_4 metabolism occurs in parallel in mesophyll and bundle sheath cells. In mesophyll cells, C_4 plants have a circular arrangement called the Kranz anatomy where CO_2 is fixed by phosphoenolpyruvate (PEP) carboxylase giving oxaloacetate which will convert into malate or aspartate (Hatch et al., 1988). The malate is transported to the bundle sheath cells where it is converted to pyruvate, releasing CO_2 to input into the Calvin-Benson Cycle in which Rubisco is responsible for the carbon fixation (Hatch et al., 1988).

CAM plants also have a specific leaf structure with thick cuticles, large vacuoles and small stomata. Besides that, the striking feature of CAM metabolism is the temporal separation between carbon fixation and stomatal opening (Gilman and Edwards, 2020). The assimilation of CO_2 from the atmosphere occurs at night via the enzyme PEP carboxylase, as the production of malic acid and its accumulation in the vacuole (Nimmo, 2000; Winter and Smith, 2022). In addition, PEP is regenerated via glycolysis. During the daytime, malic acid is released into the cytosol, with the decarboxylation of malate releasing CO_2 that will be assimilated via Rubisco into the Calvin-Benson Cycle (Schiller and Bräutigam, 2021). Storage carbohydrates are also regenerated through gluconeogenesis of pyruvate or PEP (Winter, 2019). To control the flux of PEP carboxylase during day/night shifts, plants have developed a mechanism that was first identified in *Kalanchoë fedtschenkoi*. PEP carboxylase is phosphorylated at night, being inhibited by an increase in malate levels and dephosphorylated during the day when the levels of malate decrease (Nimmo, 2000). This phosphorylation/dephosphorylation is controlled largely by the activity of PEP carboxykinase (PEPCK), an enzyme that is under circadian regulation (Carter et al., 1991; Nimmo, 2000; Nimmo et al., 2001).

In spite of these pronounced changes in metabolism, the circadian system in CAM species is broadly conserved compared to Arabidopsis. However, some changes in the timing of peak transcript abundance have been observed. For example, some orthologous genes peak at midday in *Kalanchoë fedtschenkoi* but are morning phased in Arabidopsis. Conversely, another group of *Kalanchoë fedtschenkoi* genes peaking at midday had Arabidopsis orthologues that peaked just after dusk, suggesting differential regulation of rhythmic orthologous gene pairs (Moseley et al., 2018). In addition, the evening complex genes *ELF4, ELF3* and *LUX* have a shift of expression mostly to midday in *Kalanchoë fedtschenkoi* whereas *ELF4* was altered in gene copy number (Moseley et al., 2018).

Although not as pronounced, C_4 plants also utilise circadian timing to coordinate metabolism during the day. As in CAM plants, clock genes are functionally retained between C_4 and C_3 species since studies in maize (*Zea mays*) found several orthologues of Arabidopsis circadian clock such as *PRR7* and *LUX* (Khan et al., 2010). Besides that, the genes encoding PEP carboxylase, an enzyme that fixes CO_2, are rhythmically expressed (Khan et al., 2010). Moreover, the expression of maize enzymes related to Calvin-Benson cycle and the other components of photosynthesis, like light-harvesting complex B and rubisco activase proteins, follows a coordinated expression phased preceding dawn or at dawn (Khan et al., 2010). In commercial sugar cane (*Saccharum* hybrids), the transcripts regulating photosynthesis and carbohydrate metabolism are mainly expressed during the day; meanwhile, at night the genetic machinery has a fundamental role in DNA replication, histone regulation and RNA polymerase, ribosomes and proteins synthesis (Hotta et al., 2013). Despite these differences, clock components are highly conserved in sugarcane, and about 32% of sugarcane transcripts exhibit a rhythmic expression (Hotta et al., 2013). Comparing two C_4 species, sugarcane and maize, it is possible to affirm that sugarcane has a higher number of orthologues clusters controlled by the clock (114 against 29 from maize) (Hotta et al., 2013). This is likely due to an increase in the ploidy and aneuploidy levels that come from agronomic crossings and selections to improve yield (Hotta et al., 2013).

Despite the similarity between the clock of C_4 and CAM species to that found in Arabidopsis, in CAM plants the circadian downregulation of photosynthesis is made through the regulation of PEPCK expression. This enzyme phosphorylates PEP carboxylase reducing the feedback inhibition of PEP by malate and enhancing nocturnal CO_2 fixation to stored malate (the first product from the CO_2 fixation in CAM plants). In addition, PEPCK presents a peak transcript accumulation and enzyme activity in the mid-to-late night (Carter et al., 1991; Nimmo, 2000; Nimmo et al., 2001). In *K. daigremontiana* it has been shown that *PPC1*, a gene that is responsible for controlling the expression of the enzyme PEPCK, is connected to the regulation of guard cell genes (Boxall et al., 2020).

4 Interactions between photosynthates and the circadian system

In order to synchronise with prevailing environmental conditions, the circadian system is highly responsive to environmental factors such as light and temperature. The majority of circadian rhythms comprise sinusoidal waves that vary by phase, period and amplitude. These factors can be modified in response to different environmental conditions to maximise growth and synchronise physiological responses, improving fitness and conferring competitive advantage (Dodd et al., 2005; Harmer, 2009). Indeed, the circadian period is not fixed, and the circadian phase can be advanced or delayed throughout the day in response to environmental factors (Webb et al., 2019). Although the role of photoreceptors has been well explored (Oakenfull and Davis, 2017), the contribution of metabolic signals such as sugars, ROS, and other signalling metabolites have only recently emerged (Webb et al., 2019). In this section, we will discuss how the circadian system regulates the metabolism of photosynthates and also highlight how these metabolites can feedback into the circadian system.

4.1 Starch metabolism as a consistent energy supply during the night

Photosynthesis allows plants to accumulate photosynthates during the day, but how best to utilise this harvested energy? Given the metabolic expense of gluconeogenesis, it is beneficial for plants to accumulate sufficient starch to allow respiration to continue during the night (Smith and Stitt, 2007). Conversely, unutilised starch remaining at dawn represents an inefficient use of resources. For example, plants unable to accumulate starch (or those with impaired starch catabolism) have reduced growth rates except in continuous light or very long days (Smith and Stitt, 2007). The production of starch is regulated by ADP Glucose Pyrophosphorylase (AGPase), due to an increasing ratio of glycerate-3-phosphate to inorganic phosphate (Stitt et al., 2010; Zeeman et al., 2010). Considering all carbon fixed, only 15% is turned into starch, with the remaining amount being distributed to compose cell structures and metabolites such as proteins, cell walls, lipids, pigments and waxes. This carbon exportation from source leaves to sink organs begins immediately after dawn (Kölling et al., 2015). At dusk, there is a decrease in carbon exportation from source organs coordinated with a raising in starch production (Kölling et al., 2015), which contributes to the accumulation of starch in the leaves that can be used during the night (Sulpice et al., 2014). Nevertheless, the ratio of fixed carbon/starch synthesis is pliable and will be modified in response to the photoperiod. For example, to prevent starvation Arabidopsis plants increase the proportion of fixed carbon into starch at the beginning of the diel period when grown under only 3 h of light per day (Gibon et al., 2004; Sulpice et al., 2014). This change in allocation is complemented by a reduced rate of starch degradation that prioritises plant preservation over growth enabling plants to survive under shading or changing seasons (Jones, 2017).

In order to optimise the utilisation of starch reserves during the night, plants exploit the circadian system to anticipate the timing of dawn (Graf et al., 2010). This mechanism ensures carbohydrate availability is stable until dawn by managing the resources between catabolism and metabolism. Importantly, plants with an accelerated circadian system (for example *cca1-11 lhy-21* seedlings) utilise starch at night much faster than in wildtype plants, resulting in the consumption of starch and the presentation of starvation responses in these plants (Graf et al., 2010). The most presumable connection between starch accumulation and the circadian clock is PHOSPHOGLUCAN WATER DIKINASE (PWD), which is fundamental for normal starch degradation through a cycle of phosphorylation reactions, retaining starch at dawn (Smith, 2012). The degradation of starch was not observed in plants lacking PWD following an unpredicted early dusk; moreover, the levels of phosphate bound to the granules are connected to cellular starch content during the day, suggesting that a phosphate-derived signal is sufficient to control cellular starch content (Scialdone et al., 2013). Alternately, plants could regulate starch degradation depending on the accumulation of Trehalose 6-Phospate (Tre6P) in source leaves, connecting starch metabolism to the demand for sucrose (Figueroa and Lunn, 2016). This metabolic approach is favourable to balance source/sink relationships, thereby improving plant growth.

4.2 Indicators of cellular energy status and its integration into the circadian system

Carbohydrate (triose phosphate) synthesis occurs in the chloroplast stroma during the second stage of photosynthesis via the Calvin-Benson cycle (Sharkey, 2019). The reduction of CO_2, and consequently carbon fixation, requires energy in the form of ATP and NADPH, which has been provided by the light reaction of photosynthesis. The amount of NADPH and ATP produced is regulated by photosynthetic parameters such as the efficiency of carboxylation and the regeneration rate of Rubisco, as well as by photorespiration and the rate of transport of carbohydrates out of the cell (Baker, 2008). These endogenous sugars provided by photosynthesis serve as a metabolic signal that can be integrated into the circadian system [Fig. 2, (Dodd et al., 2015; Haydon et al., 2013b)]. But how do plants monitor the accumulation of photosynthates and adjust circadian timing as required?

Tre6P acts as a global regulator of metabolism and transcription promoting plant growth and triggering developmental phase transitions in response to sugar availability (Tsai and Gazzarrini, 2014). Tre6P enables plants to control their intracellular sucrose levels, forming a negative feedback looping where sucrose induces Tre6P production, and Tre6P limits sucrose accumulation (Dos Anjos et al., 2018; Figueroa and Lunn, 2016). Sucrose and Tre6P concentrations

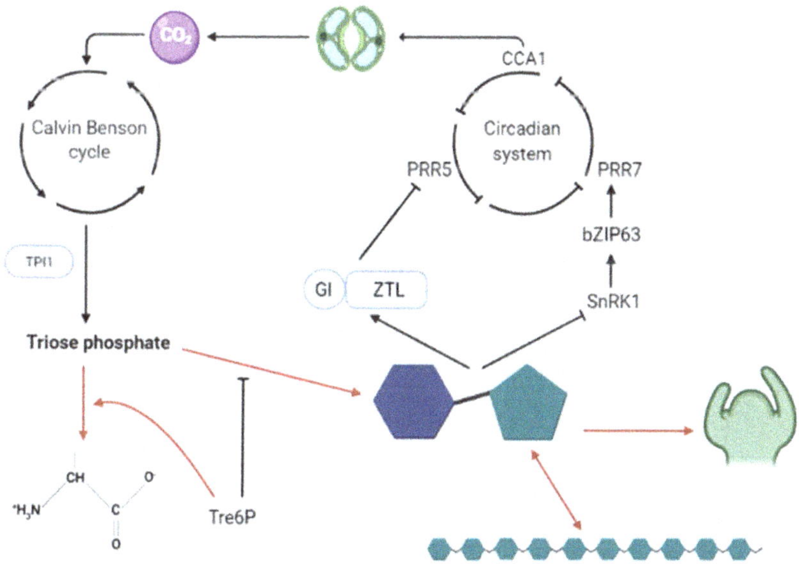

Figure 2 The circadian regulation of the products of photosynthesis. Triose phosphates generated from photosynthesis can be utilised to produce organic acids, amino acids and sugar. Sugars are utilised immediately for plant metabolism and development or can be diverted for starch synthesis; this partitioning is controlled in part by Trehalose 6-phosphate (Tre6P). Sugars feed into the circadian system via two identified pathways. SnRK1 activity is inhibited by sugar, leading to the repression of *PRR7* within the circadian system. Separately, sucrose enhances the accumulation of GIGANTEA (GI) and ZTL, leading to the repression of other circadian factors including *PRR5*. Circadian rhythms can induce stomatal opening, allowing CO_2 entry into the leaf for carbon fixation. Red lines indicate sugar partitioning pathways. Source: Created with Biorender .com.

consequently follow the same pattern, increasing during the day and decreasing at night. One of the targets of Tre6P is the SUCROSE-NON-FERMENTING1-RELATED KINASE1 (SnRK1) family, which allows cells to maintain energetic balance and homeostasis during harmful conditions (Tsai and Gazzarrini, 2014). SnRK1 is induced during starvation conditions, resulting in reduced anabolism and upregulated catabolism to limit energy expenditure (Wurzinger et al., 2018). Processes such as storage compound mobilisation and autophagy are also promoted to recover an energy deficit, while energetically demanding processes such as protein translation and cell proliferation are inhibited (Tsai and Gazzarrini, 2014). SnRK1 has consequently been a prime candidate to integrate metabolic signals into the circadian system.

Photosynthates can entrain the circadian system through two pathways: a feedback loop involving regulation of *PRR7* expression in the morning and a GI/ZTL-dependent post-translational regulation in the evening [Fig. 2, (Dalchau et al., 2011; Haydon et al., 2013b, 2017)]. *prr7* seedlings are unable to adjust circadian timing in response to light or sugar, suggesting that this gene has an essential role in integrating photosynthate-mediated signals (Haydon et al., 2013b). Indeed, SnRK1 promotes the activity of bZIP63 (a transcriptional activator that induces *PRR7* expression) thereby connecting SnRK1 and Tre6P to circadian timing (Frank et al., 2018; Mair et al., 2015). During the day, photosynthesis leads to the accumulation of sugar, inactivating SnRK1 and repressing *PRR7*, contributing to the coordination of circadian and metabolic signals (Haydon et al., 2013a; Frank et al., 2018).

Separately, the addition of sucrose to growth media is sufficient to maintain circadian rhythms during extended dark periods, with GIGANTEA having an important role (Dalchau et al., 2011). GIGANTEA is necessary to maintain this sucrose-dependent rhythmicity under dark through the stabilisation of GIGANTEA and ZTL aggregations (Dalchau et al., 2011; Haydon et al., 2017). In the absence of sucrose, GIGANTEA (along with other circadian components) are degraded, therefore reducing circadian amplitude (Haydon et al., 2017). This latter pathway may serve to limit circadian rhythms under low light (or starvation) conditions, although additional experimentation will be needed to test this hypothesis.

5 Generation of reactive oxygen species during photosynthesis

The generation of ROS is a natural process that occurs as a consequence of metabolic processes such as photosynthesis and respiration, as well as being utilised as a mechanism to dissipate excess absorbed light energy (Khorobrykh et al., 2020). Absorbed energy harvested by the photosynthetic antennae oxidises a water molecule and initiates the electron transfer pathway. The electron can follow three paths enabling cyclic, pseudocyclic and noncyclic phosphorylation. Noncyclic phosphorylation generates an electrical gradient with the production of the NADPH from $NADP^+$ as the electron acceptor, ATP and molecular oxygen ending the photochemistry stage of photosynthesis, whereas cyclic and pseudocyclic pathways are non-productive (Allen, 2003). Cyclic phosphorylation only occurs at PSI without the production and consumption of oxygen. The pseudocyclic path (water-water cycle) requires both PSII and PSI, with the production and consumption of oxygen as oxygen serves as an electron acceptor (Allen, 2003).

Due to an imbalance in O_2 reduction in the PSI, the pseudocyclic path can produce superoxide anions radical ($O_2^{\bullet-}$) through two main mechanisms: reduced plastoquinone donates electrons to O_2 via the plastid terminal oxidase (Heyno et al., 2009) and unbalance of PSI due to disparity between demand and supply of electron acceptors such as $NADP^+$, nitrite or sulphite, which transfers electrons from ferredoxin or the Fe-S-centre Fx to O_2 (Dietz, 2016).

The superoxide anion radical is converted into H_2O_2 by the action of thylakoid superoxide dismutase. Produced H_2O_2 is reduced to water by chloroplast ascorbate peroxidases and the 2-cysteine peroxiredoxins (PRXs), creating the water-water cycle to preserve the photosynthetic apparatus (Awad et al., 2015).

Plants have complex regulatory pathways to balance the accumulation of ROS as signalling and its degradation to limit potential damage (Foyer, 2018). When these compounds reach a high enough level, apoptosis is activated to regulate growth, development and eliminate damaged or infected cells (Wang et al., 2013). The imbalance between ROS production and ROS scavenging (which happens under stress conditions) causes fluctuations in the redox state of the cell, modifying the levels of reduced and oxidised forms of antioxidants like ascorbate, glutathione (GSH) and thiol-containing compounds (Foyer, 2018; Khorobrykh et al., 2020). Superoxide anion and H_2O_2 are largely generated at PSI on the stromal side of non-appressed thylakoid membranes (Wang et al., 2016), although it has been reported that H_2O_2 could be directly transferred to the nucleus from attached chloroplasts or via stromules (Caplan et al., 2015; Exposito-Rodriguez et al., 2017). In addition, several retrograde signalling pathways have been identified that report oxidative damage in the chloroplast and mitochondria via secondary messengers that accumulate in response to impaired organelle performance (Jones, 2019).

5.1 Integration of reactive oxygen species into circadian timing

The change in light availability during the day leads to changes in the rate of photosynthesis and therefore affects the generation of ROS and the redox status of the cell (del Río, 2015). Interestingly, PRX hyper-oxidation varies with a circadian rhythm in many species, leading to speculation that redox rhythms may have contributed to the ancestral circadian system (Edgar et al., 2012). Despite this possibility, it remains to be determined if rhythms in PRX oxidation contribute to the maintenance of nuclear circadian oscillations, or whether PRX rhythms simply contribute towards observed rhythms of photosynthetic capacity (Litthauer et al., 2015; Simon et al., 2019).

In addition to these chloroplastic PRX rhythms, the nuclear circadian system has been shown to contribute to cellular ROS homeostasis (Lai et al., 2012). Mutations in *CCA1* and *LHY* increased the sensitivity of these plants to ROS, emphasising the importance of a functional clock for ROS homeostasis (Lai et al., 2012). It has also been reported that sucrose initiates the accumulation of O_2^-, suggesting that sugar and ROS signals could be linked within the circadian system (Román et al., 2021). Despite these advances, it still remains unclear how ROS contribute to circadian timing, and the link between photosynthesis and these pathways remains a promising avenue for future investigation. Such work will allow us to understand how photosynthetic performance can be integrated with daily and seasonal changes.

6 Conclusion

Recent technical advances and innovations have enabled the integration of photosynthetic metabolism with the circadian system. These efforts reveal how photosynthetic products feed into the circadian system as an indicator of metabolic productivity as well as demonstrating the contribution of the circadian system towards photosynthetic regulation. Such complex interactions highlight the importance of a holistic viewpoint when considering how plants perceive and respond to changing environmental conditions. In future, multi-disciplinary approaches will be necessary to fully understand how crop production can be increased during the global heating crisis.

7 Where to look for further information

The role of circadian biology in regulating photosynthesis and gas exchange is becoming apparent. Future advances will provide better comprehension of the association between the circadian system and photosynthesis and will enable improvements in crop growth under stressful conditions. This will allow better management of natural resources and more effective exploitation of marginal land.

The School of Molecular Biosciences at the University of Glasgow provides an integrative environment aimed at driving basic and applied research of the highest standard from the molecular and cellular level to whole organisms and systems. The School hosts a strong group of 10 plant science principal investigators that maintain a broad interest in understanding how plant growth and development is regulated by environmental stimuli, and how plants cope with a changing environment. Our research focuses on identifying the molecular mechanisms underlying physiological and developmental adaptation to environmental cues, such as light quality and day length, as well as environmental stress factors such as drought, high temperatures and salinity.

8 References

Allen, J. F. (2003). Cyclic, pseudocyclic and noncyclic photophosphorylation: new links in the chain. *Trends in Plant Science* 8(1), 15–19. https://doi.org/10.1016/S1360-1385(02)00006-7.

Awad, J., Stotz, H. U., Fekete, A., Krischke, M., Engert, C., Havaux, M., Berger, S. and Mueller, M. J. (2015). 2-cysteine peroxiredoxins and thylakoid ascorbate peroxidase create a water-water cycle that is essential to protect the photosynthetic apparatus under high light stress conditions. *Plant Physiology* 167(4), 1592–1603. https://doi.org/10.1104/pp.114.255356.

Baker, N. R. (2008). Chlorophyll fluorescence: a probe of photosynthesis in vivo. *Annual Review of Plant Biology* 59, 89–113. https://doi.org/10.1146/annurev.arplant.59.032607.092759.

Black, C. C. and Osmond, C. B. (2003). Crassulacean acid metabolism photosynthesis: "working the night shift." *Photosynthesis Research* 76(1-3), 329–341. https://doi.org/10.1023/A:1024978220193.

Boxall, S. F., Kadu, N., Dever, L. V., Kneřová, J., Waller, J. L., Gould, P. J. D. and Hartwell, J. (2020). Kalanchoë PPC1 is essential for crassulacean acid metabolism and the regulation of core circadian clock and guard cell signaling genes. *The Plant Cell* 32(4), 1136–1160. https://doi.org/10.1105/tpc.19.00481.

Brenna, A. and Albrecht, U. (2020). Phosphorylation and circadian molecular timing. *Frontiers in Physiology* 11(November), 612510. https://doi.org/10.3389/fphys.2020.612510.

Caplan, J. L., Kumar, A. S., Park, E., Padmanabhan, M. S., Hoban, K., Modla, S., Czymmek, K. and Dinesh-Kumar, S. P. (2015). Chloroplast stromules function during innate immunity. *Developmental Cell* 34(1), 45–57. https://doi.org/10.1016/j.devcel.2015.05.011.

Carter, P. J., Nimmo, H. G., Fewson, C. A. and Wilkins, M. B. (1991). Circadian rhythms in the activity of a plant protein kinase. *The EMBO Journal* 10(8), 2063–2068.

Dalchau, N., Baek, S. J., Briggs, H. M., Robertson, F. C., Dodd, A. N., Gardner, M. J., Stancombe, M. A., Haydon, M. J., Stan, G.-B. B., Gonçalves, J. M. and Webb, A. A. R. R. (2011). The circadian oscillator gene GIGANTEA mediates a long-term response of the Arabidopsis thaliana circadian clock to sucrose. *Proceedings of the National Academy of Sciences of the United States of America* 108(12), 5104–5109. https://doi.org/10.1073/pnas.1015452108.

del Río, L. A. (2015). ROS and RNS in plant physiology: an overview. *Journal of Experimental Botany* 66(10), 2827–2837. https://doi.org/10.1093/jxb/erv099.

Dietz, K. J. (2016). Thiol-based peroxidases and ascorbate peroxidases: why plants rely on multiple peroxidase systems in the photosynthesizing chloroplast? *Molecules and Cells* 39(1), 20–25. https://doi.org/10.14348/molcells.2016.2324.

Dodd, A. N., Belbin, F. E., Frank, A. and Webb, A. A. R. (2015). Interactions between circadian clocks and photosynthesis for the temporal and spatial coordination of metabolism. *Frontiers in Plant Science* 6, 245. https://doi.org/10.3389/fpls.2015.00245.

Dodd, A. N., Parkinson, K. and Webb, A. A. R. (2004). Independent circadian regulation of assimilation and stomatal conductance in the ztl-1 mutant of Arabidopsis. *New Phytologist* 162(1), 63–70.

Dodd, A. N., Salathia, N., Hall, A., Kévei, E., Tóth, R., Nagy, F., Hibberd, J. M., Millar, A. J. and Webb, A. A. R. (2005). Plant circadian clocks increase photosynthesis, growth, survival and competitive advantage. *Science* 309(5734), 630–633. https://doi.org/10.1126/science.1115581.

Dos Anjos, L., Pandey, P. K., Moraes, T. A., Feil, R., Lunn, J. E. and Stitt, M. (2018). Feedback regulation by trehalose 6-phosphate slows down starch mobilization below the rate that would exhaust starch reserves at dawn in Arabidopsis leaves. *Plant Direct* 2(8), e00078. https://doi.org/10.1002/pld3.78.

Dusenge, M. E., Duarte, A. G. and Way, D. A. (2019). Plant carbon metabolism and climate change: elevated CO2 and temperature impacts on photosynthesis, photorespiration and respiration. *New Phytologist* 221(1), 32–49. https://doi.org/10.1111/nph.15283.

Edgar, R. S., Edgar, R. S., Green, E. W., Zhao, Y., van Ooijen, G., Olmedo, M., Qin, X., Xu, Y., Pan, M., Valekunja, U. K., Feeney, K. A., Maywood, E. S., Hastings, M. H., Baliga, N. S., Merrow, M., Millar, A. J., Johnson, C. H., Kyriacou, C. P., O'Neill, J. S. and Reddy, A. B. (2012). Peroxiredoxins are conserved markers of circadian rhythms. *Nature* 485, 459–464.

Exposito-Rodriguez, M., Laissue, P. P., Yvon-Durocher, G., Smirnoff, N. and Mullineaux, P. M. (2017). Photosynthesis-dependent H2O2 transfer from chloroplasts to nuclei provides a high-light signalling mechanism. *Nature Communications* 8(1), 49. https://doi.org/10.1038/s41467-017-00074-w.

Figueroa, C. M. and Lunn, J. E. (2016). A tale of two sugars: trehalose 6-phosphate and sucrose. *Plant Physiology* 172(1), 7–27. https://doi.org/10.1104/pp.16.00417.

Foyer, C. H. (2018). Reactive oxygen species, oxidative signaling and the regulation of photosynthesis. *Environmental and Experimental Botany* 154, 134–142. https://doi.org/10.1016/j.envexpbot.2018.05.003.

Frank, A., Matiolli, C. C., Viana, A. J. C., Hearn, T. J., Kusakina, J., Belbin, F. E., Wells Newman, D., Yochikawa, A., Cano-Ramirez, D. L., Chembath, A., Cragg-Barber, K., Haydon, M. J., Hotta, C. T., Vincentz, M., Webb, A. A. R. and Dodd, A. N. (2018). Circadian entrainment in Arabidopsis by the sugar-responsive transcription factor bZIP63. *Current Biology: CB* 28(16), 2597–2606.e6. https://doi.org/10.1016/j.cub.2018.05.092.

Gibon, Y., Bläsing, O. E., Palacios-Rojas, N., Pankovic, D., Hendriks, J. H. M., Fisahn, J., Höhne, M., Günther, M. and Stitt, M. (2004). Adjustment of diurnal starch turnover to short days: depletion of sugar during the night leads to a temporary inhibition of carbohydrate utilization, accumulation of sugars and post-translational activation of ADP-glucose pyrophosphorylase in the following light period. *The Plant Journal: For Cell and Molecular Biology* 39(6), 847–862. https://doi.org/10.1111/j.1365-313X.2004.02173.x.

Gilman, I. S. and Edwards, E. J. (2020). Crassulacean acid metabolism. *Current Biology* 30(2), R57–R62. https://doi.org/10.1016/j.cub.2019.11.073.

Graf, A., Schlereth, A., Stitt, M. and Smith, A. M. (2010). Circadian control of carbohydrate availability for growth in Arabidopsis plants at night. *Proceedings of the National Academy of Sciences of the United States of America* 107(20), 9458–9463. https://doi.org/10.1073/pnas.0914299107.

Harmer, S. L. (2009). The circadian system in higher plants. *Annual Review of Plant Biology* 60, 357–377. https://doi.org/10.1146/annurev.arplant.043008.092054.

Harms, E., Kivimäe, S., Young, M. W. and Saez, L. (2004). Posttranscriptional and posttranslational regulation of clock genes. *Journal of Biological Rhythms* 19(5), 361–373. https://doi.org/10.1177/0748730404268111.

Hassidim, M., Dakhiya, Y., Turjeman, A., Hussien, D., Shor, E., Anidjar, A., Goldberg, K. and Green, R. M. (2017). CIRCADIAN CLOCK ASSOCIATED1 (CCA1) and the circadian control of stomatal aperture. *Plant Physiology* 175(4), 1864–1877. https://doi.org/10.1104/pp.17.01214.

Hatch, M. D., Agostino, A. and Burnell, J. N. (1988). Photosynthesis in phosphoenolpyruvate carboxykinase-type C4 plants: activity and role of mitochondria in bundle sheath cells. *Archives of Biochemistry and Biophysics* 261(2), 357–367. https://doi.org/10.1016/0003-9861(88)90351-7.

Haydon, M. J., Mielczarek, O., Robertson, F. C., Hubbard, K. E. and Webb, A. A. R. (2013a). Photosynthetic entrainment of the Arabidopsis thaliana circadian clock. *Nature* 502(7473), 689–692. https://doi.org/10.1038/nature12603.

Haydon, M. J., Hearn, T. J., Bell, L. J., Hannah, M. A. and Webb, A. A. R. (2013b). Metabolic regulation of circadian clocks. *Seminars in Cell and Developmental Biology* 24(5), 414–421. https://doi.org/10.1016/j.semcdb.2013.03.007.

Haydon, M. J., Mielczarek, O., Frank, A., Román, Á. and Webb, A. A. R. (2017). Sucrose and ethylene signaling interact to modulate the circadian clock. *Plant Physiology* 175(2), 947–958. https://doi.org/10.1104/pp.17.00592.

Hennessey, T. L. and Field, C. B. (1991). Circadian rhythms in photosynthesis: oscillations in carbon assimilation and stomatal conductance under constant conditions. *Plant Physiology* 96(3), 831–836. https://doi.org/10.1104/pp.96.3.831.

Heyno, E., Gross, C. M., Laureau, C., Culcasi, M., Pietri, S. and Krieger-Liszkay, A. (2009). Plastid alternative oxidase (PTOX) promotes oxidative stress when overexpressed in tobacco. *Journal of Biological Chemistry* 284(45), 31174–31180. https://doi.org/10.1074/jbc.M109.021667.

Hotta, C. T., Nishiyama Jr., M. Y. and Souza, G. M. (2013). Circadian rhythms of sense and antisense transcription in sugarcane, a highly polyploid crop. *PLoS ONE* 8(8), e71847. https://doi.org/10.1371/journal.pone.0071847.

Hsu, P. Y. and Harmer, S. L. (2014). Wheels within wheels: the plant circadian system. *Trends in Plant Science* 19(4), 240–249. https://doi.org/10.1016/j.tplants.2013.11.007.

Ito, S., Matsushika, A., Yamada, H., Sato, S., Kato, T., Tabata, S., Yamashino, T. and Mizuno, T. (2003). Characterization of the APRR9 pseudo-response regulator belonging to the APRR1/TOC1 quintet in Arabidopsis thaliana. *Plant and Cell Physiology* 44(11), 1237–1245. https://doi.org/10.1093/pcp/pcg136.

Jones, M. A. (2017). Interplay of circadian rhythms and light in the regulation of photosynthesis-derived metabolism. *Progress in Botany*, 147–171. https://doi.org/10.1007/124_2017_2.

Jones, M. A. (2019). Retrograde signalling as an informant of circadian timing. *The New Phytologist* 221, 1749–1753.

Khan, S., Rowe, S. C. and Harmon, F. G. (2010). Coordination of the maize transcriptome by a conserved circadian clock. *BMC Plant Biology* 10(1), 126. https://doi.org/10.1186/1471-2229-10-126.

Khorobrykh, S., Havurinne, V., Mattila, H. and Tyystjärvi, E. (2020). Oxygen and ROS in photosynthesis. *Plants* 9(1). https://doi.org/10.3390/plants9010091.

Kim, J., Geng, R., Gallenstein, R. A. and Somers, D. E. (2013). The F-box protein ZEITLUPE controls stability and nucleocytoplasmic partitioning of GIGANTEA. *Development* 140(19), 4060–4069. https://doi.org/10.1242/dev.096651.

Kim, W. Y., Fujiwara, S., Suh, S. S., Kim, J., Kim, Y., Han, L., David, K., Putterill, J., Nam, H. G. and Somers, D. E. (2007). ZEITLUPE is a circadian photoreceptor stabilized by GIGANTEA in blue light. *Nature* 449(7160), 356–360. https://doi.org/10.1038/nature06132.

Kölling, K., Thalmann, M., Müller, A., Jenny, C. and Zeeman, S. C. (2015). Carbon partitioning in Arabidopsis thaliana is a dynamic process controlled by the plants metabolic status and its circadian clock. *Plant, Cell and Environment* 38(10), 1965–1979. https://doi.org/10.1111/pce.12512.

Lai, A. G., Doherty, C. J., Mueller-Roeber, B., Kay, S. A., Schippers, J. H. M. and Dijkwel, P. P. (2012). Circadian Clock-Associated 1 regulates ROS homeostasis and oxidative stress responses. *Proceedings of the National Academy of Sciences of the United States of America* 109(42), 17129–17134. https://doi.org/10.1073/pnas.1209148109.

Lawson, T. and Blatt, M. R. (2014). Stomatal size, speed, and responsiveness impact on photosynthesis and water use efficiency. *Plant Physiology* 164(4), 1556–1570. https://doi.org/10.1104/pp.114.237107.

Litthauer, S., Battle, M. W., Lawson, T. and Jones, M. A. (2015). Phototropins maintain robust circadian oscillation of PSII operating efficiency under blue light. *Plant Journal: For Cell and Molecular Biology* 83(6), 1034–1045. https://doi.org/10.1111/tpj.12947.

Mair, A., Pedrotti, L., Wurzinger, B., Anrather, D., Simeunovic, A., Weiste, C., Valerio, C., Dietrich, K., Kirchler, T., Nägele, T., Vicente Carbajosa, J., Hanson, J., Baena-González, E., Chaban, C., Weckwerth, W., Dröge-Laser, W. and Teige, M. (2015). SnRK1-triggered switch of bZIP63 dimerization mediates the low-energy response in plants. *eLife* 4, e05828. https://doi.org/10.7554/eLife.05828.

Malpas, K. R. and Jones, M. A. (2016). Natural variation of circadian rhythms in Kalanchoe species. *Haseltonia* 22, 35–42. https://doi.org/10.2985/026.022.0107.

Misra, B. B., Acharya, B. R., Granot, D., Assmann, S. M. and Chen, S. (2015). The guard cell metabolome: functions in stomatal movement and global food security. *Frontiers in Plant Science* 6, 334. https://doi.org/10.3389/fpls.2015.00334.

Mizuno, T., Nomoto, Y., Oka, H., Kitayama, M., Takeuchi, A., Tsubouchi, M. and Yamashino, T. (2014). Ambient temperature signal feeds into the circadian clock transcriptional circuitry through the EC night-time repressor in Arabidopsis thaliana. *Plant and Cell Physiology* 55(5), 958–976. https://doi.org/10.1093/pcp/pcu030.

Moseley, R. C., Mewalal, R., Motta, F., Tuskan, G. A., Haase, S. and Yang, X. (2018). Conservation and diversification of circadian rhythmicity Between a model crassulacean acid metabolism plant Kalanchoë fedtschenkoi and a model C3 photosynthesis plant Arabidopsis thaliana. *Frontiers in Plant Science* 9(November), 1757. https://doi.org/10.3389/fpls.2018.01757.

Murchie, E. H. and Lawson, T. (2013). Chlorophyll fluorescence analysis: a guide to good practice and understanding some new applications. *Journal of Experimental Botany* 64(13), 3983–3998. https://doi.org/10.1093/jxb/ert208.

Nimmo, H. G. (2000). The regulation of phosphoenolpyruvate carboxylase in CAM plants. *Trends in Plant Science* 5(2), 75–80. https://doi.org/10.1016/S1360-1385(99)01543-5.

Nimmo, H. G., Fontaine, V., Hartwell, J., Jenkins, G. I., Nimmo, G. A. and Wilkins, M. B. (2001). PEP carboxylase kinase is a novel protein kinase controlled at the level of expression. *New Phytologist* 151(1), 91–97. https://doi.org/10.1046/j.1469-8137.2001.00155.x.

Oakenfull, R. J. and Davis, S. J. (2017). Shining a light on the Arabidopsis circadian clock. *Plant, Cell and Environment* 40(11), 2571–2585. https://doi.org/10.1111/pce.13033.

Rascher, U., Hütt, M. T., Siebke, K., Osmond, B., Beck, F. and Lüttge, U. (2001). Spatiotemporal variation of metabolism in a plant circadian rhythm: the biological clock as an assembly of coupled individual oscillators. *Proceedings of the National Academy of Sciences of the United States of America* 98(20), 11801–11805. https://doi.org/10.1073/pnas.191169598.

Resco de Dios, V., Anderegg, W. R. L., Li, X., Tissue, D. T., Bahn, M., Landais, D., Milcu, A., Yao, Y., Nolan, R. H., Roy, J. and Gessler, A. (2020). Circadian regulation does not optimize stomatal behaviour. *Plants* 9(9), 1–10. https://doi.org/10.3390/plants9091091.

Resco de Dios, V., Gessler, A., Ferrio, J. P., Alday, J. G., Bahn, M., Del Castillo, J., Devidal, S., García-Muñoz, S., Kayler, Z., Landais, D., Martín-Gómez, P., Milcu, A., Piel, C., Pirhofer-Walzl, K., Ravel, O., Salekin, S., Tissue, D. T., Tjoelker, M. G., Voltas, J. and Roy, J. (2016a). Circadian rhythms have significant effects on leaf-to canopy scale gas exchange under field conditions. *GigaScience* 5(1), 43. https://doi.org/10.1186/s13742-016-0149-y.

Resco de Dios, V., Loik, M. E., Smith, R., Aspinwall, M. J. and Tissue, D. T. (2016b). Genetic variation in circadian regulation of nocturnal stomatal conductance enhances carbon assimilation and growth. *Plant, Cell and Environment* 39(1), 3–11. https://doi.org/10.1111/pce.12598.

Román, Á., Li, X., Deng, D., Davey, J. W., James, S., Graham, I. A., Haydon, M. J. (2021). Superoxide is promoted by sucrose and affects amplitude of circadian rhythms in the evening. *Proceedings of the National Academy of Sciences of the United States of America* 118, e2020646118.

Rugnone, M. L., Soverna, A. F., Sanchez, S. E., Schlaen, R. G., Hernando, C. E., Seymour, D. K., Mancini, E., Chernomoretz, A., Weigel, D., Mas, P. and Yanovsky, M. J. (2013). LNK genes integrate light and clock signaling networks at the core of the Arabidopsis oscillator. *Proceedings of the National Academy of Sciences of the United States of America* 110(29), 12120–12125. https://doi.org/10.1073/pnas.1302170110.

Salomé, P. A. and McClung, C. R. (2005). Pseudo-response Regulator 7 and 9 are partially redundant genes essential for the temperature responsiveness of the Arabidopsis circadian clock. *Plant Cell* 17(3), 791–803. https://doi.org/10.1105/tpc.104.029504.

Salomé, P. A., Michael, T. P., Kearns, E. V., Fett-Neto, A. G., Sharrock, R. A. and McClung, C. R. (2002). The out of phase 1 mutant defines a role for PHYB in circadian phase control in Arabidopsis. *Plant Physiology* 129(4), 1674–1685. https://doi.org/10.1104/pp.003418.

Schiller, K. and Bräutigam, A. (2021). Engineering of crassulacean acid metabolism. *Annual Review of Plant Biology* 72, 77–103. https://doi.org/10.1146/annurev-arplant-071720-104814.

Schulze, E.-D., Beck, E., Buchmann, N., Clemens, S., Müller-Hohenstein, K. and Scherer-Lorenzen, M. (2019). *Plant Ecology* (2nd edn.). Springer. https://doi.org/10.1007/978-3-662-56233-8.

Scialdone, A., Mugford, S. T., Feike, D., Skeffington, A., Borrill, P., Graf, A., Smith, A. M. and Howard, M. (2013). Arabidopsis plants perform arithmetic division to prevent starvation at night. *eLife* 2, e00669. https://doi.org/10.7554/eLife.00669.

Sharkey, T. D. (2019). Discovery of the canonical Calvin–Benson cycle. *Photosynthesis Research* 140(2), 235–252. https://doi.org/10.1007/s11120-018-0600-2.

Simon, N. M. L., Graham, C. A., Comben, N. E., Hetherington, A. M. and Dodd, A. N. (2020). The circadian clock influences the long-term water use efficiency of Arabidopsis1. *Plant Physiology* 183(1), 317–330. https://doi.org/10.1104/pp.20.00030.

Simon, N. M. L., Litthauer, S., Jones, M. A. and Dodd, A. N. (2019). Interactions between circadian rhythms, ROS and redox. In: Panda, S. K. and Yamamoto, Y. Y. (eds.). *Redox Homeostasis in Plants: From Signalling to Stress Tolerance*. Series: Signaling and communication in plants. Springer: Cham, pp. 67-84. ISBN: 9783319953144. DOI: 10.1007/978-3-319-95315-1_4.

Smith, A. M. (2012). Starch in the Arabidopsis plant. *Starch - Stärke* 64(6), 421-434. https://doi.org/10.1002/star.201100163.

Smith, A. M. and Stitt, M. (2007). Coordination of carbon supply and plant growth. *Plant, Cell and Environment* 30(9), 1126–1149. https://doi.org/10.1111/j.1365-3040.2007.01708.x.

Sorkin, M. L. and Nusinow, D. A. (2021). Time will tell: intercellular communication in the plant clock. *Trends in Plant Science* 26(7), 706–719. https://doi.org/10.1016/j.tplants.2020.12.009.

Stitt, M., Lunn, J. and Usadel, B. (2010). Arabidopsis and primary photosynthetic metabolism - more than the icing on the cake. *The Plant Journal: For Cell and Molecular Biology* 61(6), 1067–1091. https://doi.org/10.1111/j.1365-313X.2010.04142.x.

Sulpice, R., Flis, A., Ivakov, A. A., Apelt, F., Krohn, N., Encke, B., Abel, C., Feil, R., Lunn, J. E. and Stitt, M. (2014). Arabidopsis coordinates the diurnal regulation of carbon allocation and growth across a wide range of photoperiods. *Molecular Plant* 7(1), 137–155. https://doi.org/10.1093/mp/sst127.

Tsai, A. Y. L. and Gazzarrini, S. (2014). Trehalose-6-phosphate and SnRK1 kinases in plant development and signaling: the emerging picture. *Frontiers in Plant Science* 5, 119. https://doi.org/10.3389/fpls.2014.00119.

Voet, D., Voet, J. G. and Pratt, C. W. (2016). *Fundamentals of Biochemistry: Life at the Molecular Level* (5th edn.). Wiley.

Wang, L., Kim, C., Xu, X., Piskurewicz, U., Dogra, V., Singh, S., Mahler, H. and Apel, K. (2016). Singlet oxygen- and EXECUTER1-mediated signaling is initiated in grana margins and depends on the protease FtsH2. *Proceedings of the National Academy of Sciences of the United States of America* 113(26), E3792–E3800. https://doi.org/10.1073/pnas.1603562113.

Wang, Y., Loake, G. J. and Chu, C. (2013). Cross-talk of nitric oxide and reactive oxygen species in plant programed cell death. *Frontiers in Plant Science* 4. https://www.frontiersin.org/article/10.3389/fpls.2013.00314.

Wang, Z. Y. and Tobin, E. M. (1998). Constitutive expression of the CIRCADIAN CLOCK ASSOCIATED 1 (CCA1) gene disrupts circadian rhythms and suppresses its own expression. *Cell* 93(7), 1207–1217. https://doi.org/10.1016/S0092-8674(00)81464-6.

Webb, A. A. R., Seki, M., Satake, A. and Caldana, C. (2019). Continuous dynamic adjustment of the plant circadian oscillator. *Nature Communications* 10(1), 550. https://doi.org/10.1038/s41467-019-08398-5.

Winter, K. (2019). Ecophysiology of constitutive and facultative CAM photosynthesis. *Journal of Experimental Botany* 70(22), 6495–6508. https://doi.org/10.1093/jxb/erz002.

Winter, K. and Smith, J. A. C. (2022). CAM photosynthesis: the acid test. *New Phytologist* 233(2), 599–609. https://doi.org/10.1111/nph.17790.

Wurzinger, B., Nukarinen, E., Nägele, T., Weckwerth, W. and Teige, M. (2018). The snrk1 kinase as central mediator of energy signaling between different organelles. *Plant Physiology* 176(2), 1085–1094. https://doi.org/10.1104/pp.17.01404.

Xie, Q., Wang, P., Liu, X., Yuan, L., Wang, L., Zhang, C., Li, Y., Xing, H., Zhi, L., Yue, Z., Zhao, C., McClung, C. R. and Xu, X. (2014). LNK1 and LNK2 are transcriptional coactivators in the Arabidopsis circadian oscillator. *The Plant Cell* 26(7), 2843–2857. https://doi.org/10.1105/tpc.114.126573.

Zeeman, S. C., Kossmann, J. and Smith, A. M. (2010). Starch: its metabolism, evolution, and biotechnological modification in plants. *Annual Review of Plant Biology* 61(1), 209-234. https://doi.org/10.1146/annurev-arplant-042809-112301.

Chapter 4

Modifying photosystem antennas to improve light harvesting for photosynthesis in crops

Min Chen, The University of Sydney, Australia; and Robert E. Blankenship, Washington University in St Louis, USA

1 Introduction

Photosynthesis is the biological process by which plants, algae and other photosynthetic organisms use sunlight to synthesize carbohydrates (sugars, $(CHO)_n$) from carbon dioxide (CO_2) and water (H_2O). Photosynthesis is a chemical reaction that takes place inside of plants and is driven by sunlight to make nutrients (food) for plants, eventually providing food and energy that sustain life on Earth.

The light-dependent CO_2 fixation process can be summarized as the following chemical equation, which indicates that the amount of light captured by plants poses a physical limit for photosynthesis.

$$CO2 + H2O \xrightarrow[\text{Light}]{} (CHO)n + O2$$

Efficiently harvesting sunlight is an essential aspect of improving the biomass production of photosynthetic organisms. If the sun is in the acme

http://dx.doi.org/10.19103/AS.2022.0119.08

of intensity, the amounts of ~1.4 kW.m^{-2} impinging on the Earth's surface vastly surpasses the biological energy needs of the plant (Chen and Scheer, 2013). However, sunlight is also a diffuse and diluted source of energy. Most photosynthetic organisms are unable to efficiently use midday light intensity and are unable to reach the maximum efficiency at low intensity of light as well. The temporal and spatial variations of the sunlight flux and its spectral composition pose considerable problems to photosynthetic organisms, for example, the sun breaks through clouds or a forest canopy. Photosynthetic organisms must compete for light but, at the same time, must avoid damage by excess light. In addition, only a few photopigments can perform photochemical reactions in the reaction center; instead, many photopigment molecules cooperate to provide light energy to ensure the photochemical reactions in the reaction centers. This balance between starvation and being overloaded is mainly maintained by a modular composition of the photosynthetic apparatus and regulatory networks, most importantly the balance between the primary processes of light-harvesting complexes and the energy transduction in the reaction centers. Maximizing the potential for light harvesting per unit or per area will play significant roles in improving biomass. In this chapter, we will discuss the possibilities of maximizing light-harvesting efficiency to meet various competitive situations, including light, nutrients, water use, etc.

2 Photopigments and their functions in light-harvesting complexes

The light quality and quantity reaching the Earth's surface are modified by lower solar altitudes and scattering of the atmosphere (Fig. 1). Oxygenic photosynthetic organisms can only use light in the spectral region of 300–760 nm setting by narrow absorption of photopigments. Individual photopigment captures only a narrow range of the spectrum, e.g. carotenoids will absorb blue light (Fig. 1). The diversity of photopigments is the mechanism enabling photosynthetic organisms to capture more of the sun's energy to maximize light-harvesting efficiency (Fig. 1). There are three basic classes of photosynthetic pigments: chlorophylls, carotenoids and bilins.

Chlorophylls are greenish pigments functioning as the key photosynthetic pigments, contributing to the green color of plants. The other two classes of pigments play important roles in light harvesting and photoprotection. Chlorophylls are the first defined photopigments used by plants. Since the first reported chlorophyll *a* in the early twentieth century, five natural types of chlorophylls have been discovered to date. Chl *a* and Chl *b* are the main chlorophylls found in plants, Chl *c* is only detected in some algae and Chl *d* and Chl *f* are present in some cyanobacteria.

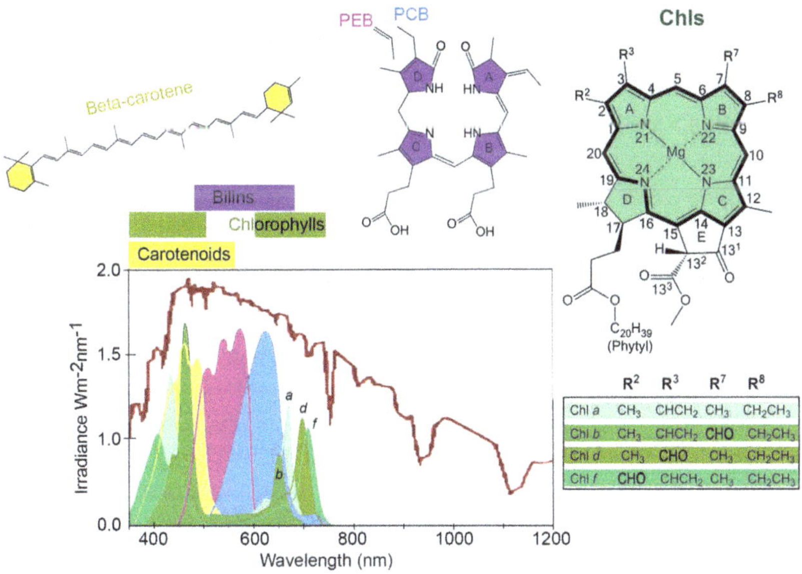

Figure 1 Solar spectrum on the Earth's surface, spectra of several photosynthetic pigments in solution and chemical structure of pigments. The red curve is the solar energy output spectrum. Yellow-colored spectral coverage represents carotenoids; pink- and blue-colored spectral coverage is from bilins; and different green-colored coverage represents spectra of different chlorophylls. The different chlorophylls have variations at positions C-2, C-3 and C-7 that are summarized in the inserted table. Structures of phycocyanobilin (PCB), phycoerythrobilin (PEB) and beta-carotene are also given.

Chemically, all chlorophylls have a Mg-containing tetrapyrrole macrocycle, like the Fe-containing hemes of our blood; therefore, the central role of chlorophylls in primary photosynthesis is reflected in its name as 'blood of photosynthesis'. Chlorophylls differ from hemes by their unique fifth (E) ring structure and a phytol chain connected at ring D of porphyrin (Fig. 1). Chl *c* is the only exception which has no phytol chain.

Chl *b* is distinguished from Chl *a* by a formyl substitution at the C-7 position of Mg-containing macrocycle and this substitution results in a blue shift of the Qy band (Fig. 1). Chl *d* and Chl *f* are also distinguished from Chl *a* by a formyl substitution, but at C-3 and C-2 position of Mg-containing macrocycle, respectively (Chen, 2014).

Chlorophylls have two transition dipole moments of π-π electronic transitions, polarized along the x and y axes of the macrocycle. They have two absorption bands, Soret bands in the blue spectral region and long-wavelength absorption band (Qy) in the red spectral region (Fig. 1). The significant absorption changes of chlorophylls are caused by small chemical modification and formyl group substitution. The peaks of the Qy maximum are located at 647 nm, 665 nm, 696 nm and 706 nm in methanolic solvents for Chl *b*, *a*, *d* and *f*, respectively (Table 1) (Sawicki et al., 2019). Such observations of the variations found in natural Chls stimulate the ingenious idea of expanding the light-harvesting efficiency via extending the absorption of photosynthetic pigments into additional spectral regions (Chen and Blankenship, 2011).

All chlorophylls have weak absorption at green wavelengths, making plants appear green. Chemical modification of chlorophylls manipulates the π-π electron conjugation resulting in changed spectral properties; therefore, photosynthetic organisms with different chlorophyll contents may be much better matched to the solar spectrum than those containing only one kind of chlorophylls (Fig. 1). Plants growing in the shaded environments have a higher content of Chl *b* to capture more light (Hikosaka and Terashima, 1995). Even within the same plant, the position of leaves shows a changed ratio of chlorophylls to match the changing light environments. On the top of the canopy, less chlorophyll content per area is observed; in contrast, a higher ratio of Chl *b* to Chl *a* is present in the leaves at the lower level or shaded parts of the plant (Gu et al., 2017; Hikosaka and Terashima, 1995; Lichtenthaler et al., 2007). Additionally, introducing vertically oriented leaves to change the canopy structure demonstrated enhanced yield of rice (*Oryza sativa*) (Long et al., 2006).

The photosynthetic spectral limit (red-edges/red-drops at ~700 nm) of photosynthetic active radiation (PAR) is set by the red-absorbing (Qy) band position of Chl *a* at around 700 nm. The integrated 400–700 nm PAR photon flux is 1.05×10^{21} photons $m^{-2} S^{-1}$ [1,740 $\mu E\ m^{-2} S^{-1}$, E (Einstein) is a mole of photons] (Chen et al., 2011). The PAR region has expanded with the discovery of Chl *d*

Table 1 Chlorophyll molecules' optical properties, distribution and their functions

		λ nm			
	Solvent	Soret	Qy	Type of organisms	Functions
Chl *a*	Methanol [ab]	432	665	Cyanobacteria, Algae, Plants	Light harvesting and photochemistry in reaction centers
	Diethyl ether [b]	428	660		
	Acetone [b]	430	662		
Chl *b*	Methanol [b]	469	652	Cyanobacteria, Algae, Plants	Light harvesting
	Diethyl ether [b]	452	642		
	Acetone [b]	456	645		
Chl *d*	Methanol [a]	456	697	Cyanobacterium: *Acaryochloris marina* spp.	Light harvesting and photochemistry in reaction centers
	Diethyl ether [a]	446	686		
	Acetone [a]	447	688		
Chl *f*	Methanol [a]	407	707	Cyanobacteria	Light harvesting
	Diethyl ether [a]	396	695		
	Acetone [a]	397	698		

[a] Li et al. (2012) Biochimica et Biophysica Acta (BBA) - Bioenergetics, 1857, 107–114.
[b] Lichtenthaler (1987) Methods in Enzymology, 148, 350–382.

(Miyashita et al., 1996) and Chl *f* (Chen et al., 2010) in cyanobacteria. Compared with Chl *a*, Chl *d* absorbs in solvents and in photosynthetic complexes ~30 nm to the red of Chl *a* because of a formyl substitution at C-3 position (Fig. 1; Table 1). Although Chl *d* was first reported in 1943, the reliance for photosynthesis using Chl *d* was discovered in cyanobacterium, *Acaryochloris marina*, where Chl *d* is major chlorophyll and replaces Chl *a* not only in the light-harvesting complexes but also in reaction centers (Miyashita et al., 1996; Loughlin et al., 2013). Chl *d*-binding protein complexes in *Acaryochloris marina* shift the established PAR energetic limits for oxygenic photosynthesis to ~ 740 nm, suggesting ~15% increase in solar energy input (Chen and Blankenship, 2011). Chl *f*, which was identified in 2010, demonstrates the most red-shifted absorption feature, 706 nm in methanol, having a formyl substitution at the C-2 position (Table 1) (Li et al., 2013). If the PAR region covers the wavelength portion of 400–760 nm, the photon flux is 1.25×10^{21} photons $m^{-2} S^{-1}$, or 2070 $\mu E\ m^{-2} S^{-1}$, with increased available PAR photons by 19% (Chen and Blankenship, 2011).

The lesson from a natural photosynthetic community is that different photosynthetic organisms do exploit regions of the solar spectrum not harvested by those organisms that live spatially above them, such as the *Acaryochloris marina* thriving in an ascidian-coral bio-community (Kühl et al., 2005) and *Halomicronima hongdechloris* thriving inside of stromatolite (Chen et al., 2012). In one sense, the community can be viewed as a stacked optical series that maximizes the energy captured using different photopigments. Such a strategy of stacked optical series in a bio-community could be applied to intelligently organize a multilayered agricultural crop environment.

Altering Chl contents *via* bioengineering could be possible, utilizing the knowledge of the pathway of Chl biosynthesis. The basic biosynthetic pathway to tetrapyrrole molecules (Chls and hemes) is the same in almost all organisms. It begins with glutamate and produces protoporphyrin IX after nine enzymatic reactions (Chen, 2014). Protoporphyrin IX is at the branch point of biosynthetic pathways for hemes and chlorophylls. The heme pathway initiated by Fe-chelatase leads to the biosynthesis of hemes, biliverdin and phycobilins. Mg-chelatase catalyzes the insertion of Mg into protoporphyrin IX to produce Mg-protoporphyrin IX and then the production of chlorophyllide *a* via multiple enzymatic steps. Biochemically, all chlorophylls share a common biosynthetic pathway to the production of chlorophyllide *a*, and the chemical modification of Chl happens at the later stage of biosynthesis either from chlorophyllide *a* or Chl *a* (Chen, 2014). Chl *b* is synthesized from Chl *a* (or chlorophyllide a) by a single enzyme chlorophyll *a* oxygenase (CAO) that oxygenates the 7-methyl to formyl group (Oster et al., 2000). The CAO is also able to introduce the second formyl group into Chl *d* at the C-7 position (Tsuchiya et al., 2012), indicating the specificity of CAO, formyl substitution at the C-7 position of Chls. Chl *d* is also derived from Chl *a* (Schliep et al., 2010), but by a different reaction that involves the oxidation of the 3-vinyl group with the concomitant loss of one C-atom (Schliep et al., 2010). The aerobic oxidative reduction of a vinyl group into a formyl group at the C-3 position is a reaction that may occur with several different types of chemical catalysis. *In vitro*, Chl *a* is readily converted to Chl *d* in an aqueous solution if heme, oxygen and thiols are present (Loughlin et al., 2014). However, the enzyme(s) catalyzing this step have not been identified yet. The formyl substitution in Chl *f* is biosynthesized from oxidation of Chl *a* and directly uses molecular oxygen as a substrate (Garg et al., 2017). A paralogous gene of the photosystem II protein, PsbA (D1), carries out biosynthesis of Chl *f* under far-red light conditions (Ho et al., 2016). Incorporation of Chl *f* into Chl *a*-containing photosystems demonstrated substantially decreased photosystem II, and insignificant impacts on the efficiency of photosystem I advances photosynthetic efficiency in far-red light conditions as an efficiency trade-off strategy for extending PAR and photo-adaptation (Mascoli et al., 2020). Interestingly, the production of Chl *f* is tightly regulated by near-infrared light conditions, and all reported cyanobacteria producing Chl *f* have a specific gene cluster, the Far-Red Light Photoacclimation gene cluster (FaRLiP) (Gao et al., 2015).

Carotenoids are linear tetraterpenoids containing a polyene chain that absorbs light mainly in the blue-light region. As the second class of pigments required for photosynthesis, they are involved in light harvesting and in the absorption of excessive light energy for photo protection. There are approximately 30 different carotenoid structures reported in photosynthetic organisms belonging to two main groups of carotenoids: carotenes and xanthophylls. Carotenes have nonpolar oxygen-lacking structure (Fig. 1) and xanthophylls are polar oxygen-containing carotenoids. Having carotenoids in the photosystem allows the organism to capture light energy from about 400 nm to 550 nm and transfer the energy to chlorophylls. Carotenoid contents and compositions generally change in response to light-stress conditions, for example, zeaxanthin concentration is increased under high-light irradiation following a rapid epoxidation and de-epoxidation xanthophyll cycle (Demmig-Adams, 1990; Goss and Jakob, 2010). Carotenoids also can desensitize long-lived Chl triplet states that can lead to the production of reactive oxygen species (ROS).

Some algae and cyanobacteria produce a class of water-soluble proteins known as phycobiliproteins. Bilins are linear tetrapyrroles and are chromophores for phycobiliproteins. Phycobiliproteins are basic subunits of phycobilisomes (PBS), peripheral light-harvesting complexes mainly in cyanobacteria and some algae. Phycocyanobilin (PCB) and phycoerythrobilin (PEB) are common chromophores in PBS and capture different wavelengths, filling in the 'green window' of Chls (Fig. 1). Most bilin-type chromophores share the biosynthesis pathway from heme as mentioned above, suggesting potential regulatory functions shared with chlorophyll biosynthesis. In addition, bilin-type chromophores are also important in phytochrome, a class of photoreceptor in plants, involving control plant development including the synthesis of chlorophyll and the germination of seeds (Hikosaka et al. 2016).

Photosynthetic organisms never contain free photopigments *in vivo* because they are likely to become a harmful species after capturing light energy with no pathway to either deliver it to the reaction center or safely dissipate the energy. Photopigments are bound by protein and form different pigment-protein complexes. Protein provides structural scaffolds that maintain the pigments in an optimized orientation that prevents non-productive decay of the excited state. Free pigment aggregates are also subject to concentration-dependent quenching that would negate their use for the light-harvesting function. There are two components: antenna pigment-protein complexes for photon capture and energy transfer and reaction centers for photochemical charge separation and energy transduction.

3 Photosynthetic light-harvesting protein complexes

There are many natural elements changing light environments beyond the scattering of the atmosphere and cloudy days. Light quality (color) and quantity (irradiance) reaching a photosynthetic organism vary in time and space. Light-harvesting complexes establish the gate for manipulating photons used by photosynthesis. The diversification of photopigments with different light-absorption properties and the properties of antenna complexes could increase the number of photons absorbed. Expanding the spectral coverage of the antenna system could be used to optimize the photochemistry in reaction centers. Therefore, antenna complexes are extremely important in determining the ecophysiology of photosynthetic organisms. Many antenna systems can be found coexisting in complex microbial communities to facilitate the ability of these organisms to occupy specific photic environments, in other words, to improve light-harvesting efficiency along myriad light-harvesting paths that are dictated by evolutionary selection. In this chapter, we will introduce the myriad light-harvesting systems and discuss the structural and functional constraints to define the possibilities of enhancing light-harvesting efficiency.

The diversity of photopigments, light-harvesting protein complexes and their different regulatory mechanisms contrasts with the high homology of the photosynthetic RCs within the phylogenetic classification (Blankenship, 2010). Light-harvesting or antenna systems have evolved to adapt to variable intensity and spectral qualities of light in diverse environments and are divided into inner and outer antenna systems (Table 2). The inner antenna is an essential component of RCs. In contrast, the outer antenna systems include multiple photopigment-bound protein families having different pigments and pigment-protein complexes. The light-harvesting complexes (LHCs) in plants and other eukaryotic photosynthetic organisms belong to a single protein family and bind

Table 2 Types of photosynthetic antenna systems

		Integral membrane Chl-binding antenna systems		
	Outer-membrane antenna	3-helixes antenna	6-helixes antenna	
	Phycobilisomes in cyanobacteria and red algae	LHCI and LHCII	Inner antenna (CP43/CP47)	Chl-binding proteins in cyanobacteria
				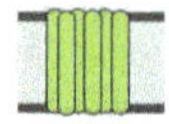
Pigments	1-2 Bilins per phycobilin protein subunit	15-22 Chl *a*+*b*/ or *c*	32-36 Chl *a* + *b* (*d* or *f*)*	32-36 Chl *a*, Chl *d*, Chl *f*
Functions	Mainly for PSII	PSI and PSII	PSII	PSI and PSII

*Chl *d*-containing cyanobacterium, Acaryochloris marina spp. or Chl *f*-producing cyanobacteria.

15-22 chlorophylls. LHCs make up the outer antenna system and are composed of three membrane-spanning proteins. Cyanobacteria lack members of the LHC family and instead use a soluble antenna, comprising phycobiliproteins or six-transmembrane Chl-binding proteins (CBP) as their main light-harvesting system (Chen et al., 2008; Mirkovic et al., 2017). Three important parameters decide the optimization and efficiency of antenna systems: (1) the distance between photopigments, (2) the orientation of photopigments and (3) spectral overlap between the pigments.

3.1 Chl-based light-harvesting protein complexes

In addition to the role of Chl in the reactions performing photochemistry, Chl is an important pigment in the light-harvesting antenna to ensure that the photochemical activities in reaction centers occur at maximal rates. There are inner Chl-binding antenna complexes as a part of photosystems and intrinsic membrane-bound Chl-binding antenna associated with photosystems. The inner antenna proteins CP43 and CP47 are indispensable subunits of photosystem II complexes (Eaton-Rye and Putnam-Evans, 2005). They function in the same manner as that of the PSII system in plants. The N-terminal domain of photosystem I reaction center proteins, PsaA and PsaB, serves as the inner antenna and has a homologous sequence similar to CP43 and CP47 (Jordan et al., 2001). The inner antenna binds 15-30 Chls and directly transfers the captured energy to the core of reaction centers.

In plants and algae, the intrinsic Chl-binding antenna is three helical transmembrane Chl *a*/*b*-binding LHCs, except for Chl *c*-containing alga, which have Chl *a*/*c*-binding LHC. All three-helix light-harvesting protein complexes belong to a single protein family and are found in all eukaryotic photosynthetic organisms. The typical LHCs bind approximately 15–22 Chls (Kühlbrandt et al., 1994). Two main classes of LHC, LHCI and LHCII, are defined following their roles serving for photosystem I or photosystem II. Commonly, LHC forms trimers, and multiple LHC subunits associated with photosystems forming supercomplexes expand the light-capture capability and increase energy to drive the photochemical reaction in reaction centers (Caffarri et al., 2014; Umena et al., 2011).

3.2 CP43-like protein family, chlorophyll-binding proteins

Most cyanobacteria have two antenna systems, PBS and Chl-binding antenna systems. The Chl-binding light-harvesting systems are six-helical transmembrane Chl-binding protein complexes, structurally different from the LHCs in eukaryotic photosynthetic organisms, including Pcb and IsiA, also named as CBP (Chen et al., 2008; La Roche et al., 1996). Most CBP will bind available Chls and function as an intrinsic Chl-binding light-harvesting protein in cyanobacteria (Table 2). Two main groups are identified under this CBP family: IsiA (iron-stress-induced protein A) and Pcb (prochlorophyte Chl *a/b* protein), which are phylogenetically related to CP43 and contain six-transmembrane helices. They differ from CP43 in not having the long extrinsic domain joining helices V and VI (Table 2).

The IsiA gene owes its name to its induction during iron limitation, even though it also accumulates under other stresses. The light-harvesting function of the IsiA protein was confirmed after the discovery of an 18-mer IsiA ring around cyanobacterial trimeric photosystem I (Bibby et al., 2001; Boekema et al., 2001). Several functions have been reported for IsiA, from light-harvesting for PSI (Andrizhiyevskaya et al., 2002; Melkozernov et al., 2003) to a role in non-photochemical quenching (NPQ) (van der Weij-de Wit et al., 2007) or to serve as a temporal reservoir of Chl under iron-limiting conditions (Ma et al., 2017).

In contrast, PCB proteins are constitutively expressed and form 18-mers around trimeric PSI. In prochlorophytes (*Prochloron didemni*, *Prochlorothrix hollandica* and *Prochlorococcus marinus*) Pcb-photosystem supra-complexes are observed for Pcb-PSI and Pcb-PSII system (Matthijs et al., 1994; Bibby et al., 2003; Murray et al., 2006). In the Chl *d*-containing cyanobacterium *A. marina*, Pcb and IsiA proteins bind Chl *d* instead of Chl *a* (Chen et al., 2005a,2005b; Li et al., 2017).

3.3 Phycobilin-based antennas

Phycobilisomes (PBS) are very large multiprotein complexes assembled from phycobiliproteins and linker polypeptides (Adir et al., 2020). Phycobiliproteins are the basic unit of phycobilisomes and are covalently bound to one or more chromophores, bilins. Phycobiliproteins absorb light wavelengths that are complementary to those absorbed by Chl *a*, commonly having the absorption maxima in the range of 500–670 nm (Fig. 1). The PBPs absorb light in the blue-green region that transmits into deeper water than red light. Consequently, PBPs are commonly observed in endolithic photosynthetic algae and cyanobacteria. PBS can transfer light energy directly or indirectly to photosystems depending on PBS-specific structure and composition, which are tightly regulated by light environments.

Naturally, three main structural classes of PBS have been described: (1) Hemi-ellipsoida/hemidiscodial PBS; (2) bundle-shaped PBS; (3) rod-shaped PBS. For example, the simple rod structure of PBP in *Acaryochloris marina sp.* MIBC11017 is tightly associated with the photosynthetic membranes and provides a shorter energy transfer pathway to photosystems (Chen et al., 2009). The bundle-shaped PBSs are reported from *Gloeobacter violaceus*, a cyanobacterium lacking thylakoid membranes (Koyama et al., 2006; Guglielmi et al., 1981).

Photoacclimation processes of PBP enable organisms to change the absorbance of PBP in response to the changing light environment to maximize the light-harvesting efficiency. Rearrangement of the assembled PBS structure plays an important role in changing the absorption properties of PBS, including structural changes. According to recent -omics studies, PBSs can be remodeled depending on the light conditions (Wiltbank and Kehoe, 2019). For example, the production of PBS that extends the absorbance into the red/far-red region of 710 nm is regulated by far-red light conditions in *H hongdechloris*, which rearrange the PBS by assembling the dicylindrical core structure only (Li et al., 2016). The red-shifted PBS in *Synechococcus* PCC 7335 uses a different strategy, the random reorganization of PBP structural blocks contributes to the changed bound-bilins position and interactions and results in red-shifted absorbance properties of PBS under far-red light conditions (Herrera-Salgado et al., 2018; Ho et al., 2017). In general, most bilins in the PBS are spatially evenly distributed around the core structure. However, the disordered structure of PBS can be achieved by adding new protein subunits with different orientation and position of chromophores.

Another PBS acclimation process is known as 'state transitions', which switch the coupling of PBS to photosystem II and/or photosystem I, hence controlling the distribution of excitation energy from the PBS to the two photosystems. Understanding the 'state transition' of PBS could provide the mechanism of PBS acclimation and their functions subject to an intrinsic quenching and recovery process of photosystems in response to the changed light conditions.

Therefore, engineering PBS could expand the PAR wavelength range substantially and increase the energy transferring to photosystems. The distances and orientations among bilins in PBS are determined by individual PBP subunits. The local protein environment surrounding each of the bilins is important for PBS absorption properties. Protein engineering technologies could accomplish the specified tasks in LHCs by designing proteins to tune to the different absorption properties.

3.4 Light acclimation and adaptation

Variant light-harvesting protein complexes reflect the complexity of the light environment and the evolutionary history of photosynthetic organisms. Individual photosynthetic organisms have developed their own light-harvesting systems and developed different acclimation strategies to adapt to their specific light conditions. Photoacclimation represents a series of reversible changes reflecting short- (minutes) to long-term (days) changes in spectral properties. Light acclimation changes include altering the pigment composition, pigment-binding protein structure, re-constituting the interactional relationship among the light-harvesting components and reaction centers. The high-light-grown leaves commonly have the following features: (1) less chlorophyll per unit nitrogen; (2) a higher Chl *a*:Chl *b* ratio; (3) increased photosynthetic electron transport rate; and (4) an increased ratio of electron transport rate to Rubisco activity. The reversible changes in the ratio of Chl *a*:Chl *b* play key roles in photoacclimation in plants (Hikosaka and Terashima, 1995) and chromatic acclimation of PBS in cyanobacteria and red algae (Sanfilippo et al., 2019). Altering the size of LHCs is another photoacclimation process. For example, in clear oceanic environments, the maximal spectral transmission occurs around 475 nm, with short-wavelength light reaching deeper water layers than long-wavelength light. At ~200-m depth, the light intensity of ~0.05 $\mu mol.m^{-2}.s^{-1}$ is only <0.005% when compared with that at the surface, and specific cyanobacteria with multiple LHCs thrive there. Multiple copies of Chl-binding light-harvesting protein-encoded genes have been reported from low-light-adapted *Prochlorococcus sp.* SS120, thriving at ~120 m below sea level, in contrast to high-light-adapted *Prochlorococcus sp.* MED4 isolated from the surface of the ocean that have only one copy of CBP (Garczarek et al., 2000). The Chl *d*-containing cyanobacterium, *A. marina* MBIC11017, was isolated from a very low-light intensity niche and its genome sequence revealed 11 copies of genes encoding CBP proteins (Kühl et al., 2005; Swingley et al., 2008). The total RNAseq comparison revealed that 6 out of 11 CBP-encoded genes are upregulated under FRL conditions, while another 2 CBP-encoded genes were upregulated under white light conditions (Hernandez-Prieto et al., 2018).

Antenna systems are comprised of various photopigments and a different number of photopigments per photosystem, which reflects cellular nutrient resources. Commonly, antenna systems contain much less protein per bound photopigment. The extended antenna system increases the light absorption cross-section of the photosystems, but at biosynthetic cost in an adjustable fashion (Chen et al., 2013). At moderate or high-light conditions, the rate of photosynthesis will decrease as the introduction of additional metabolic costs for photoprotection and photoinhibitions. The reduced antenna size leads to cellular economic optimization and potentially enhances light transmittance and reflection within the canopy (Beckman et al., 2009). For example, uneven distribution of nitrogen content is reported in many plant canopies, an adaptive strategy for the efficient use of nitrogen. Theoretical studies have shown that the nitrogen allocation has a strong influence on canopy photosynthetic rates and is thus one of the primary factors determining vegetation productivity (Hikosaka et al., 2016).

There are variations in accessible light energy for a plant. The leaves of plant canopies are directly exposed to full sunlight and lead to the wasteful situation of saturated and dissipation of light, while the shaded leaves are light-limited. Introducing vertically oriented leaves to change the canopy structure demonstrated enhanced yield of rice (*Oryza sativa*) (Long et al., 2006). Reducing the photopigment content in the top level of the canopy or top layer of microalgae is another strategy to enhance light transmittance and reflectance within the plant canopy or microalgae mass cultures (Blankenship and Chen, 2013).

Some cyanobacteria retain one or two helices of transmembrane domain with similarity to the membrane-spanning region of plant chlorophyll *a-/b-*binding proteins (LHCs), in addition to the presence of CBP (Dolganov et al., 1995). The function of those small chlorophyll-binding proteins mainly contributes to photoprotection. For example, high-light-induced protein, found in *Synechocystis sp.* PCC6803, binds Chls and carotenes and associates with PSII to protect the D1 protein from degradation under high-light stress conditions (Knoppov et al., 2014).

Most Chl *f*-producing cyanobacteria demonstrate the capability to reorganization their photosynthetic protein complexes, including rearrangement of far-red light absorbing phycobilisomes (Chen et al., 2019; Gan et al., 2015; Li et al., 2016). The switchable photosynthetic apparatus regulated by far-red light empowers an agile mechanism for developing a regulatory mechanism of assembly of photosynthetic apparatus to meet the requirements of changed environments. An additional set of paralogous photosystem genes in Chl *f*-producing cyanobacteria appears. They are silenced under white light and upregulated immediately under far-red light, including red-shifted chlorophyll biosynthesis and the formation of red-shifted phycobilisomes. The wider distribution of Chl *f*-producing cyanobacteria indicates the importance of flexible photosystem organization enabling photosynthetic organisms to thrive in extreme light environments. Could we build such switchable photosystems in plants to achieve higher photosynthetic efficiency?

4 Photoinhibition

Under full sunlight (high-light intensity) photosynthetic energy supply (ATP) and reducing power (NADPH) exceed the demand for the metabolic processes in carbon fixation, resulting in the accumulation of ROS. At low concentrations, ROS act as a signal to induce defense responses, while at high concentrations ROS induces lipid peroxidation in the cell membranes and causes oxidative damage to cellular components (Dat et al., 2000; Khorobrykh et al., 2020). Photosynthesis suppression due to exposure to high-light intensities is known as photoinhibition. The photosystem II (PSII) complex is considered the primary target of photoinhibition (Vass, 2012; Roach and Krieger-Liszkay, 2014). There are multiple mechanisms to dissipate the excess energy protecting from the damage of photoinhibition. Thermal dissipation is known as NPQ of Chl fluorescence, including energy-dependent quenching by the involvement of the xanthophyll cycle, conformational changes in light-harvesting complex II (LHCII), which are known as state transitions and photoinhibition resulting in the reduction of quantum yield because of light-induced damage (Müller et al., 2001; Zhao et al., 2016).

The process of rebalancing the energy distribution between the two photosystems to divert the excess energy away from one photosystem to another is known as state transitions, which take place on short timescales (seconds to minutes). For example, if an excess of light energy captured by LHCII is delivered to photosystem I instead of photosystem II, an opposite adjustment can occur to shift some energy captured by LHCI to photosystem II. Cyanobacteria, without LHCI and LHCII, shift energy transfer pathways of PBS between the photosystems. Photosynthetic state transition helps to reduce photodamage under stressed conditions and also helps to offset the imbalance in the rate of generating ATP and reductant NADPH.

The trigger for the induction of NPQ might cause changes in LHC positions (state transition) and newly synthesized minor LHCs, such as PsbS antenna protein, high-light-induced CBP. When NPQ is activated, the number of carotenoids is increased, especially the xanthophylls. Xanthophylls serve as NPQ agents to dissipate overcharged Chl via xanthophyll cycle. The xanthophyll cycle is catalyzed by epoxidase and de-epoxidase enzymes to interconvert violaxanthin and zeaxanthin in response to changed light conditions. Under the low-light condition, epoxidase enzyme is active and produces an increased amount of violaxanthin, a double-epoxide compound; in contrast, excess light triggers the action of the de-epoxidase enzyme and converts violaxanthin to zeaxanthin, without the epoxide functional group. The xanthophyll cycle is one main strategy against photoinhibition in plants.

5 Photosynthesis efficiency

The overall efficiency of photosynthesis is quite low (~1%) according to the biomass production (Zhu et al., 2010; Ort et al., 2015). It relies firstly on the efficiency of solar light collection by the LHCs and secondly on the efficiency of primary energy transduction, and dark reactions. The quantum efficiency of the first two is near 100% at a stable light flux or a stable spectral composition in the laboratory. This instantaneous efficiency would only be achievable under low-light conditions, where essentially every incident photon of the appropriate wavelength can be absorbed and utilized for productive electron transfers. Under full sunlight, natural photosynthesis uses only a portion of these incident photons. Downstream inefficiencies during carbon fixation further reduce the attainable efficiency. Additionally, many photosynthetic organisms have seasonal variations in their photosynthetic rates. Thus, on an annual basis, photosynthesis is remarkably inefficient, with energy conversion efficiencies <1% for land-based organisms, and <5% for microalgae. Even with these low efficiencies, solar energy converted into fuels via photosynthesis provides essentially 100% of the energy input for life on this planet.

One physical limit of photosynthetic processes in plants is the narrow window of PAR by Chl *a* and Chl *b*. The red-shifted chlorophylls, Chl *d* and Chl *f*, in cyanobacteria can expand their PAR into the far-red light region. The high absorptivity provides effective capture at low-light intensities, such as dawn, dusk, cloudy days and unique ecological niches. The challenge here is to balance the absorption of more photons, which can be used to drive useful photochemistry (as described in the reaction center section). Excess energy needs to be carefully extinguished to avoid photoinhibition and photodamages. Ideally, if plants had fewer LHC pigments per photosystem and fewer photosystems in their upper canopy leaves, light energy can be used more efficiently (Ort et al., 2010; Blankenship and Chen, 2011; Negi et al., 2020). Genetically engineering higher plants for expanding spectral features will potentially increase photosynthetic light-capture efficiency under low-light conditions (Wolf and Blankenship, 2019).

For field crops, the upper part of the canopy receives the full solar spectrum and the ratio of red light to far-red light is almost 1:1 (Fig. 1), but the floor part of the field receives only light that was not absorbed above. Therefore, different strategies might be applied to increase the efficiency. The filtered light reaching the floor part of the field is enriched in far-red light, and the ratio of red to far-red light could be 0.3: 1, supporting red-shifted chlorophyll-driving photosynthesis. The increased antenna systems will increase the amount of energy that can be absorbed compared with a single photopigment. However, under high-light conditions or other forms of stresses, captured light energy can be more than can be used, resulting in photodamage. Photoprotection mechanisms could be activated rapidly, and light-harvesting antenna systems can play important roles in extensive photo-regulation, protection and regulation processes.

6 Challenges and future trends in research

The factors that contribute to the inefficiency of photosynthesis (light-dependent reactions: light energy capture, transfer and transduction) are:

1. absorbed light: PAR comprises only ~48% of the incident solar energy (Fig. 1);
2. excitation energy transfer, including maximized antenna pigments, absorbs and transfers energy to the photosynthetic reaction center;
3. high energy absorbance by antenna and reemitted as heat and fluorescence, (photoprotection and photoinhibition); and
4. carbon fixation.

Proposed approaches to improving light-harvesting transferring and conversion efficiency:

1. Changing pigment composition: (a) increased variety of light-harvesting pigments (increasing cross area of light-harvesting); (b) increased variety of LHCs (introduction of PBS, additional capability for photoacclimation); (c) increased photoprotective pigments (regulation of xanthophylls and the xanthophyll cycles); and (d) altering pigment composition (ratio of Chl *a*/Chl *b*).
2. Light-harvesting strategy: reduced antenna size; controlling state transition for balanced photosynthetic light reactions.

An ideal modified plant would expand the available solar spectrum to harvest more of the available solar spectrum and would distribute the energy throughout the canopy so that upper parts of the canopy have a reduced antenna size, while the lower parts of the canopy have an increased antenna size. These changes would permit increased light absorption but not increase photodamage due to excess photon flux. Additional changes in the dark reactions such as described in Kromdijk et al. (2016) could increase the ability of the plant to process the increased energy input without activating dissipative mechanisms.

7 Conclusion

Photosynthesis is an essential biological process, which is responsible for all of our food and most of our energy resources on Earth. However, the efficiency of conversion of solar energy into biomass is not very high compared to some types of photoconversion devices. Research has identified a number of factors that limit the efficiency of photosynthesis, and numerous proposals that may improve the efficiency have been proposed. Current research is focused on implementing these proposals and searching for additional factors that might be explored.

8 Where to look for further information

Additional information can be found in general books on photosynthesis (Blankenship, 2021; Ruban et al., 2022; Shevala et al., 2018) or in specialized books on energy transfer in photosynthesis (Croce et al., 2018; Ruban 2012; Valkunas et al., 2000) and the computer models on enhancing crop growth and yield (Wu et al. 2022).

9 References

Adir, N., Bar-Zvi, S. and Harris, D. (2020). The amazing phycobilisome, *Biochim. Biophys. Acta Bioenerg.* 1861(4):148047.

Andrizhiyevskaya, E. G., Schwabe, T. M. E., Germano, M., D'Haene, S., Kruip, J., van Grondelle, R. and Dekker, J. P. (2002). Spectroscopic properties of PSI-IsiA supercomplexes from the cyanobacterium *Synechococcus* PCC 7942, *Biochim. Biophys. Acta* 1556(2-3):265-272.

Bibby, T. S., Mary, I., Nield, J., Partensky, F. and Barber, J. (2003). Low-light-adapted Prochlorococcus species possess specific antennae for each photosystem, *Nature* 424(6952):1051-1054.

Bibby, T. S., Nield, J. and Barber, J. (2001). Iron deficiency induces the formation of an antenna ring around trimeric photosystem I in cyanobacteria, *Nature* 412(6848):743-745.

Blankenship, R. E. (2010). Early evolution of photosynthesis, *Plant Physiol.* 154(2):434-438.

Blankenship, R. E. (2021). *Molecular Mechanisms of Photosynthesis*, Chichester: Wiley.

Blankenship, R. E. and Chen, M. (2013). Spectral expansion and antenna reduction can enhance photosynthesis for energy production, *Curr. Opin. Chem. Biol.* 17(3):457-461.

Boekema, E. J., Hifney, A., Yakushevska, A. E., Piotrowski, M., Keegstra, W., Berry, S., Michel, K. P., Pistorius, E. K. and Kruip, J. (2001). A giant chlorophyll-protein complex induced by iron deficiency in cyanobacteria, *Nature* 412(6848):745-748.

Caffarri, S., Tibiletti, T., Jennings, R. C. and Santabarbara, S. (2014). A comparison between plant photosystem I and photosystem II architecture and functioning, *Curr. Protein Pept. Sci.* 15(4):296-331.

Chen, M. (2014). Chlorophyll modifications and their spectral extension in oxygenic photosynthesis, *Annu. Rev. Biochem.* 83:317-340.

Chen, M., Bibby, T. S., Nield, J., Larkum, A. W. D. and Barber, J. (2005a). Iron deficiency induces a chlorophyll d-binding Pcb antenna system around photosystem I in *Acaryochloris marina*, *Biochim. Biophys. Acta* 1708(3):367-374.

Chen, M., Bibby, T. S., Nield, J., Larkum, A. W. D. and Barber, J. (2005b). Structure of a large photosystem II supercomplex from *Acaryochloris marina*, *FEBS Lett.* 579(5): 1306-1310.

Chen, M. and Blankenship, R. E. (2011). Expanding the solar spectrum used by photosynthesis, *Trends Plant Sci.* 16(8):427–431.

Chen, M., Floetenmeyer, M. and Bibby, T. S. (2009). Supramolecular organization of phycobiliproteins in the chlorophyll d-containing cyanobacterium *Acaryochloris marina*, *FEBS Lett.* 583(15):2535–2539.

Chen, M., Hernandez-Prieto, M. A., Loughlin, P. C., Li, Y. and Willows, R. D. (2019) Genome and proteome of the chlorophyll f-producing cyanobacterium Halomicronema hongdechloris: adaptative proteomic shifts under different light conditions, *BMC Genomics* 20(1):207.

Chen, M., Li, Y., Birch, D. and Willows, R. D. (2012). A cyanobacterium that contains chlorophyll *f* - a red-absorbing photopigment, *FEBS Lett.* 586(19):3249–3254.

Chen, M. and Scheer, H. (2013). Extending the limit of natural photosynthesis and implications of technical light harvesting, *J. Porphyrins Phthalocyanines* 17(01n02):1–15.

Chen, M., Schliep, M., Willows, R. D., Cai, Z. L., Neilan, B. A. and Scheer, H. (2010). A red-shifted chlorophyll, *Science* 329(5997):1318–1319.

Chen, M., Zhang, Y. and Blankenship, R. E. (2008). Nomenclature for membrane-bound light-harvesting complexes of cyanobacteria, *Photosynth. Res.* 95(2–3):147–154.

Croce, R., van Grondelle, R., van Amerongen, H. and van Stokkum, I. (Eds). (2018). Light Harvesting in Photosynthesis. Boca Raton, FL: CRC Press.

Dat, J., Vandenabeele, S., Vranová, E., Van Montagu, M., Inzé, D. and Van Breusegem, F. (2000). Dual action of the active oxygen species during plant stress responses, *Cell. Mol. Life Sci.* 57(5):779–795.

Demmig-Adams, B. (1990). Carotenoids and photoprotection in plants: a role for the xanthophyll zeaxanthin, *Biochim. Biophys. Acta Bioenerg.* 1020(1):1–24.

Dolganov, N. A., Bhaya, D. and Grossman, A. R. (1995). Cyanobacterial protein with similarity to the chlorophyll a/b binding proteins of higher plants: evolution and regulation, *Proc. Natl Acad. Sci. U. S. A.* 92(2):636–640.

Eaton-Rye, J. J. and Putnam-Evans, C. (2005). The CP47 and CP43 core antenna components. In: Wydrzynski, T. J., Satoh, K. and Freeman, J. A. (Eds), *Photosystem II: The Light-DrivenWater:Plastoquinone Oxidoreductase*, pp. 45–70. Dordrecht: Springer Netherlands.

Gan, F., Shen, G. and Bryant, D. A. (2015). Occurrence of far-red light photoacclimation (FaRLiP) in diverse cyanobacteria, *Life (Basel)* 5(1):4–24.

Garczarek, L., Hess, W. R., Holtzendorff, J., Van Der Staay, G. W. and Partensky, F. (2000). Multiplication of antenna genes as a major adaptation to low light in a marine prokaryote, *Proc. Natl Acad. Sci. U. S. A.* 97(8):4098–4101.

Garg, H., Loughlin, P. C., Willows, R. D. and Chen, M. (2017). The C2^1-formyl group in chlorophyll *f* originates from molecular oxygen, *J. Biol. Chem.* 292(47):19279–19289.

Goss, R. and Jakob, T. (2010). Regulation and function of xanthophyll cycle-dependent photoprotection in algae, *Photosynth. Res.* 106(1–2):103–122.

Gu, J., Zhou, Z., Li, Z., Chen, Y., Wang, Z., Zhang, H. and Yang, J. (2017). Photosynthetic properties and potentials for improvement of photosynthesis in pale green leaf rice under high light conditions, *Front. Plant Sci.* 8:1082.

Guglielmi, G., Cohen-Bazire, G.and Bryant, D. A. (1981). The structure of *Gloeobacter violaceus* and its phycobilisomes, *Arch. Microbiol.* 129(3):181–189.

Hernandez-Prieto, M. A., Li, Y., Postier, B. L., Blankenship, R. E. and Chen, M. (2018). Far-red light promotes biofilm formation in the cyanobacterium Acaryochloris marina, *Environ. Microbiol.* 20(2):535–545.

Herrera-Salgado, P., Leyva-Castillo, L. E., Ríos-Castro, E. and Gómez-Lojero, C. (2018). Complementary chromatic and far-red photoacclimations in Synechococcus ATCC 29403 (PCC 7335). I: the phycobilisomes, a proteomic approach, *Photosynth. Res.* 138(1):39–56.

Hikosaka, K., Anten, N. P., Borjigidai, A., Kamiyama, C., Sakai, H., Hasegawa, T., Oikawa, S., Iio, A., Watanabe, M., Koike, T., Nishina, K. and Ito, A. (2016). A meta-analysis of leaf nitrogen distribution within plant canopies, *Ann. Bot.* 118(2):239–247.

Hikosaka, K. and Terashima, I. (1995). A model of the acclimation of photosynthesis in the leaves of C-3 plants to sun and shade with respect to nitrogen use, *Plant Cell Environ.* 18(6):605–618.

Ho, M. Y., Gan, F., Shen, G. and Bryant, D. A. (2017). Far-red light photoacclimation (FaRLiP) in *Synechococcus* sp. PCC 7335. II. Characterization of phycobiliproteins produced during acclimation to far-red light, *Photosynth. Res.* 131(2):187–202.

Ho, M. Y., Shen, G., Canniffe, D. P., Zhao, C. and Bryant, D. A. (2016). Light-dependent chlorophyll *f* synthase is a highly divergent paralog of PsbA of photosystem II, *Science* 353(6302):aaf9178.

Jordan, P., Fromme, P., Witt, H. T., Klukas, O., Saenger, W. and Krauss, N. (2001). Three-dimensional structure of cyanobacterial photosystem I at 2.5A resolution, *Nature* 411(6840):909–917.

Khorobrykh, S., Havurinne, V., Mattila, H. and Tyystjarvi, E. (2020). Oxygen and ROS in photosynthesis, *Plants (Basel)* 9(1):91.

Knoppov., J., Sobotka, R., Tichý, M., Yu, J., Konik, P., Halada, P., Nixon, P. J. and Komenda, J. (2014). Discovery of a chlorophyll binding protein complex involved in the early steps of photosystem II assembly in *Synechocystis*, *Plant Cell* 26(3):1200–1212.

Koyama, K., Tsuchiya, T., Akimoto, S., Yokono, M., Miyashita, H. and Mimuro, M. (2006). New linker proteins in phycobilisomes isolated from the cyanobacterium *Gloeobacter violaceus* PCC 7421, *FEBS Lett.* 580(14):3457–3461.

Kromdijk, J., Głowacka, K., Leonelli, L., Gabilly, S. T., Iwai, M., Niyogi, K. K. and Long, S. P. (2016). Improving photosynthesis and crop productivity by accelerating recovery from photoprotection, *Science* 354(6314):857–861.

Kühl, M., Chen, M., Ralph, P. J., Schreiber, U. and Larkum, A. W. D. (2005). Ecology: a niche for cyanobacteria containing chlorophyll d, *Nature* 433(7028):820.

Kühlbrandt, W., Wang, D. N. and Fujiyoshi, Y. (1994). Atomic model of plant light-harvesting complex by electron crystallography, *Nature* 367(6464):614–621.

La Roche, J., Van Der Staay, G. W., Partensky, F., Ducret, A., Aebersold, R., Li, R., Golden, S. S., Hiller, R. G., Wrench, P. M., Larkum, A. W. and Green, B. R. (1996). Independent evolution of the prochlorophyte and green plant chlorophyll a/b light-harvesting proteins, *Proc. Natl Acad. Sci. U. S. A.* 93(26):15244–15248.

Li, Y., Cai, Z. L. and Chen, M. (2013). Spectroscopic properties of chlorophyll *f*, *J. Phys. Chem. B* 117(38):11309–11317.

Li, Y., Lin, Y., Garvey, C. J., Birch, D., Corkery, R. W., Loughlin, P. C., Scheer, H., Willows, R. D. and Chen, M. (2016). Characterization of red-shifted phycobilisomes isolated from the chlorophyll *f*-containing cyanobacterium *halomicronema hongdechloris*, *Biochim. Biophys. Acta* 1857(1):107–114.

Li, Z.-K., Yin, Y.-C., Zhang, L.-D., Zhang, Z.-.C., Dai, G.-Z., Chen, M. and Qiu, B.-S. (2017). The identification of isiA proteins binding chlorophyll d in the cyanobacterium *Acaryochloris marina*, *Photosynth. Res.* 135:1650175.

Lichtenthaler, H. K., Ac, A., Marek, M. V., Kalina, J. and Urban, O. (2007). Differences in pigment composition, photosynthetic rates and chlorophyll fluorescence images of sun and shade leaves of four tree species, *Plant Physiol. Biochem.* 45(8):577-588.

Long, S. P., Zhu, X. G., Naidu, S. L. and Ort, D. R. (2006). Can improvement in photosynthesis increase crop yield?, *Plant Cell Environ.* 29(3):315-330.

Loughlin, P., Lin, Y. and Chen, M. (2013). Chlorophyll d and *Acaryochloris marina*: current status, *Photosynth. Res.* 116(2-3):277-293.

Loughlin, P. C., Willows, R. D. and Chen, M. (2014). *In vitro* conversion of vinyl to formyl groups in naturally occurring chlorophylls, *Sci. Rep.* 4:6069.

Ma, F., Zhang, X., Zhu, X., Li, T., Zhan, J., Chen, H., He, C. and Wang, Q. (2017). Dynamic changes of IsiA-containing complexes during long-term iron deficiency in *Synechocystis* sp. PCC 6803, *Mol. Plant* 10(1):143-154.

Matthijs, H. C. P., van der Staay, G. W. M. and Mur, L. R. (1994). 'Prochlorophytes: the 'other' cyanobacteria? In: Bryant, D. A. (Ed.), *The Molecular Biology of Cyanobacteria*. Dordrecht: Springer Netherlands, pp. 49-64.

Melkozernov, A. N., Bibby, T. S., Lin, S., Barber, J. and Blankenship, R. E. (2003). 'Time-resolved absorption and emission show that the CP43'antenna ring of iron-stressed *Synechocystis* sp. PCC6803 is efficiently coupled to the photosystem I reaction center core, *Biochemistry* 42(13):3893-3903.

Mirkovic, T., Ostroumov, E. E., Anna, J. M., van Grondelle, R., Govindjee and Scholes, G. D. (2017). Light absorption and energy transfer in the antenna complexes of photosynthetic organisms, *Chem. Rev.* 117(2):249-293.

Miyashita, H., Ikemoto, H., Kurano, N., Adachi, K., Chihara, M. and Miyachi, S. (1996). Chlorophyll *d* as a major pigment, *Nature* 383(6599):402.

Müller, P., Li, X. P. and Niyogi, K. K. (2001). Non-photochemical quenching. A response to excess light energy, *Plant Physiol.* 125(4):1558-1566.

Murray, J. W., Duncan, J. and Barber, J. (2006). CP43-like chlorophyll binding proteins: structural and evolutionary implications, *Trends Plant Sci.* 11(3):152-158.

Negi, S., Perrine, Z., Friedland, N., Kumar, A., Tokutsu, R., Minagawa, J., Berg, H., Barry, A. N., Govindjee, G. and Sayre, R. (2020). Light regulation of light-harvesting antenna size substantially enhances photosynthetic efficiency and biomass yield in green algae†, *Plant J.* 103(2):584-603.

Ort, D. R., Merchant, S. S., Alric, J., Barkan, A., Blankenship, R. E., Bock, R., Croce, R., Hanson, M. R., Hibberd, J. M., Long, S. P., Moore, T. A., Moroney, J., Niyogi, K. K., Parry, M. A., Peralta-Yahya, P. P., Prince, R. C., Redding, K. E., Spalding, M. H., van Wijk, K. J., Vermaas, W. F., von Caemmerer, S., Weber, A. P., Yeates, T. O., Yuan, J. S. and Zhu, X. G. (2015). Redesigning photosynthesis to sustainably meet global food and bioenergy demand, *Proc. Natl Acad. Sci. U. S. A.* 112(28):8529-8536.

Ort, D. R., Zhu, X. and Melis, A. (2010). Optimizing antenna size to maximize photosynthetic efficiency, *Plant Physiol.* 155:79-85.

Oster, U., Tanaka, R., Tanaka, A. and Rüdiger, W. (2000). Cloning and functional expression of the gene encoding the key enzyme for chlorophyll b biosynthesis (CAO) from *Arabidopsis thaliana*, *Plant J.* 21(3):305-310.

Roach, T. and Krieger-Liszkay, A. (2014). Regulation of photosynthetic electron transport and photoinhibition, *Curr. Protein Pept. Sci.* 15(4):351-362.

Ruban, A. V. (2012). The Photosynthetic Membrane: Molecular Mechanisms and Biophysics of Light Harvesting. Chichester: Wiley.

Ruban, A. V., Foyer, C. and Murchie, E. (2022). *Photosynthesis in Action: Harvesting Light, Generating Electrons, Fixing Carbon*. London: Academic Press.

Sanfilippo, J. E., Garczarek, L., Partensky, F. and Kehoe, D. M. (2019). Chromatic acclimation in cyanobacteria: a diverse and widerspead process for optimizing photosynthesis, *Annu. Rev. Microbiol.* 73:407–433.

Sawicki, A., Willows, R. D. and Chen, M. (2019). Spectral signatures of chlorophyll derivatives derived by non-toxic methods, *Photosynth. Res.* 140(1):115–127.

Schliep, M., Crossett, B., Willows, R. D. and Chen, M. (2010). ^{18}O-labelling of chlorophyll d in *Acaryochloris marina* reveal chlorophyll a and molecular oxygen are precursors, *J. Biol. Chem.* 285(37):28450–28456.

Shevala, D., Björn, L. O. Govindjee. (2018). *Photosynthesis: Solar Energy for Life*. Singapore: World Scientific.

Swingley, W. D., Chen, M., Cheung, P. C., Conrad, A. L., Dejesa, L. C., Hao, J., Honchak, B. M., Karbach, L. E., Kurdoglu, A., Lahiri, S., Mastrian, S. D., Miyashita, H., Page, L., Ramakrishna, P., Satoh, S., Sattley, W. M., Shimada, Y., Taylor, H. L., Tomo, T., Tsuchiya, T., Wang, Z. T., Raymond, J., Mimuro, M., Blankenship, R. E. and Touchman, J. W. (2008). Niche adaptation and Genome expansion in the chlorophyll d-producing cyanobacterium *Acaryochloris marina*, *Proc. Natl Acad. Sci. U. S. A.* 105(6):2005–2010.

Tsuchiya, T., Mizoguchi, T., Akimoto, S., Tomo, T., Tamiaki, H. and Mimuro, M. (2012). Metabolic engineering of the Chl *d*-dominated cyanobacterium *Acaryochloris marina*: production of a novel Chl species by the introduction of the chlorophyllide *a* oxygenase gene, *Plant Cell Physiol.* 53(3):518–527.

Umena, Y., Kawakami, K., Shen, J. R. and Kamiya, N. (2011). Crystal structure of oxygen-evolving photosystem II at a resolution of 1.9 Å, *Nature* 473(7345):55–60.

van der Weij-de Wit, C. D., Ihalainen, J. A., van de Vijver, E., D'Haene, S., Matthijs, H. C., van Grondelle, R. and Dekker, J. P. (2007). Fluorescence quenching of IsiA in early stage of iron deficiency and at cryogenic temperatures, *Biochim. Biophys. Acta* 1767(12):1393–1400.

Valkunas, L., van Amerongen, H. and van Grondelle, R. (2000). *Photosynthetic Excitons*. Singapore: World Scientific.

Vass, I. (2012). Molecular mechanisms of photodamage in the photosystem II complexes, *Biochim. Biophys. Acta Bioenerg.* 1817(1):209–217.

Wiltbank, L. B. and Kehoe, D. M. (2019). Diverse light responses of cyanobacteria mediated by phytochrome superfamily photoreceptors, *Nat. Rev. Microbiol.* 17(1):37–50.

Wolf, B. M. and Blankenship, R. E. (2019). Far-red light acclimation in diverse oxygenic photosynthetic organisms, *Photosynth. Res.* 142(3):349–359.

Wu, A., Brider, J., Busch, F. A., Chen, M., Chenu, K., Clarke, V. C., Collins, B., Ermakova, M., Evans, J. R., Farquhar, G. D., Forster, B., Furbank, R. T., Gorszmann, M., Hernandez, M. A., Long, B. M., Mclean, G., Potgieter, A., Price, G. D., Sharwood, R. E., Stower, M., van Oosterom, E., von Caemmerer, S., Whitney, S. M., Hammer, G. L. (2022). A cross-scale analysis to understand and quantify effects of photosynthetic enhancement on crop growth and yield across environments, *Plant Cell Environ.* doi: 10.1111/pce.14453

Zhao,X., Chen, T., Feng, B., Zhang, C., Peng, S., Zhang, X., Fu, G. and Tao, L. (2016). Non-photochemical quenching plays a key role in light acclimation of rice plants differing in leaf color, *Front. Plant Sci.* 7:1968.

Zhu, X. G., Long, S. P. and Ort, D. R. (2010). Improving photosynthetic efficiency for greater yield, *Annu. Rev. Plant Biol.* 61:235–261.

Chapter 5

Relaxing non-photochemical quenching (NPQ) to improve photosynthesis in crops

Johannes Kromdijk, University of Cambridge, UK and Carl R. Woese Institute for Genomic Biology, University of Illinois at Urbana-Champaign, USA; and Julia Walter, University of Cambridge, UK

1 Introduction

The conversion of energy from the sun into biochemical energy by plants and other photosynthetically competent organisms drives and sustains life on the earth. While the ability to perform photosynthesis provides autotrophic growth, it can be a double-edged sword. Even for non-photosynthetic organisms, exposure to sunlight can be treacherous, as anyone who has spent a little too much time in the sun will know (the global market for after-sun products currently amounts to ~US$2 billion – Statista, 2021). However, in photosynthetic organisms, the light-harvesting antennae, which are used to collect and concentrate the radiant energy for photochemistry, aggravate the risk of overexposure, precisely because of their function.

http://dx.doi.org/10.19103/AS.2022.0119.09

While low-intensity energy fluxes can be processed via photochemistry in a hazard-free manner, performing photosynthesis under high-intensity sunlight can be risky business. Given that plants depend on light to grow and survive, yet are liable to suffer damage to proteins, nucleic acids and membranes as a result of high light exposure, they have a clear need for high light avoidance strategies, as well as ways to deal with energy absorbed in excess. For example, light absorption can be reduced via paraheliotropic movements (Koller, 1990). These leaf movements position the leaf lamina in parallel to the solar beam, thereby reducing energy load. Alternatively, or in addition, some plants can modulate leaf reflectance, such as via trichome density on the leaf surface (Ehleringer, 1984), which helps to adjust the fraction of incident irradiance that gets absorbed by the leaf. However, such avoidance strategies are typically too slow to respond to rapid fluctuations in light intensity and thus come with a significant penalty to photosynthetic CO_2 fixation, while still allowing overexposure upon sharp increases in light. This explains why more dynamic mitigation strategies to safely dissipate excessively absorbed light energy are universally found across plants and other photosynthetic organisms. The focus in this chapter will be on the protective energy dissipation strategies at the level of the thylakoid, the chloroplast internal membrane system, which contains the proteins and pigments for light harvesting and photochemistry. At this level, non-photochemical quenching (NPQ) processes in the antennae of photosystem II can dynamically adjust light-harvesting efficiency in response to the relative excess of intercepted irradiance, thereby offering tailored photoprotection to the photochemical machinery downstream. In the following paragraphs, the mechanism of photoprotection via NPQ as well as the opportunities to improve photosynthetic efficiency and crop yield via adjustment of NPQ will be laid out.

2 Light harvesting and photochemistry

In higher plants, the photosynthetic light-dependent reactions are organised into two photosystems, which harvest light energy to drive redox chemistry in their reaction centres. The light-absorbing cross-section of these photosystems is greatly enlarged by pigment-binding antenna proteins surrounding the core. These antenna proteins are members of the light-harvesting complex superfamily (named a or b for photosystem I or II, respectively) and bind light-absorbing pigments in configurations that allow absorbed energy to efficiently flow from the site of absorption to the reaction centre in the core of each photosystem (Fig. 1).

In the reaction centres, a special pair of chlorophylls forms an energy trap to concentrate the energy flux from the surrounding antennae. This energy is used to start electron transfer. Photosystem II (PSII) forms a homodimeric

Published by Burleigh Dodds Science Publishing Limited, 2023.

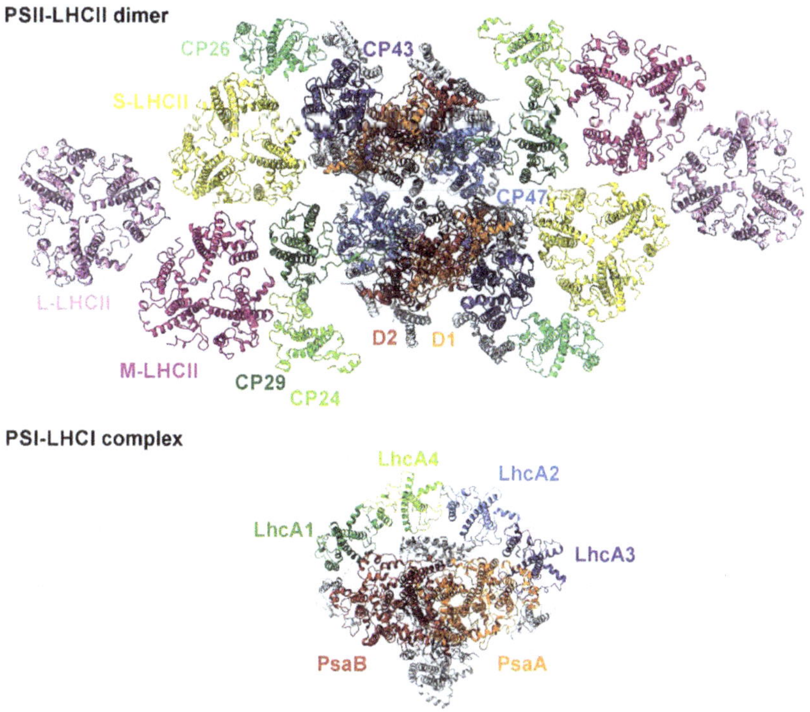

Figure 1 Pigment-protein supercomplexes PSII-LHCII and PSI-LHCI in photosynthetic thylakoid membranes of higher plants (top view). Trimeric light-harvesting complexes (LHC) are strongly (S-LHCII), moderately (M-LHCII) and loosely (L-LHCII) connected to the PSII reaction centre via the monomeric LHCII proteins CP29/CP24 and CP26 and the core proteins CP43 and CP47. Absorbed light energy is channelled from the periphery towards the PSII reaction centre, which is composed of the D1 and D2 proteins. The PSI-LHCI supercomplex contains the reaction centre proteins PsaA and PsaB, which are linked to two heterodimers, consisting of the LhcA1/A4 and LhcA2/A3 proteins. PDB entries 7OUI (PSII-LHCII from *Arabidopsis thaliana*) and 7DKZ (PSI-LHCI from *Pisum sativum*) were downloaded into ChimeraX and protein residues were coloured by amino acid chains. Grey subunits indicate PSII/PSI proteins adjacent to the reaction centres and core proteins.

complex and, moving from the reaction centre proteins D1/D2 outwards, consists of the CP47 and CP43 core proteins, Lhcb4, Lhcb5 and Lhcb6 (CP29, CP26 and CP24) minor distal antennae, and the major distal antennae Lhcb1, Lhcb2 and Lhcb3 (Jansson, 1994), which are involved in light harvesting. Whereas the core and minor antennae are present as monomers, the major antennae form homotrimers of Lhcb1 or heterotrimers of Lhcb1/2 and Lhcb1/3 (Jansson, 1999), which are classified as strongly (S), moderately (M) or loosely (L) bound, corresponding to specific binding positions around the PSII core. In contrast to PSII, photosystem I (PSI) in higher plants is formed by a monomeric core complex, which coordinates most chlorophyll and carotenoid pigments of

PSI. The absorption cross-section of PSI is further enhanced by four peripheral antenna pigment-protein complexes, Lhca1, Lhca2, Lhca3 and Lhca4, which are arranged in a half-moon shape of two heterodimers (Lhca1/4 and Lhca2/3) around the core (Qin et al., 2015). In addition, a mobile pool of LHCII trimers can flexibly (dis-)associate to either photosystem in a process called state transitions (Allen, 2003), which helps to balance the excitation energy across both photosystems upon changing light conditions.

The chlorophyll and carotenoid pigments used in light harvesting absorb light in the spectral region of 400–700 nm, which is termed the Photosynthetically Active Radiation (PAR) part of the solar spectrum. PSII absorbs at slightly different wavelengths than PSI, with the latter having a broader range of absorption into the far-red region. Absorption of the reaction centre pigments peak at approximately 680 nm (P680) and 700 nm (P700) for PSII and PSI, respectively. Whole-chain electron transport can be neatly summarised using the Z-scheme (Fig. 2, reviewed by Govindjee et al., 2017)

Electron transfer is initiated by charge separation in P680 and reduction of the first electron acceptor pheophytin. This very unstable $Pheo^-$ intermediate rapidly passes on the electron to the first stable electron acceptor plastoquinone (PQ) bound to the Q_A site in D2, leaving oxidised $P680^+$. $P680^+$ in the PSII reaction centre is reduced back to P680 via tyrosine 161 (Y_Z) in D1 by electrons liberated from water-splitting by the manganese cluster in the oxygen-evolving complex (OEC). For every four oxidation events, the OEC generates molecular oxygen (O_2) and four protons (H^+) from two water (H_2O) molecules in the thylakoid lumen. Electrons accepted by PQ_A are then passed on from Q_A to the

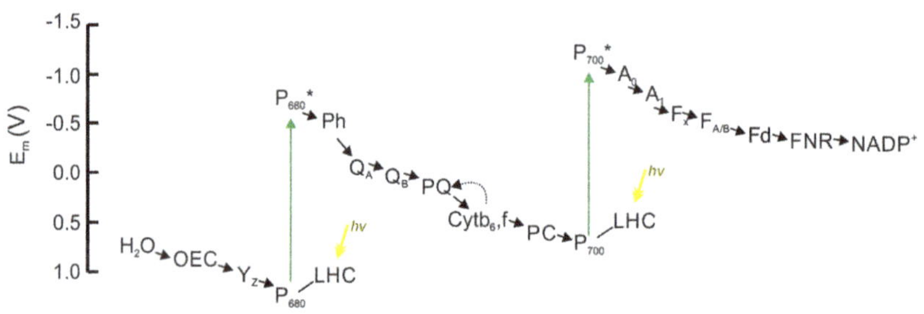

Figure 2 Linear electron transfer according to the Z-scheme. Black arrows indicate electron flow with electron carriers plotted according to their mid-point potential (E_m) at pH 7; green arrows indicate charge separation in PSII (P680) and PSI (P700) reaction centres and yellow arrows (hv) represent excitation energy. LHC, light-harvesting complex; OEC, oxygen-evolving complex; Yz, tyrosine 161; P680, PSII special chlorophyll pair; Ph, pheophytin; Q_A, quinone A; Q_B, quinone B; PQ, plastoquinone/quinol; $Cytb_6$,f, cytochrome b_6,f complex; PC, plastocyanin; P700, PSI special chlorophyll pair; A_0, chlorophyll a; A_1, phylloquinone; $F_{x/A/B}$, iron-sulphur (Fe_4S_4) clusters; Fd, ferredoxin; FNR, ferredoxin-$NADP^+$ reductase; $NADP^+$, nicotinamide adenine dinucleotide phosphate.

quinone bound to Q_B on D1, which upon acceptance of two electrons recruits two protons from the stroma to form the mobile electron carrier plastoquinol (PQH_2). PQH_2 undocks from Q_B, gets replaced by PQ from the inner-membrane pool, and diffuses to the lumen side of the thylakoid to become oxidised at the cytochrome b_6,f complex. Here, the path of electron transfer bifurcates. Electrons are either donated to the Rieske iron-sulphur (Fe-S) complex, leading to linear electron flow towards cytochrome f and plastocyanin, or passed on to cytochromes bc_i, inducing the Q-cycle around cytochrome $b_{6,f}$. In this cyclic route, electrons are transferred back to the PQ pool via several cytochromes, which leads to the reduction of PQ to PQH_2 at the stromal side of the membrane. The resulting PQH_2 then diffuses back through the membrane to the *p*-side (electrochemically positive side; Baniulis et al., 2013) of cytochrome b_6,f at the lumenal side of the membrane. In summary, this PQ/PQH_2 shuttle displaces two protons for every two electrons from the stromal to the lumenal side of the thylakoid membrane for each turn. In the other (linear) pathway, electron flow from PQH_2 to the Rieske Fe-S protein and cytochrome f reduces a second mobile electron carrier, plastocyanin, which carries electrons through the lumen to reduce $P700^+$ to P700 in PSI, following charge separation in the P700 reaction centre. Charge separation in PSI generates electron flow from P700 via phylloquinone to a series of three Fe-S clusters: F_X, F_A and F_B. Electrons from F_B are accepted by ferredoxin, a stromal electron carrier which can donate electrons to reduce $NADP^+$ to NADPH via the ferredoxin-$NADP^+$-reductase (FNR). Although useful for visualisation, it is important to remember that the Z-scheme shown in Fig. 2 does not accurately reflect the stoichiometry or localisation of the photosynthetic machinery. Photosystem II is predominantly found in the appressed regions of the thylakoid membranes in so-called grana stacks, whereas photosystem I is more prominent in the lumen-exposed margins.

The accumulation of protons in the lumen in conjunction with the electron transfer pathways, either liberated by water-splitting or displaced from the stroma to the lumen via the Q cycle, leads to the establishment of an electro-chemical proton gradient across the thylakoid membrane. This gradient drives the formation of adenosine triphosphate (ATP) via the activity of the chloroplastic F_1F_o ATP synthase (Hahn et al., 2018). The cF_1F_o ATPase conducts protons from the thylakoid lumen to the stroma through two aqueous channels, each spanning half of the membrane (Hahn et al., 2018). The resulting proton motive force drives rotation of the central stalk of the ATPase leading to ATP synthesis from adenosine diphosphate (ADP) and inorganic phosphate in the stroma-positioned catalytic F_1-head. Both NADPH and ATP are essential energy carriers which are used for downstream metabolic processes, such as the Calvin-Benson-Bassham cycle for photosynthetic carbon fixation.

3 Non-photochemical quenching: dynamic regulation of light-harvesting efficiency in the PSII antennae

Under high-intensity light, the electrons liberated from water-splitting can become too plentiful for downstream biochemistry to keep up. As a result, the components of the photosynthetic electron transport chain, that is the photosystems and mobile electron carriers that form the light-dependent reactions in photosynthesis, get more reduced. This enhances the probability of electron donation to alternative, potentially less favourable acceptors. A reduced electron transfer chain blocks efficient electron flow at the acceptor side of PSII, which increases the lifetime of singlet excited chlorophyll. This in turn enhances the probability of triplet chlorophyll formation via inter-system crossing. Energy transfer from triplet chlorophyll to molecular oxygen (which has a triplet ground state) produces singlet oxygen, a highly reactive molecule which can create damage to proteins, membranes and nucleic acids (Khorobrykh et al., 2020). In addition to this reactive oxygen species (ROS), an imbalance between the photosynthetic light reactions and utilisation of NADPH downstream also increases the chances of electron transfer to molecular oxygen at the acceptor side of PSI, leading to the formation of superoxide radical anions, another type of ROS. Superoxide damages DNA indirectly via the production of hydroxyl radicals, as well as Fe–S clusters via direct oxidation.

As explained above, the large antenna complexes surrounding each photosystem help to enhance light absorption per reaction centre and thereby improve light use efficiency of the nitrogen-costly photosynthetic machinery downstream. When light energy is absorbed in the light-harvesting antennae, the energy contained by the chlorophyll excited state can be quenched via photochemistry. In order to drive photochemistry, the absorbed energy needs to be transferred to the reaction centre, which can be done at very high efficiency via resonance energy transfer due to the proximity and orientation of neighbouring light-harvesting pigments in the light-harvesting complexes. However, when light energy is in excess, these large antennae instead enhance the risk of photoinhibition, defined as the light-induced decline of photosynthetic efficiency. For most plants, the availability of light is not constant but instead varies across orders of magnitude on timescales ranging from seconds to seasons. Hence, dynamic adjustment is crucial to leverage the trade-off between damage and efficiency. For this reason, NPQ processes are ubiquitous to photosynthetic eukaryotes (Müller et al., 2001), where they form part of an elaborate feedback de-excitation mechanism to dissipate excess energy harmlessly as heat. NPQ induction is triggered by the acidification of the thylakoid lumen, whose pH was estimated to stay between pH of 6.5 and 5.8 under most conditions (Kramer et al., 1999). Accumulation of protons in the lumen, deriving from several steps of the light-dependent reactions, thus provides an early signal that harvested light energy is in excess. This stress signal is further amplified via slowing down of ATP synthesis to match demand, or via stimulation of cyclic electron flow around PSI, which occurs due to the decreased abundance of electron acceptors downstream of PSI (high NADPH/$NADP^+$ ratio). To alleviate photoinhibition of PSI, electrons from PSI are instead recycled back to the PQ pool via alternative pathways to reduce PQ to PQH_2, aggravate lumen acidification and enhance NPQ. NPQ prevents over-reduction of the electron transport chain by funnelling energy away as heat, which in turn decreases the probability of ROS formation and concomitant damage. However, because NPQ competes with photochemistry, it can lead to a reduced light use efficiency of photochemistry, which is especially detrimental when light levels are non-saturating. To avoid this, NPQ is dynamically controlled in response to prevailing light levels. When light levels are low, dissipation of energy via NPQ is kept low and photochemistry can take place at high efficiency. On the other hand, when light levels are high enough, such that NADPH and ATP synthesis exceed the demands of the reactions downstream, NPQ is induced.

4 Assessing non-photochemical quenching via fluorescence measurements

While photochemical and non-photochemical quenching can de-excite chlorophyll, excited chlorophylls can also be quenched via the re-emission of a red-shifted photon, termed fluorescence. Although the fluorescence pathway is not meaningful for photoprotection, it offers a very useful, non-destructive mode to measure the energy transfer through the photochemical and non-photochemical routes; namely, both pathways competitively reduce fluorescence yield. While fluorescence yield can be retrieved as an absolute emission intensity as a function of absorbed excitation intensity, a more commonly used technique is pulse amplitude modulated (PAM) fluorescence, whereby a weak modulated measuring pulse elicits modulated fluorescence emission. Since the amplitude of the measuring pulse remains constant, the amplitude of the fluorescence modulation provides a relative fluorescence yield proxy, which can be easily determined in the background of actinic light (reviewed in Baker, 2008). A typical fluorescence quenching analysis is shown in Fig. 3.

Using brief saturating flashes of actinic light, photochemical quenching can be temporarily overwhelmed, which allows resolving the PSII operating efficiency from the fluorescence yield during the flash (F_m, F_m') and immediately prior to the flash (F_o, F'). Measurements of maximal fluorescence yield after full dark-adaptation and prior to illumination (F_m) can be used together with the maximal fluorescence yields during subsequent flashes (F_m') to determine NPQ at any point during a time series of fluorescence measurements. In this way, induction and relaxation kinetics of NPQ can be determined from the time series of NPQ in response to moving from darkness to actinic illumination or vice versa.

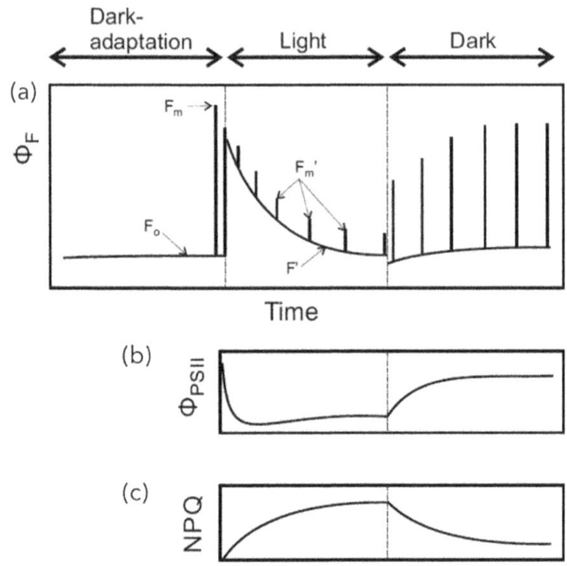

Figure 3 Typical PAM fluorescence experiment to estimate NPQ kinetics. (a) Fluorescence yield of minimal (F_o) and maximal (F_m) fluorescence after dark-adaptation are determined at the start of the protocol. Subsequently, actinic light is turned on at high intensity and saturating flashes are spaced out across the light period to measure F′ and F_m' which are used to resolve induction kinetics. Finally, light is turned off again and recovery of maximal fluorescence yield over time is probed with saturating flashes to resolve relaxation kinetics. (b) Time series of PSII operating efficiency (Φ_{PSII}) corresponding to fluorescence trace in panel A. (c) Time series of NPQ corresponding to fluorescence trace in panel A.

5 PsbS and zeaxanthin: important factors controlling non-photochemical quenching formation and relaxation in higher plants

The identification of major molecular factors for NPQ induction has greatly benefited from the power of forward genetics. Using a fluorescence imaging screen on a population of mutagenised *A. thaliana* plants, Li et al. (2000) discovered a mutant with severely reduced levels of NPQ upon high light exposure. The corresponding mutation in this *npq4* mutant was found to be a complete deletion of the photosystem II subunit S (PsbS) sequence. PsbS is a member of the light-harvesting complex superfamily and contains four transmembrane helices, rather than the common three helices found in the major and minor LHCII antennae (Kim et al., 2020). PsbS was initially hypothesised to be the quenching site in LHCII (see review in Niyogi et al., 2005). PsbS is found in PSII preparations as a 22 kDa protein (Berthold et al., 1981) and immunoprecipitates with the 33 and 23 kDa subunits of the OEC

(Ljungberg et al., 1984), but also with several LHC proteins (Bergantino et al., 2003). This suggests that it associates both with the core and periphery of PSII, perhaps as a function of the prevailing conditions. If PsbS was the quenching site one would have expected to find a strong interaction with chlorophyll and carotenoid pigments. However, although purified PsbS was reported to bind exogenous zeaxanthin *in vitro* (Aspinall-O'Dea et al., 2002), the protein was stable in the absence of chlorophyll (Funk et al., 1995) and no clear evidence for chlorophyll-binding by PsbS could be found (reviewed in Niyogi et al., 2005). Instead, results indicated that the role of PsbS might be to sense the acidification of the lumen (ΔpH) in response to excess light and relay this signal to induce quenching in LHCII. In this context, protonation of two lumen-exposed glutamate-residues seems to be important for NPQ induction, as shown by site-directed mutational analysis (Li et al., 2002a). In addition, the effect of PsbS on NPQ induction could be restored in the *psbs* deficient *npq4* mutant, using the ΔpH enhancer diaminodurene (Johnson and Ruban, 2011). It is further proposed that upon sensing a decrease in lumen pH PsbS might interact with the LHCII components and initiate the dissociation of certain major antennae from the PSII core. Thus, the structural organisation of PSII-LHCII supercomplexes might be rearranged and the qE-dependent NPQ site activated in detached LHCII trimers (Miloslavina et al., 2011; Jahns and Holzwarth, 2012; Correa-Galvis et al., 2016).

Whereas PsbS seems to signal a decrease in lumen pH to induce NPQ, the reversible xanthophyll cycle also strongly affects NPQ. When the lumen acidifies, enhanced activity of violaxanthin de-epoxidase (VDE) drives the formation of de-epoxidated xanthophylls antheraxanthin and zeaxanthin (Yamamoto and Kamite, 1972). The process is reversed under darkness or low light via the action of zeaxanthin epoxidase (ZEP), which is expressed in the stroma. Early seminal work had already established an important role for the xanthophyll cycle (Demmig-Adams and Adams, 1996) and zeaxanthin especially (Demmig-Adams, 1990) in photoprotection of photosynthesis. Thus, it came as no surprise that fluorescence imaging screening for mutants with altered levels of NPQ (Niyogi et al., 1998) yielded several mutant alleles for violaxanthin de-epoxidase (*npq1*). The *npq1* plants are unable to accumulate antheraxanthin and zeaxanthin and exhibit severely reduced NPQ. In contrast, *npq2* mutants identified on the same screen display similar levels of NPQ compared to wildtype but respond differently to changes in light intensity. In the *npq2* mutants, the ZEP enzyme is deficient, resulting in overaccumulation of zeaxanthin and no detectable levels of antheraxanthin or violaxanthin. Although this mutation also creates significant deficiencies downstream, most notably in abscisic acid synthesis (*aba1* identified by Koornneef et al. (1982) is a ZEP mutant allele), it is clear that NPQ induction in the *npq2* plants is much more rapid and more sustained, taking significantly longer to switch off under dark-adapted conditions. Although the role of zeaxanthin in thermal energy dissipation was initially met with scepticism (see the personal account of Demmig-Adams, 2003), it is now undisputed. But how and where zeaxanthin exerts its influence on NPQ and how it interacts with PsbS is still subject to debate. The lowest excited singlet states (S_1) of antheraxanthin and zeaxanthin lie just below that of singlet chlorophyll, permitting downhill excitation transfer from chlorophyll to these xanthophylls, but not to violaxanthin which has a slightly higher S_1 state, lying above that of chlorophyll (Frank et al., 1994). Additionally, NPQ may result from charge transfer from chlorophyll to zeaxanthin, via the transient formation of a radical zeaxanthin cation (Holt et al., 2005). In both cases, the reversible xanthophyll cycle would control the conversion between accessory light-harvesting violaxanthin pigments to energy-dissipating antheraxanthin and zeaxanthin pigments. Alternatively, the effect of zeaxanthin on NPQ may work via allosteric regulation, whereby binding of zeaxanthin to LHCs leads to a change in the pH sensitivity of NPQ induction towards higher pH values, effectively priming NPQ for repeat exposure (Nocter et al., 1991). In this case, the induction of NPQ is not a result of zeaxanthin as a quencher, but rather of the protonation and aggregation of LHCs, both of which are promoted by the presence of zeaxanthin (Horton et al., 1991; Ruban et al., 2012). Although these mechanisms are not necessarily specific about which LHC is involved, nor mutually exclusive, they all rely on zeaxanthin to interact and associate with LHCs. All LHCs (majors and minors) have a conserved structure with three transmembrane helices and a maximum of 14 chlorophylls per monomer. In addition, up to four carotenoid binding sites, L1, L2, N1 and V1, coordinate xanthophylls in each LHC (Jahns and Holzwarth, 2012). The N1 and V1 sites may not be present in all LHCs although their peripheral locations make this harder to assess, since any bound xanthophylls are easily removed during purification. From these four binding sites, experimental data seem to be most consistent with a putative role in NPQ for the xanthophylls bound to the V1 site in the LHCII trimers (Ruban et al., 1997, 1999) and the L2 site of the monomeric minor antennae (Dall'Osto et al., 2005; Ahn et al., 2008), although it is also possible that zeaxanthin interacts with chlorophyll at other peripheral positions between the protein-protein interface of adjacent LHCs (Xu et al., 2015).

Whereas PsbS and zeaxanthin are both required for the full induction of NPQ, PsbS seems to be specifically associated with energy-dependent quenching or qE (Wraight and Crofts, 1970), named to indicate a form of NPQ which directly depends on low lumen pH and thus readily switches off in 10-30 seconds in darkness. The accumulation of PsbS protein strongly determines the amplitude of qE (Li et al., 2002b) with saturation reached approximately 5 PsbS per D1 protein. In contrast to PsbS, zeaxanthin seems to affect multiple forms of NPQ. A threshold level of zeaxanthin seems to be required for the complete formation of qE (Niyogi et al., 1998) and when zeaxanthin levels are elevated, qE induces more rapidly and relaxes more slowly. In addition, zeaxanthin is also associated with a form of NPQ called qZ (Nilkens et al., 2010) which does not directly depend on the lumen pH, but instead depends on the turnover kinetics of the reversible xanthophyll cycle. Thus, through its involvement in qE and qZ, zeaxanthin provides a memory effect to NPQ, to maintain higher levels after repeated or prolonged exposure to high light.

6 Manipulating non-photochemical quenching to improve photosynthetic efficiency

The previous paragraphs have introduced NPQ as an important photoprotective mechanism to harmlessly dissipate excess energy. Exposure to light in the absence of adequate photoprotection leads to a decrease in the light use efficiency of photosynthesis, which is clearly observable in the shape of light response curves of electron transport rate or CO_2 fixation. These light response curves can be simplified as non-rectangular hyperbolae (NRH; Fig. 4).

Under low light, photosynthesis is determined by the initial slope, which represents the maximal efficiency with which light energy can be used to drive electron transfer ($\varphi PSII_{max}$) or CO_2 fixation ($\varphi CO_{2,max}$). In contrast, at high light, electron transport and CO_2 fixation rates are no longer limited by available light energy and plateau at their maximum capacities J_{max} and A_{sat}. The inflection from the initial slope towards the asymptote values is determined by a curvature factor θ, which ranges between zero and one for very gradual or sharp transitions, respectively. When photosynthesis is inhibited due to high light exposure, this can be seen in a decrease in the initial slope of the light response, as well as a decrease in θ and eventually in A_{sat} when the inhibition is particularly severe. Decreases in any of the NRH parameters thus signify that electron transport and CO_2 fixation rates fall short of the uninhibited rates in part or across the full light response.

Photoinhibition can either result from reversible downregulation of light-harvesting via NPQ processes as described in previous paragraphs or via photoinactivation, often involving photodamage of the reaction centre protein D1 (Aro et al., 1993; Keren and Krieger-Liszkay, 2011) which requires a dedicated repair cycle to recover uninhibited efficiency, taking hours or even longer. In contrast, photoinhibition due to NPQ is usually reversible in timescales ranging from seconds to minutes (qE) or minutes to hours (qZ) although some sustained forms of NPQ such as qH (Malnoë, 2018) can also take much longer. Using a modelling approach to simulate the decrease in maximal PSII quantum yield as a function of light history, Long et al. (1994) predicted the impact of transient photoinhibition due to reversible photoprotection on a clear-sky day in southern England. Based on their calculations, canopy CO_2

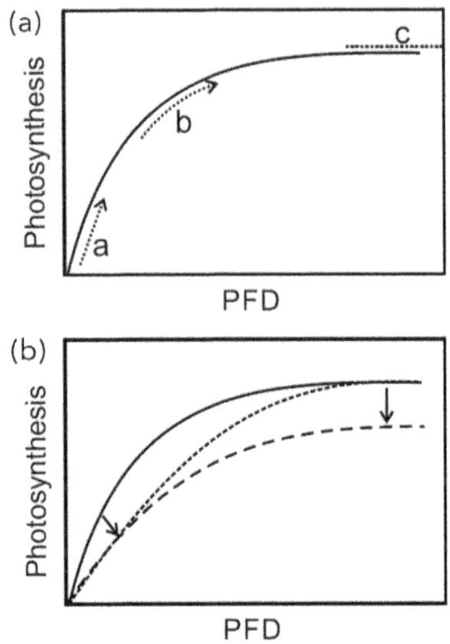

Figure 4 Non-rectangular hyperbolae to describe the light response of photosynthesis. Photosynthesis in non-stressed leaves (a) can be depicted by hyperbolic response to light intensity, which can be characterised by a non-rectangular hyperbole with parameters a (initial slope), b (curvature factor) and c (asymptote). (b) Photoprotective downregulation of light-harvesting efficiency or photoinhibition due to reaction centre damage both can lead to a decrease in the initial slope (dotted line), as well as the curvature factor and asymptote parameters (intermitted line) in more severe cases.

fixation of a hypothetical canopy with a leaf area index of 3 m^2 leaf/m^2 soil could be decreased by approximately 9%, with more pronounced losses at a higher leaf area index. Using a 3D model parameterised for Mediterranean oak, Werner et al. (2001) calculated a hypothetical loss of 6.1% carbon gain as a lower bound of photoinhibitory losses. In both cases, the light environment and concomitant loss of efficiency were predicted at hourly resolution. Realising that the averaging of light intensity profiles over hourly timesteps would remove the sharp transitions in light occurring within a plant canopy, Zhu et al. (2004) performed simulations with a ray-tracer algorithm at one second time resolution for a clear-sky spring day at 44°N at a range of temperatures. Reversible photoprotection was predicted to cause greater carbon loss in chilling-susceptible species (maize) compared to chilling-tolerant species (willow), ranging between 17–32% and 13–24%, respectively. The increase in predicted losses associated with the higher time resolution of the simulated light environment emphasised the importance of being able to recover more rapidly from the photoinhibited state.

Experiments in which PsbS was overexpressed to enhance qE were done for a range of plant species, such as the model plant *A. thaliana* (Li et al., 2002b), as well as the crop species rice (Hubbart et al., 2012) and tobacco

(Głowacka et al., 2018). The effect on NPQ induction and recovery is shown in Fig. 5a. An increase in PsbS levels results in the enhancement of the level of qE, which potentially decreases the extent of photoinhibitory damage. If so, this could enhance the recovery rate of photosynthetic efficiency after high light exposure since recovery from qE is much more rapid than from photodamage. The effects on crop productivity were surprisingly different between rice and tobacco. PsbS overexpressing rice plants showed higher biomass accumulation and yield under greenhouse conditions (Hubbart et al., 2018), whereas field-grown tobacco plants overexpressing PsbS were instead slightly smaller in size and weight than the control plants (Głowacka et al., 2018). These differences may be explained by the differences in plant architecture between these two species, leading to vastly different light environments. In addition, the PsbS overexpressing tobacco plants were found to have a dampened stomatal opening response to light, which enhanced photosynthetic water use efficiency but may have impacted their growth characteristics.

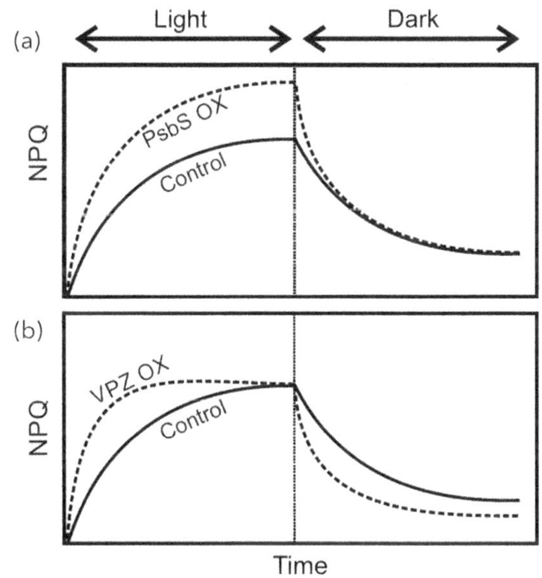

Figure 5 NPQ induction and recovery as a result of different genetic manipulations. Using the experimental protocol shown in Fig. 3, NPQ induction and relaxation can be quantified for genotypes with contrasting characteristics. The plots show example responses to illustrate the main effects associated with specific transgenic manipulations. (a) Overexpression of photosystem II subunit S (PsbS) increases the amplitude of qE and thereby total NPQ. Relaxation kinetics is similar to the control. (b) Combined overexpression of violaxanthin de-epoxidase (VDE), PsbS and zeaxanthin epoxidase (ZEP) increases induction and relaxation kinetics of NPQ, while maintaining the same amplitude of the NPQ response.

Since the overexpression of PsbS raises the amplitude of NPQ, this is unlikely to decrease the losses due to slowly reversible photoprotection, such as qZ, substantially. Experiments with plants overexpressing β-carotene hydroxylase 1 (Johnson et al., 2008), which have an enlarged xanthophyll cycle pool size, demonstrated that the cycle turnover rate is controlled by the VDE and ZEP enzyme abundance relative to the size of the pigment pool. Thus, in an attempt to speed up NPQ responses without affecting the amplitude of the response, Kromdijk et al. (2016) overexpressed the xanthophyll-converting enzymes VDE and ZEP in conjunction with PsbS in transgenic tobacco plants. This strategy aimed to speed up the xanthophyll cycle to decrease qZ amplitude and speed up qZ recovery, as well as complement the decrease in total NPQ levels specifically with qE, by overexpressing PsbS. The resulting VPZ overexpressing plants showed NPQ responses with similar amplitude to the control plants at high light, but a faster recovery rate leading to lower NPQ than the control plants at low light (Fig. 5b). Interestingly, these differences could only be observed under fluctuating light conditions, whereas under constant light, NPQ levels were indistinguishable from the control plants. In line with the model predictions by Zhu et al. (2004), this acceleration of NPQ responses improved photosynthetic efficiency, as well as biomass productivity under greenhouse and field conditions by 15%. However, when the same approach was used in *Arabidopsis*, productivity was negatively affected (Garcia-Molina and Leister, 2020). Thus, similar to the effects of overexpression of PsbS, the effects of VPZ overexpression on photosynthetic efficiency and productivity seem to be specific to species and growth conditions. In fact, the large impact of NPQ relaxation kinetics on photosynthetic efficiency is likely to be enhanced by an ongoing change in cultivation conditions. Namely, the trend in increasing plant density across many staple crops (e.g. Assefa et al., 2018) has led to an increasing proportion of leaves routinely subjected to sharp fluctuations in light intensity and may also make yield more dependent on photosynthetic efficiency (Sales et al., 2021). Adjusting the photosynthetic and photoprotective responses to these changing production conditions may offer opportunities to enhance photosynthetic efficiency and crop productivity.

7 Conclusion

Plants routinely absorb light energy strongly in excess of potential utilisation via photochemistry and have therefore developed elegant feedback mechanisms to harmlessly dissipate this energy. Modification of these NPQ mechanisms in the chloroplast thylakoid membrane system can have a profound effect on plant performance under fluctuating field conditions and could offer a trajectory towards enhanced photosynthetic efficiency and crop productivity.

8 Where to look for further information

The following articles provide a good overview of the subject:

- Murchie, E. H. and Niyogi, K. K. (2011). 'Manipulation of photoprotection to improve plant photosynthesis', *Plant Physiol.*, 155(1), 86-92.
- Kromdijk, J., Głowacka, K., Leonelli, L., Gabilly, S. T., Iwai, M., Niyogi, K. K. and Long, S. P. (2016). 'Improving photosynthesis and crop productivity by accelerating recovery from photoprotection', *Science*, 354(6314), 857-861.
- Murchie, E. H. and Ruban, A. V. (2019). 'Dynamic non-photochemical quenching in plants: from molecular mechanism to productivity', *Plant J.*, 101(4), 885-896.

9 Acknowledgements

This work was supported by the Realizing Increased Photosynthetic Efficiency (RIPE) project at the University of Illinois via a subaward to J.K. RIPE and is possible through support from the Bill & Melinda Gates Foundation, Foreign, Commonwealth & Development Office, and the Foundation for Food and Agriculture Research Grant No. OPP1172157.

10 References

Ahn, T. K., Avenson, T. J., Ballottari, M., Cheng, Y. C., Niyogi, K. K., Bassi, R. and Fleming, G. R. (2008). Architecture of a charge-transfer state regulating light harvesting in a plant antenna protein, *Science* 320(5877), 794-797.

Allen, J. F. (2003). Botany: state transitions–a question of balance, *Science* 299(5612), 1530-1532.

Aro, E. M., Virgin, I. and Andersson, B. (1993). Photoinhibition of photosystem II. Inactivation, protein damage and turnover, *Biochim. Biophys. Acta* 1143(2), 113-134.

Aspinall-O'Dea, M., Wentworth, M., Pascal, A., Robert, B., Ruban, A. V. and Horton, P. (2002). In vitro reconstitution of the activated zeaxanthin state associated with energy dissipation in plants, *Proc. Natl. Acad. Sci. U. S. A.* 99(25), 16331-16335.

Assefa, Y., Carter, P., Hinds, M., Bhalla, G., Schon, R., Jeschke, M., Paszkiewicz, S., Smith, S. and Ciampitti, I. A. (2018). Analysis of long term study indicates both agronomic optimal plant density and increase maize yield per plant contributed to yield gain, *Sci. Rep.* 8(1), 4937.

Baker, N. R. (2008). Chlorophyll fluorescence: a probe of photosynthesis in vivo, *Annu. Rev. Plant Biol.* 59, 89-113.

Baniulis, D., Hasan, S. S., Stofleth, J. T. and Cramer, W. A. (2013). Mechanism of enhanced superoxide production in the cytochrome b(6)f complex of oxygenic photosynthesis, *Biochemistry* 52(50), 8975-8983.

Bergantino, E., Segalla, A., Brunetta, A., Teardo, E., Rigoni, F., Giacometti, G. M. and Szabò, I. (2003). Light- and pH-dependent structural changes in the PsbS subunit of photosystem II, *Proc. Natl. Acad. Sci. U. S. A.* 100(25), 15265-15270.

Berthold, D. A., Babcock, G. T. and Yocum, C. F. (1981). A highly resolved, oxygen-evolving photosystem II preparation from spinach thylakoid membranes: EPR and electron-transport properties, *FEBS Lett.* 134(2), 231–234.

Correa-Galvis, V., Poschmann, G., Melzer, M., Stühler, K. and Jahns, P. (2016). PsbS interactions involved in the activation of energy dissipation in *Arabidopsis*, *Nat. Plants* 2, 15225.

Dall'Osto, L., Caffarri, S. and Bassi, R. (2005). A mechanism of nonphotochemical energy dissipation, independent from PsbS, revealed by a conformational change in the antenna protein CP26, *Plant Cell* 17(4), 1217–1232.

Demmig-Adams, B. (1990). Carotenoids and photoprotection in plants: a role for the xanthophyll zeaxanthin, *Biochim. Biophys. Acta Bioenerg.* 1020(1), 1–24.

Demmig-Adams, B. (2003). Linking the xanthophyll cycle with thermal energy dissipation, *Photosynth. Res.* 76(1–3), 73–80.

Demmig-Adams, B. and Adams III, W. W. (1996). The role of xanthophyll cycle carotenoids in the protection of photosynthesis, *Trends Plant Sci.* 1(1), 21–26.

Ehleringer, J. (1984). Ecology and ecophysiology of leaf pubescence in North American desert plants. In: Rodriguez, E., Healey, P. and Mehta, I. (Eds), *Biology and Chemistry of Plant Trichomes*, Plenum Press, New York, pp. 113–132.

Frank, H. A., Cua, A., Chynwat, V., Young, A., Gosztola, D. and Wasielewski, M. R. (1994). Photophysics of the carotenoids associated with the xanthophyll cycle in photosynthesis, *Photosynth. Res.* 41(3), 389–395.

Funk, C., Adamska, I., Green, B. R., Andersson, B. and Renger, G. (1995). The nuclear-encoded chlorophyll-binding photosystem II-S protein is stable in the absence of pigments, *J. Biol. Chem.* 270(50), 30141–30147.

Garcia-Molina, A. and Leister, D. (2020). Accelerated relaxation of photoprotection impairs biomass accumulation in *Arabidopsis*, *Nat. Plants* 6(1), 9–12.

Głowacka, K., Kromdijk, J., Kucera, K., Xie, J., Cavanagh, A. P., Leonelli, L., Leakey, A. D. B., Ort, D. R., Niyogi, K. K. and Long, S. P. (2018). Photosystem II Subunit S overexpression increases the efficiency of water use in a field-grown crop, *Nat. Commun.* 9(1), 868.

Govindjee, Shevela, D. and Björn, L. O. (2017). Evolution of the Z-scheme of photosynthesis: a perspective, *Photosynth. Res.* 133(1–3), 5–15.

Hahn, A., Vonck, J., Mills, D. J., Meier, T. and Kühlbrandt, W. (2018). Structure, mechanism, and regulation of the chloroplast ATP synthase, *Science* 360(6389), 620.

Holt, N. E., Zigmantas, D., Valkunas, L., Li, X. P., Niyogi, K. K. and Fleming, G. R. (2005). Carotenoid cation formation and the regulation of photosynthetic light harvesting, *Science* 307(5708), 433–436.

Horton, P., Ruban, A. V., Rees, D., Pascal, A. A., Noctor, G. and Young, A. J. (1991). Control of the light-harvesting function of chloroplast membranes by aggregation of the LHCII chlorophyll-protein complex, *FEBS Lett.* 292(1–2), 1–4.

Hubbart, S., Ajigboye, O. O., Horton, P. and Murchie, E. H. (2012). The photoprotective protein PsbS exerts control over CO_2 assimilation rate in fluctuating light in rice, *Plant J.* 71(3), 402–412.

Hubbart, S., Smillie, I. R. A., Heatley, M., Swarup, R., Foo, C. C., Zhao, L. and Murchie, E. H. (2018). Enhanced thylakoid photoprotection can increase yield and canopy radiation use efficiency in rice, *Commun. Biol.* 1, 22.

Jahns, P. and Holzwarth, A. R. (2012). The role of the xanthophyll cycle and of lutein in photoprotection of photosystem II, *Biochim. Biophys. Acta* 1817(1), 182–193.

Published by Burleigh Dodds Science Publishing Limited, 2023.

Jansson, S. (1994). The light-harvesting chlorophyll ab-binding proteins, *Biochim. Biophys. Acta Bioenerg.* 1184(1), 1-19.

Jansson, S. (1999). A guide to the Lhc genes and their relatives in *Arabidopsis*, *Trends Plant Sci.* 4(6), 236-240.

Johnson, M. P., Davison, P. A., Ruban, A. V. and Horton, P. (2008). The xanthophyll cycle pool size controls the kinetics of non-photochemical quenching in *Arabidopsis thaliana*, *FEBS Lett.* 582(2), 262-266.

Johnson, M. P. and Ruban, A. V. (2011). Restoration of rapidly reversible photoprotective energy dissipation in the absence of PsbS protein by enhanced delta pH, *J. Biol. Chem.* 286(22), 19973-19981.

Keren, N. and Krieger-Liszkay, A. (2011). Photoinhibition: molecular mechanisms and physiological significance, *Physiol. Plant.* 142(1), 1-5.

Khorobrykh, S., Havurinne, V., Mattila, H. and Tyystjärvi, E. (2020). Oxygen and ROS in photosynthesis, *Plants (Basel)* 9(1), 91.

Kim, E., Watanabe, A., Duffy, C. D. P., Ruban, A. V. and Minagawa, J. (2020). Multimeric and monomeric photosystem II supercomplexes represent structural adaptations to low- and high-light conditions, *J. Biol. Chem.* 295(43), 14537-14545.

Koller, D. (1990). Light-driven leaf movements, *Plant Cell Environ.* 13(7), 615-632.

Koornneef, M., Jorna, M. L., Brinkhorst-van der Swan, D. L. and Karssen, C. M. (1982). The isolation of abscisic acid (ABA) deficient mutants by selection of induced revertants in non-germinating gibberellin sensitive lines of *Arabidopsis thaliana* (L.) heynh, *Theor. Appl. Genet.* 61(4), 385-393.

Kramer, D. M., Sacksteder, C. A. and Cruz, J. A. (1999). How acidic is the lumen?, *Photosynth. Res.* 60(2/3), 151-163.

Kromdijk, J., Głowacka, K., Leonelli, L., Gabilly, S. T., Iwai, M., Niyogi, K. K. and Long, S. P. (2016). Improving photosynthesis and crop productivity by accelerating recovery from photoprotection, *Science* 354(6314), 857-861.

Li, X. P., Björkman, O., Shih, C., Grossman, A. R., Rosenquist, M., Jansson, S. and Niyogi, K. K. (2000). A pigment-binding protein essential for regulation of photosynthetic light harvesting, *Nature* 403(6768), 391-395.

Li, X. P., Phippard, A., Pasari, J. and Niyogi, K. K. (2002a). Structure-function analysis of photosystem II subunit S (PsbS) in vivo, *Funct. Plant Biol.* 29(10), 1131-1139.

Li, X. P., Müller-Moulé, P., Gilmore, A. M. and Niyogi, K. K. (2002b). PsbS-dependent enhancement of feedback de-excitation protects photosystem II from photoinhibition, *Proc. Natl. Acad. Sci. U. S. A.* 99(23), 15222-15227.

Ljungberg, U., Åkerlung, H.-E., Larsson, C. and Andersson, B. (1984). Identification of polypeptides associated with the 23 and 33 kDa proteins of photosynthetic oxygen evolution, *Biochim. Biophys. Acta Bioenerg.* 767(1), 145-152.

Long, S. P., Humphries, S. and Falkowski, P. G. (1994). Photoinhibition of photosynthesis in nature, *Annu. Rev. Plant Physiol. Plant Mol. Biol.* 45(1), 633-662.

Malnoë, A. (2018). Photoinhibition or photoprotection of photosynthesis? Update on the (newly termed) sustained quenching component qH, *Environ. Exp. Bot.* 154, 123-133.

Miloslavina, Y., de Bianchi, S., Dall'Osto, L., Bassi, R. and Holzwarth, A. R. (2011). Quenching in *Arabidopsis thaliana* mutants lacking monomeric antenna proteins of photosystem II, *J. Biol. Chem.* 286(42), 36830-36840.

Müller, P., Li, X. P. and Niyogi, K. K. (2001). Non-photochemical quenching: a response to excess light energy, *Plant Physiol.* 125(4), 1558-1566.

Murchie, E. H. and Niyogi, K. K. (2011). Manipulation of photoprotection to improve plant photosynthesis, *Plant Physiol.* 155(1), 86–92.

Murchie, E. H. and Ruban, A. V. (2019). Dynamic non-photochemical quenching in plants: from molecular mechanism to productivity, *Plant J.* 101(4), 885–896.

Nilkens, M., Kress, E., Lambrev, P., Miloslavina, Y., Müller, M., Holzwarth, A. R. and Jahns, P. (2010). Identification of a slowly inducible zeaxanthin-dependent component of non-photochemical quenching of chlorophyll fluorescence generated under steady-state conditions in *Arabidopsis*, *Biochim. Biophys. Acta* 1797(4), 466–475.

Niyogi, K. K., Grossman, A. R. and Björkman, O. (1998). *Arabidopsis* mutants define a central role for the xanthophyll cycle in the regulation of photosynthetic energy conversion, *Plant Cell* 10(7), 1121–1134.

Niyogi, K. K., Li, X. P., Rosenberg, V. and Jung, H. S. (2005). Is PsbS the site of non-photochemical quenching in photosynthesis?, *J. Exp. Bot.* 56(411), 375–382.

Nocter, G., Rees, D., Young, A. and Horton, P. (1991). The relationship between zeaxanthin, energy-dependent quenching of chlorophyll fluorescence, and trans-thylakoid pH gradient in isolated chloroplasts, *Biochim. Biophys. Acta Bioenerg.* 1057(3), 320–330.

Qin, X., Suga, M., Kuang, T. and Shen, J. R. (2015). Photosynthesis. Structural basis for energy transfer pathways in the plant PSI-LHCI supercomplex, *Science* 348(6238), 989–995.

Ruban, A. V., Phillip, D., Young, A. J. and Horton, P. (1997). Carotenoid-dependent oligomerization of the major chlorophyll a/b light harvesting complex of photosystem II of plants, *Biochemistry* 36(25), 7855–7859.

Ruban, A. V., Lee, P. J., Wentworth, M., Young, A. J. and Horton, P. (1999). Determination of the stoichiometry and strength of binding of xanthophylls to the photosystem II light harvesting complexes, *J. Biol. Chem.* 274(15), 10458–10465.

Ruban, A. V., Johnson, M. P. and Duffy, C. D. P. (2012). The photoprotective molecular switch in the photosystem II antenna, *Biochim. Biophys. Acta* 1817(1), 167–181.

Sales, C. R. G., Wang, Y., Evers, J. B. and Kromdijk, J. (2021). Improving C4 photosynthesis to increase productivity under optimal and suboptimal conditions, *J. Exp. Bot.* 72(17), 5942–5960.

Statista (2021). Revenue of the sun protection skin care market in the U.S. 2012–2025. Available at: https://www.statista.com/forecasts/1244392/united-states-us-revenue-sun-protection-market (accessed October 12, 2021).

Werner, C., Ryel, R. J., Correia, O. and Beyschlag, W. (2001). Effects of photoinhibition on whole-plant carbon gain assessed with a photosynthesis model, *Plant Cell Environ.* 24(1), 27–40.

Wraight, C. A. and Crofts, A. R. (1970). Energy-dependent quenching of chlorophyll a fluorescence in isolated chloroplasts, *Eur. J. Biochem.* 17(2), 319–327.

Xu, P., Tian, L., Kloz, M. and Croce, R. (2015). Molecular insights into zeaxanthin-dependent quenching in higher plants, *Sci. Rep.* 5, 13679.

Yamamoto, H. Y. and Kamite, L. (1972). The effects of dithiothreitol on violaxanthin deepoxidation and absorbance changes in the 500-nm region, *Biochim. Biophys. Acta Bioenerg.* 267(3), 538–543.

Zhu, X. G., Ort, D. R., Whitmarsh, J. and Long, S. P. (2004). The slow reversibility of photosystem II thermal energy dissipation on transfer from high to low light may cause large losses in carbon gain by crop canopies: a theoretical analysis, *J. Exp. Bot.* 55(400), 1167–1175.

Chapter 6

Modifying mesophyll conductance to optimise photosynthesis in crops

Coralie E. Salesse-Smith, University of Illinois at Urbana-Champaign, USA; Steven M. Driever, Wageningen University and Research, The Netherlands; and Victoria C. Clarke, The Australian National University, Australia

1 Introduction

The need to feed a growing global population, combined with a changing climate and diminishing resources, has driven researchers towards improving crop productivity through a variety of approaches (Bailey-Serres et al., 2019; Ray et al., 2013). Recently, this effort has begun to focus on improving photosynthesis as a key driver of crop yields. The process of photosynthesis is dependent upon acquisition of carbon dioxide (CO_2) from the atmosphere, which is converted into carbohydrates by the enzyme Rubisco within the chloroplast. In C_3 plants,

http://dx.doi.org/10.19103/AS.2022.0119.10

the diffusion of atmospheric CO_2 into leaf chloroplasts is a limiting step in this process, and improving the conductance to CO_2 diffusion within leaves, termed mesophyll conductance (g_m), is predicted to increase photosynthetic rates and improve water use efficiency (Long et al., 2015; Lundgren and Fleming, 2020).

Modelling has suggested that improving the conductance of CO_2 into chloroplasts would increase photosynthetic assimilation rates due to several mechanisms (Wu et al., 2019). Rubisco efficiency would be improved as higher concentration of available CO_2 increases the carboxylation rates of Rubisco, resulting in less photorespiration. With more CO_2 available for fixation, the nitrogen cost of synthesising and maintaining photosynthetic machinery is also reduced due to increased efficiency in the system (Evans and Clarke, 2019), allowing for grain protein content to be maintained under higher assimilation rates without additional fertilisation.

An increase in g_m would also allow higher photosynthetic rates to be achieved without greater water loss through an adjustment of stomatal conductance. At the surface of the leaf, the boundary layer and stomatal properties influence both water vapour transpiration and gas diffusion into the sub-stomatal cavity. Thus, a trade-off exists between preventing water loss through transpiration and maintaining CO_2 diffusion to allow for photosynthesis, which is regulated dynamically by stomatal conductance. With higher g_m, stomatal conductance could be reduced while still maintaining the same chloroplastic CO_2 concentration, reducing water losses due to transpiration through the stomata.

The partial pressure of CO_2 within the chloroplast is highly variable due to the rate of CO_2 assimilation by Rubisco and the total conductance to CO_2. Higher assimilation rates consume more CO_2, tending to decrease CO_2 concentration in the chloroplast, while increases in g_m tend to increase it. Therefore, the combination of improved g_m with other complementary photosynthetic improvements will be essential to offset reductions in chloroplast CO_2 partial pressure, further enhancing the impact of increased g_m.

The path of CO_2 to the chloroplast is different in C_3 and C_4 photosynthesis plants, with the specialised anatomy of C_4 plants creating a spatial separation that concentrates CO_2. This impacts the way in which mesophyll conductance is estimated and interpreted. In this chapter, we will focus on strategies to improve mesophyll conductance in C_3 plants only. For a comprehensive discussion of recent advances in mesophyll conductance in C_4 (and CAM photosynthesis) plants see Cousins et al. (2020).

On its journey from the atmosphere into chloroplasts in C_3 plants, CO_2 faces several barriers that impose resistance to diffusion, including the cell wall, plasma membrane, cytosol, chloroplast envelopes and chloroplast stroma (Fig. 1, Clarke et al., 2021). These resistances together determine the overall mesophyll conductance. This book chapter will describe each factor contributing

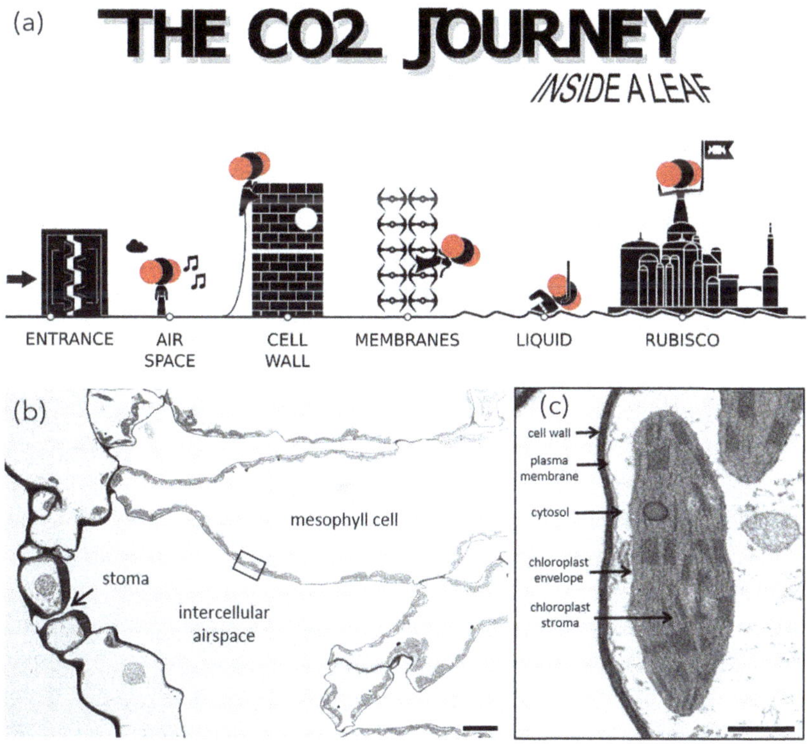

Figure 1 A Star Wars-inspired cartoon depicting the CO_2 diffusion journey inside the leaf with Princess Leia representing a CO_2 molecule (a) and the corresponding electron micrographs illustrating the actual CO_2 diffusion pathway within a tobacco leaf (b and c). For photosynthetic CO_2 assimilation to occur, CO_2 has to diffuse through stomata (entrance), intercellular airspace, cell wall, membranes (plasma membrane and chloroplast envelope) and the liquid phase in the cytosol and chloroplast stroma. Once inside the chloroplast, CO_2 is then fixed by Rubisco into energy-rich sugars as part of the photosynthesis process. Bar in *b* = 10 μm, bar in *c* = 1 μm. Adapted from Clarke et al. (2021).

to the resistance to CO_2 diffusion within C_3 leaves and discuss research efforts to overcome these resistances and improve the CO_2 concentration at the site of carboxylation.

2 Points of resistance to diffusion of CO_2

The CO_2 diffusion path within the leaf and its encompassing sequence of barriers directly determines mesophyll conductance. These barriers to CO_2 movement can be described as a series of resistances in the leaf, referred to as mesophyll resistance (r_m), which is inverse to g_m ($r_m = 1/g_m$). Although existing experimental methods cannot measure the individual components of r_m or g_m, these components can be studied separately using computational models, and understanding each component is critical to understand g_m as a whole.

Mesophyll resistance is generally divided as follows (Lundgren and Fleming, 2020; Ren et al., 2019):

- gas phase resistance;
- liquid phase resistance.

Gas phase resistance encompasses CO_2 movement through intercellular airspaces of the leaf, while liquid phase resistance is made up of the following series of resistances (Cousins et al., 2020; Evans et al., 2009):

- cell wall;
- plasma membrane;
- cytosol;
- chloroplast envelope;
- chloroplast stroma.

Liquid phase resistance accounts for up to 90% of mesophyll resistance to CO_2 diffusion through the leaf (Lu et al., 2016; Tosens et al., 2012) and is the primary focus of the following sections which discuss the roles of cell anatomy and biochemical processes in determining g_m.

3 The interaction between mesophyll cell anatomy, light and g_m

The constraints on CO_2 diffusion within the mesophyll are determined to a large extent by the shape, size and density (packing) of mesophyll cells and the position and orientation of the chloroplasts within, since these factors determine the liquid path length that CO_2 must travel from the intercellular air spaces to the chloroplasts. First, the arrangements of cells within the mesophyll in relation to CO_2 diffusion and light are discussed. In particular, the mesophyll can be considered a porous medium, where CO_2 diffuses through intercellular airspace and into the cells containing the chloroplasts. Some of the largest physical determinants of g_m are the surface area of the cells exposed to the intercellular airspace (S_m), the surface area of the chloroplasts exposed to the intercellular airspace (S_c) and the ratio S_c/S_m. S_c/S_m expresses the relative fraction of available area that is occupied by chloroplasts through which CO_2 can diffuse towards the site of Rubisco.

The most straightforward improvement of g_m would be through an increase of S_m and S_c. The presence of many small cells within the mesophyll

Published by Burleigh Dodds Science Publishing Limited, 2023.

increases both S_m and S_c, which has been proposed as a potential target for improving g_m. Indeed, several studies have shown that this strategy increases g_m, with a positive effect on photosynthetic capacity. In *Arabidopsis*, increased cell number and packing increased g_m (Lehmeier et al., 2017) and in rice, leaves with lobed cells had an increased S_m, increased g_m and photosynthetic capacity (Sage and Sage, 2009). The lobed cell shape of rice leaves is an example of a mesophyll cell anatomy that can increase cell surface to volume ratio and enhance S_m as well as light absorption. Cell shape as well as cell number and packing are therefore both important to consider when improving g_m and the effects on S_m and S_c are reviewed below.

Increasing mesophyll cell number and packing does increase S_m but cannot be increased indefinitely, as it comes at a higher construction cost (cell wall material) and subsequently smaller leaves (Lundgren and Fleming, 2020; Ren et al., 2019). Increasing cell number and packing would logically increase S_m, but there is a trade-off between the number, packing and size of cells, since cell size is negatively related to cell number within a given space. Following the simplified model of Ren et al. (2019), S_m is related to the circumference of the cell when considered as a spherical object. If a single large cell has an S_m of π, then nine smaller cells within the same space would have an S_m of 3π. However, the nine cells have more areas where chloroplasts overlap and this would decrease S_c. Therefore, an increase in S_m by higher cell density is only beneficial for increasing g_m if S_c is not reduced proportionally due to cell surface overlapping.

S_c is also influenced by chloroplast number, movement and distribution in the mesophyll. A larger number of chloroplasts per cell can enhance S_c (Miyazawa and Terashima, 2001). But, as for a larger cell number, this cannot be increased without limit. Some species have a ratio of S_c to S_m (S_c/S_m) that is far below 1, where there could be benefit in an increased S_c. Still, a larger effect is expected from greater S_m to allow for further increases of S_c. Therefore S_c and S_m should be increased simultaneously to realise the greatest enhancement of g_m (Ren et al., 2019). The re-distribution and movement of chloroplasts is also known to change S_c (Hanba et al., 2002; Oguchi et al., 2003). Under high light intensity, the chloroplasts can be arranged vertically along the anticlinal cell walls (profile position) to avoid damage by excess light intensity. Under low light intensity, they can move in a perpendicular position (face position) to maximise light capture. The possibility of chloroplasts to move and still face the mesophyll airspace is limited when chloroplast number is high. For example, in rice, the chloroplasts are very tightly packed in the lobed mesophyll cells (Sage and Sage, 2009). However, under high light intensity, tightly packed chloroplasts and limited chloroplast movement can increase the risk of high light damage and may be more disadvantageous than gains in g_m.

The distribution of light strongly depends on the 3D anatomy and arrangement of mesophyll cells and has an influence on the so-called 'apparent' g_m: the effective g_m at leaf level that can vary with light intensity and CO_2 concentration, even though the underlying conductance at cell level remains the same (Evans, 2021). This is because within the mesophyll, the assimilation of CO_2 is distributed among many chloroplasts. These chloroplasts are all spatially distributed at different depths and consequently experience different light environments depending on the vertical attenuation of light in the mesophyll (Earles et al., 2019). Light attenuation is affected by the colour and intensity of the incoming light as well as the shape of the mesophyll cells. For instance, vertical columnar pallisade cells allow light to penetrate deeper into the leaf (Terashima and Saeki, 1983; Vogelmann and Evans, 2002; Vogelmann and Martin, 1993). Irregularly shaped spongy mesophyll cells and air spaces increase reflection and refraction of light, causing optical path lengthening (Terashima et al., 2009; Vogelmann and Martin, 1993). This is also known as the detour effect and, apart from cell shape, is due to the higher refractive index of cells versus air. Thus, the spatial distribution and shape of cells and chloroplasts influence light attenuation within the mesophyll. This becomes evident at low light where the chloroplasts in the lower layers of mesophyll cells may operate at the light compensation point and do not contribute much to the total g_m, while at high light, this contribution becomes substantial (Théroux-Rancourt and Gilbert, 2017). Indeed, 'apparent' g_m has been shown to respond to light intensity at both steady state (Busch et al., 2020; Xiong et al., 2018; Yin et al., 2020) and under dynamic conditions (Carriquí et al., 2019; Liu et al., 2022; Sakoda et al., 2021). So although cell wall and membrane properties like S_m and S_c are important, the 3D structure and arrangement of mesophyll cells determine to a large extent how the 'apparent' g_m reacts to environmental changes such as light.

4 Leaf age and g_m

An example of the link between leaf anatomy and g_m can be observed in leaves of different age and canopy position. Several studies have examined the modulation of g_m by the mesophyll cell architecture in relation to leaf age, canopy position and light intensity. During the development of a leaf, it has been shown that g_m can limit photosynthesis considerably (Carriquí et al., 2021; Clarke et al., 2021; Hanba et al., 2001; Hoshika et al., 2020; Marchi et al., 2008; Miyazawa et al., 2003; Miyazawa and Terashima, 2001; Niinemets et al., 2012; Tosens et al., 2012). In most species, g_m increases with leaf expansion and reaches a maximum with full expansion and then declines with leaf age (Loreto et al., 1994; Niinemets et al., 2012). The increase in g_m during expansion is largely due to the growth and expansion of mesophyll cells and the increase of leaf thickness both of which increase S_m and S_c. The subsequent decline in g_m after expansion has been attributed to the increase in cell wall thickness (Miyazawa et al., 2003; Miyazawa and Terashima, 2001; Sugiura et al., 2020; Tosens et al., 2012) and reduction of cell wall effective porosity (Niinemets et al., 2012). In tobacco, g_m reductions with leaf age were accompanied by both cell wall thickening and a decrease in S_c/S_m (Clarke et al., 2021).

Another observed effect of leaf age was that of reduction of the thickness of chloroplasts. The thickness of the chloroplast determines the diffusion path length of CO_2 inside the chloroplast stroma and, in theory, can influence the conductance of CO_2 in the liquid phase; an important component of g_m. A decrease in chloroplast stroma thickness with leaf age has been observed in *Arabidopsis* (Carriquí et al., 2021), while other studies have not seen a decrease with age (Clarke et al., 2021) or have not found a correlation between chloroplast stroma thickness and g_m (Tosens et al., 2012). It is noteworthy that chloroplast stroma thickness was negatively affected by the growth light intensity in some species (Tosens et al., 2012) but not in others (Oguchi et al., 2003; Vidal et al., 1990). Still, chloroplast stroma thickness can potentially be an important factor for g_m. This was best illustrated in a study on transgenic tobacco with reduced Rubisco content that had lower chloroplast stroma thickness compared with the wild type and was accompanied by a lower CO_2 drawdown (Evans et al., 1994).

The observed changes in g_m with leaf age highlight that an increase in g_m can be realised by:

- an increase in S_m and S_c by mesophyll cell expansion and increased leaf thickness;
- a reduction in cell wall thickness.

However, this will need to be accomplished at a cell density that allows maximal exposure of cell walls to the intercellular airspace. In that regard, an arrangement of mesophyll cells that are more loosely packed together may allow for a more homogenous distribution of light throughout the mesophyll. This would not only increase S_m and S_c but also the light level for chloroplasts at lower layers in the mesophyll that would enhance the 'apparent g_m' (Fig. 2).

The effects of leaf age on g_m illustrate that mesophyll anatomy influences g_m to a large extent and that it can be exploited. There is a variation in the extent to which these leaf age effects are observed not only between species, but also in response to growth light intensity. Especially in canopies with steep light gradients, the effect of light intensity may potentially be stronger than that of leaf age but has not been extensively studied in relation to g_m in C_3 plants.

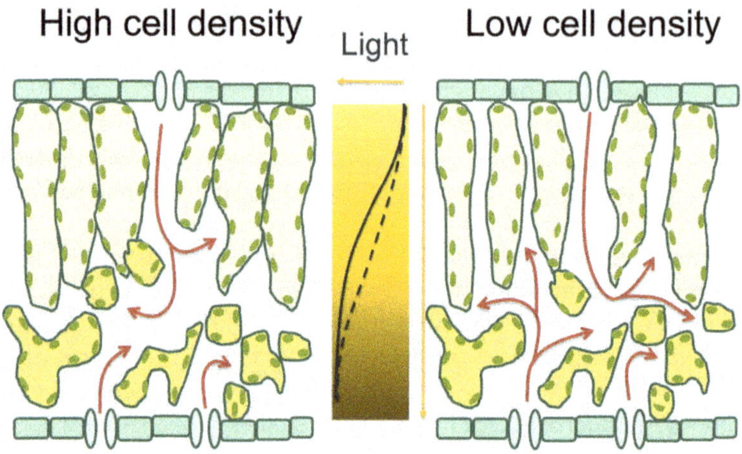

Figure 2 A diagram of mesophyll tissue shown in cross section through the leaf. The arrangement of mesophyll cells can be more closely packed (high cell density) or more loosely packed (low cell density), influencing both the gas diffusion pathways (red lines) and light distribution throughout the mesophyll (black lines).

5 Cell wall diffusion

Diffusion through the cell wall is the first step in the liquid phase of CO_2 diffusion through the leaf. Cell walls are structurally complex barriers that function to protect plants from biotic and abiotic stresses, maintain structural integrity and support cell division and plant growth (Cosgrove, 2005; Houston et al., 2016). The walls contain pores filled with apoplastic fluid; these pores form the interface between the gas and liquid phases of diffusion and provide routes for CO_2 to move from the intercellular airspaces into mesophyll cells. Several model studies of g_m have suggested that the cell wall is one of the main constraints on mesophyll conductance (Evans et al., 1994; Gago et al., 2020; Terashima et al., 2011; Tomas et al., 2013; Xiao and Zhu, 2017; Yin and Struik, 2017).

The path through the cell wall is complicated, and the overall resistance depends on the number of channels or pores, the effective length of each channel and the rate of diffusion along each channel (see Fig. 3 in Evans et al., 2009). These considerations can be expressed mathematically using four components: cell wall thickness, tortuosity of the path through cell wall pores, cell wall porosity and diffusivity of CO_2 through the apoplastic fluid (Evans, 2021). Although one cannot definitively calculate cell wall resistance and each of its components, it can be estimated with the help of plant models. The impact of cell wall thickness on g_m is relatively well studied being that it is the only component affecting cell wall conductance to CO_2 that can be easily and reliably measured. Cell wall thickness has been measured in hundreds of species using transmission electron microscopy (Gago et al., 2020). Several studies have reported that cell wall thickness is negatively correlated with mesophyll conductance, suggesting selection or genetic modification for thinner cell walls could be a good target to enhance mesophyll conductance (Clarke et al., 2021; Onoda et al., 2017; Ren et al., 2019; Terashima et al., 2011; Tosens et al., 2012). However, one must consider how thin a mesophyll cell wall can be made without jeopardising its structural integrity (Cosgrove, 2018). The ideal solution will likely be complicated and depend on factors such as the specific plant in question, and its mesophyll cell size and cell wall composition (Lundgren and Fleming, 2020). Additionally, the lack of data on the regulation of cell wall thickness and genes that may be involved make targeted changes quite challenging. Moreover, much uncertainty exists with the other three components of cell wall resistance partly due to the fact that there are currently no good methods available to experimentally measure these parameters (Evans, 2021).

The composition of the cell wall is another element which may influence g_m by modifying CO_2 diffusion. Primary cell walls are composed of a complex carbohydrate matrix of cellulose, hemicellulose and pectin (Cosgrove and Jarvis, 2012; Cosgrove, 2005; Tenhaken, 2015). It is not difficult to envision that altering the absolute amounts or ratios of these components will affect CO_2 diffusion through the cell wall. These changes could potentially modify cell wall thickness or alter porosity and the liquid path length through the cell wall (Ellsworth et al., 2018). However, how changes in cell wall composition affect tortuosity and porosity remains relatively unexplored (Carriquí et al., 2020). Pectins are known to have a role in regulating cell wall porosity but little is known about their direct effect on CO_2 diffusion (Fleischer et al., 1999; Rondeau-Mouro et al., 2008). The ability to directly measure these parameters is essential in order to answer these questions.

A small number of recently published studies have started to examine the relationships between cell wall composition and mesophyll conductance using cell wall mutants, stress treatments and interspecific variation. Ellsworth et al. (2018) observed that rice mutants unable to synthesise mixed-linkage glucans (a cell wall hemicellulose) had reduced photosynthetic rates, g_m, S_c and leaf mass per area. In addition, the authors calculated the effective porosity of the mesophyll cell walls, which was estimated to be significantly reduced in the mutants relative to the control. From this it was concluded that mixed-linkage glucans in cell walls improve effective porosity and CO_2 diffusivity in rice (Ellsworth et al., 2018). Another study looking at pectin methylesterase and pectin acetylesterase mutants in Arabidopsis reported reductions in CO_2 assimilation and g_m associated with reduced galacturonic acid content (the main component of pectin; Roig-Oliver et al., 2021). Two other studies looking at water stress in grapevines and tobacco found that decreases in photosynthesis and g_m were associated with changes in cellulose and pectin content in the cell wall (Clemente-Moreno et al., 2019; Roig Oliver et al., 2019). Furthermore, to assess whether cell wall composition plays a role in existing differences in g_m between species from common environments, Carriquí et al. (2020) measured cell wall composition, CO_2 assimilation and g_m in seven gymnosperm species. The study found g_m to be correlated with cell wall composition, likely signifying a change in the diffusivity of the cell wall. The authors conclude that higher ratios of pectin to hemicellulose and/or pectin to cellulose was correlated with increased g_m, suggesting pectin has a role in regulating CO_2 diffusivity in the cell wall of gymnosperms.

These novel studies have started to provide critical information about the relationship between g_m and cell wall properties. However, additional work is needed to explore new variations on cell wall composition such as increasing or decreasing the amounts of individual primary cell wall components. Analyses comparing species in other land plant groups besides gymnosperms will also be helpful since differences in main cell wall components exist between plant types (Popper et al., 2011). The information gained from such studies can be used to directly target changes in cell wall composition through genetic engineering or selection. Based off our current knowledge, we propose some potential targets for modification with the goals of gleaning a better understanding of and enhancing CO_2 diffusivity across the cell wall:

- Genes coding for enzymes that catalyse cellulose, hemicellulose or pectin biosynthesis
- Genes involved in cell wall loosening and cell expansion
- Genes that affect methylation and/or crosslinking of primary cell wall components

Overall, altering cell wall properties such as composition and thickness is a promising route for reducing CO_2 diffusion resistance across the cell wall and increasing mesophyll conductance. However, true engineering remains elusive because many fundamental questions about the genetic control of cell wall properties and the relationship between those properties and g_m remain unanswered. Future exploratory work, as well as the development of new measurement techniques for measuring diffusion and relevant cell wall parameters, will be crucial for successful genetic engineering of mesophyll conductance.

After crossing the mesophyll cell wall, a series of liquid phase resistances remains between CO_2 and the site of carboxylation. These are the plasma membrane, cell cytosol, chloroplast envelopes and the liquid stroma inside the chloroplasts. The resistance imposed by the membrane components (plasma and chloroplast envelope) are discussed together, followed by liquid cytosol and stroma components.

6 Cellular membranes and CO_2 diffusion

Cellular membranes are composed of a lipid bilayer and function to modulate movement of molecules in and out of the cell. CO_2 is hydrophobic and non-polar, and initial studies on membrane permeability in artificial lecithin-cholesterol bilayers concluded that biological lipid membranes offered little to no resistance to CO_2 (Gutknecht et al., 1977). These membrane models, however, did not account for the dense packing of proteins within the membrane and differing lipid composition. Unlike animal cells, plant membranes do not contain cholesterol but instead have a suite of related sterols (Reszczyńska and Hanaka, 2020).

Alteration of lipid content in plant membranes could be a target for improving g_m; however, as the membranes are very protein-dense, the available free lipid membrane for CO_2 diffusion is likely quite limited. The lipid and protein composition of membranes can nevertheless be significantly remodelled in response to environmental cues, suggesting optimisation of this process could potentially improve overall g_m (Uemura et al., 1995).

Cellular membranes impose up to 50% of the resistance to CO_2 diffusion within mesophyll cells (Endeward et al., 2017; Evans, 2021). These membranes include the plasma membrane (PM) and the dual membranes of the chloroplast envelope (CE). The CO_2 permeability of the CE is assumed to be half that of the PM, as the envelope is composed of dual membranes, thus increasing the resistance (Evans, 2021). While some argue the Meyer–Overton rule (also known as the solubility-diffusion model) accounts for all gaseous CO_2 transfer across membranes (Missner and Pohl, 2009), the PM and CE contain channels and transporters that facilitate CO_2 transfer across the membrane and manipulation of these has become a target for understanding and improving mesophyll conductance in plants.

7 Improving g_m using aquaporins as CO_2 channels

Aquaporins are a prominent gene family involved in CO_2 transfer across membranes. While initially named for their water transfer capabilities, further studies have reported aquaporins to be multifunctional proteins, capable of facilitating the transfer of various solutes, ions and gases, including CO_2 (Groszmann et al., 2017). The functional aquaporin unit is composed of four monomer channels coming together to form a tetramer, which creates a fifth central pore. Solute specificity is influenced by individual monomer properties and their combinations in the tetramer (Otto et al., 2010). Aquaporins are dynamic proteins that can move in and out of the membrane in response to environmental stimuli, and the pores within the tetramer can also be gated, allowing close regulation of channel activity (see Groszmann et al., 2017 for a detailed discussion of plant aquaporin structure, function and regulation).

Early work in mammalian systems first identified aquaporin proteins as CO_2 channels. The human red blood cell aquaporin, AQP1, when expressed in *Xenopus* oocytes, increased membrane permeability to CO_2 (Cooper and Boron, 1998; Nakhoul et al., 1998). In plants, members of the PIP aquaporin family are homologous to mammalian AQP1, and generally localise to the plasma membrane, with some isoforms also detected in chloroplast envelopes by Western blot and proteomic analysis (Beebo et al., 2013; Uehlein et al., 2008).

Expression of a tobacco AQP1 homologue, NtAQP1 (also known as NtPIP1;5s, De Rosa et al., 2020), in oocytes also conferred increased CO_2 permeability (Uehlein et al., 2003). A function in mesophyll conductance was predicted and *in vivo* ectopic or antisense expression study of NtAQP1 in tobacco altered mesophyll conductance, with ectopic expression increasing g_m by up to 20% (Flexas et al., 2006). Additional studies targeting aquaporins from other species and across environmental conditions have had mixed effects on mesophyll conductance, with some studies reporting effects on g_m (Ermakova et al., 2021; Hanba et al., 2004; Heckwolf et al., 2011; Mori et al., 2014; Uehlein et al., 2008; Xu et al., 2019), while in others studies, no effects were observed (Clarke et al., 2022; Kromdijk et al., 2020).

Questions remain about why manipulation of aquaporins does not produce consistent effects on mesophyll conductance, and several factors have been suggested that may be impacting the g_m phenotype in these studies. The underlying g_m capacity of a species may affect the ability to detect improvements in g_m. Many crop species already have relatively high g_m, driven in part by their thin cell walls (see 'Cell Walls' above). Modelling has suggested that when basal g_m is higher, increases will be subtle and harder to detect (Clarke et al., 2022). A study in tomato reported that ectopic expression of *NtAPQ1* did not increase g_m until basal g_m levels had been sufficiently reduced through simultaneous ectopic expression of hexokinase (Kelly et al., 2014). Recent studies have also highlighted the dynamic effects of environmental factors on g_m. Conditions such as photoperiod, light intensity, day/night temperature, water availability, humidity and nutrient supply can all alter anatomical and biochemical components that influence g_m. The conditional effects of growth environment and growth stage between aquaporin loss-of-function mutants and g_m have been reported in rice (Huang et al., 2021), and in tomato, an aquaporin knockout mutant was only shown to reduce g_m when mutants were grown under high CO_2 levels (Zhang et al., 2021). Further studies to better understand these complex relationships will be essential to ensure manipulation of aquaporins can provide a consistent impact on g_m, and aquaporins remain a strong engineering target for enhancing g_m.

Published by Burleigh Dodds Science Publishing Limited, 2023.

To further efforts to improve the membrane conductivity to CO_2, it may also be beneficial to look beyond aquaporins to find candidates from other systems where CO_2 transfer across membranes occurs. Red blood cells (RBCs), for example, require the ability to rapidly uptake CO_2 from the blood stream, and indeed AQP1 is present on these membranes (see Michenkova et al., 2021 for a comprehensive review). However, another protein, the Rhesus-AG protein (part of the Rhesus complex) is also present on the RBC membrane and also acts as a CO_2 channel, contributing half of CO_2 uptake in these cells (Endeward et al., 2008; Ripoche et al., 2006). Plants do not contain Rhesus proteins but do contain the somewhat homologous AmtB protein which is an ammonium channel (Marini et al., 1997). Rhesus and AmtB proteins are bidirectional channels, moving CO_2 or NH_3 respectively along a concentration gradient. Expression of a mammalian protein such as Rhesus-AG into plant membranes could be a strong target for increasing mesophyll conductance if the protein can be assembled and targeted correctly. No studies have reported success in expressing the related mammalian AQP1 *in planta* however, so this may not be a feasible approach. Identification of further CO_2 channels or transporters from plants or more closely related non-plant systems would provide new targets for engineering improved g_m. Expression of genes from non-plant photosynthesising organisms, such as green or red algae, may have more compatibility. Beyond CO_2 channels, several bicarbonate transporters have been identified in plants that may also support increased g_m if targeted correctly (Poschenrieder et al., 2018).

8 CO_2 solubility in liquids

On its journey to Rubisco, CO_2 must transverse two liquid compartments:

- the cell cytosol;
- the stroma within the chloroplasts themselves.

CO_2 gas can dissolve into liquids, but it can also be reversibly converted to bicarbonate which has a much higher solubility (Badger and Price, 1994). This reaction is catalysed by the enzyme carbonic anhydrase (CA), making the interconversion almost instantaneous. CAs are the second most abundant enzyme in plants (after Rubisco) and most of this activity is thought to support the diffusion of CO_2 through the cytosol and chloroplast stroma (Ogée et al., 2018). In plants, CAs have three forms, α, β and γ, that differ in their activity, expression and localisation patterns (Momayyezi et al., 2020). β-CA is the dominant isoform in plants and is found either free in the cytosol and stroma (Fabre et al., 2007) or bound to organelle membranes via interaction with aquaporin proteins (Wang et al., 2016).

The localisation of CA within the cell is critical for facilitating carbon concentrating mechanisms such as cyanobacterial carboxysomes or in C_4 photosynthesis, where CA localisation determines what form of CO_2 exists within each compartment. Here we discuss the role of CA in C_3 plants with respect to its impact on mesophyll conductance, and how it may be engineered for improvement.

9 Improving g_m with carbonic anhydrases

CA activity in the stroma is likely to have a large impact on g_m, with models predicting that removal of stromal CAs would decrease g_m by up to 44% and photosynthetic assimilation by 7% (Tholen and Zhu, 2011). Experimental work using anti-sense CAs confirmed carbon discrimination is altered (Price et al., 1994; Williams et al., 1996), or that photosynthesis was reduced (Sasaki et al., 1998), though g_m itself was not altered. More recent studies have shown evidence that manipulation of CA can alter g_m (Gillon and Yakir, 2001; Perez-Martin et al., 2014). With many forms of CA present within plants, it is possible that functional redundancy is compensating for CA knock-down effects, and more research is needed to determine what levels of CA are optimal in various compartments.

A key development in our understanding of the role CAs have in mesophyll conductance has been the discovery of CA interaction with aquaporin proteins (Perez-Martin et al., 2014; Wang et al., 2016). By coupling CA to a CO_2 channel, diffusing CO_2 is instantaneously converted to bicarbonate, ensuring a strong pull down effect for CO_2 into the cytosol. This relationship could potentially be further exploited through an engineered complex (metabolon) of aquaporin and CA at the plasma membrane that ensures their interaction remains constant and optimised. A similar metabolon of a CA coupled to a bicarbonate transporter at either the apoplast/plasma membrane boundary or chloroplast envelope/stroma boundary could also improve CO_2 delivery through the mesophyll cell to the stroma. Overall, more research is needed to better understand the relationships between CA isoforms, localisation and mesophyll conductance to help guide targeted improvements in CO_2 diffusion within mesophyll cells.

10 Estimating g_m

Although mesophyll conductance cannot be measured directly, a handful of techniques have been developed to indirectly estimate g_m and thorough reviews are available (Cousins et al., 2020; Pons et al., 2009). Two techniques have proven to be most reliable and are based on gas exchange measurements either (i) in conjunction with carbon isotope discrimination ($\Delta^{13}C$); or (ii) in combination with chlorophyll fluorescence (Evans et al., 1986; Harley et al., 1992).

Carbon isotope discrimination methods are based on measuring the $^{13}C/^{12}C$ ratio of CO_2 remaining after photosynthetic CO_2 fixation. The heavier isotope of carbon, ^{13}C is discriminated against during diffusion through the leaf and the biochemical processes of photosynthesis (Farquhar et al., 1982). This is because $^{13}CO_2$ has lower diffusivity in air and liquids relative to the lighter isotope ($^{12}CO_2$) and the carboxylating enzyme Rubisco preferentially binds $^{12}CO_2$. Correspondingly, photosynthetic products are enriched in ^{12}C, while gas flowing over a photosynthesizing leaf is enriched in ^{13}C and exhibits an increased $\Delta^{13}C$ compared to the ambient air. Carbon isotope discrimination has generally been measured one of two ways: by the cryogenic trapping of CO_2 and subsequent measurement of $\Delta^{13}C$ using isotope ratio mass spectrometry (Evans et al., 1986) or by using tunable diode laser (TDL) absorption spectrometry (Barbour et al., 2007) which can obtain simultaneous measurements of carbon isotope discrimination and gas exchange.

Estimating mesophyll conductance from gas exchange combined with chlorophyll fluorescence relies on the modelled relationship between CO_2 assimilation (A), the electron transport rate (J) and the concentration of CO_2 at Rubisco (C_c; FvCB model Farquhar et al., 1980). Two main modelling methods are used to estimate g_m: the constant J method and variable J method (Harley et al., 1992). In the constant J method, gas exchange measurements are made across a range of C_i values where J is assumed to be constant. In this method, fluorescence measurements are used solely to verify the assumption that J is constant under the measured range. Alternatively, the variable J method uses chlorophyll fluorescence to estimate J and then g_m (Harley et al., 1992). This method can be applied for any value of C_i, even when J is not constant.

Each of the aforementioned methods relies on different models and assumptions for the calculation of g_m and has corresponding limitations. They are all considered suitable for the determination of g_m values; however, several factors can and should be considered when determining which method is best suited for the planned experiment (Table 1). It is also common for values of g_m to vary depending on the estimation method applied, with the $\Delta^{13}C$ method often yielding higher estimates then chlorophyll fluorescence methods (Kromdijk et al., 2020). When possible, more than one method should be used to estimate g_m to increase confidence in the results due to uncertainties associated with each method.

Table 1 Factors to consider when choosing the most suitable method for estimating mesophyll conductance with regards to individual research questions

	Method			
	Carbon isotope discrimination and gas exchange		Chlorophyll fluorescence and gas exchange	
Factors to consider	Cryogenic trapping and isotope ratio mass spectrometry	Tunable Diode Laser (TDL) absorption spectrometry	Constant *J*	Variable *J*
Equipment cost	$$$	$$$$	$$	$$
Suitable for field measurements	No	No	Yes	Yes
Suitable for measuring at any CO_2 concentration	Yes	Yes	No	Yes
Suitable for measuring at non steady-state conditions	No	Yes	Yes	Yes
Measurement throughput	+	++	++++	++++
Accuracy	+++	++	+	+
Important requirements for accurate measurements	• large CO_2 draw down • large gas exchange chamber when possible • make measurements under non-photorespiratory conditions (2% O_2)		• more suitable for small leaves with low CO_2 draw down • avoid measurements under conditions of very low stomatal conductance such as drought stress	
Major sources of error	• uncertainties with assumed values of fractionation factors associated with carboxylation, respiration and photorespiration		• assumes that *J* and g_m are constant across the CO_2 concentration range	• uncertainties associated with estimating *J* from fluorescence measurements

11 Strategies for altering g_m

Strategies aimed at understanding and improving crop performance have changed over the years. During the Green Revolution, yields were increased by optimising light interception and carbon partitioning; however, these traditional methods for crop improvements are reaching a plateau, highlighting the need for novel approaches to ensure food production can keep up with the growing population (Ort et al., 2015). One promising target is the process of photosynthesis, which has not been fully optimised in food crops. Here, we discuss the use of selective breeding and synthetic biology via genetic engineering as possible approaches to enhancing photosynthesis by increasing mesophyll conductance.

Selective breeding has successfully been used in agriculture to produce crops with desirable traits, such as disease resistance and larger grains, fruits and vegetables. For artificial selection of a trait to be successful, the trait must exhibit genetic variation and that variation must be heritable. Several studies have been published looking at existing interspecific and intraspecific natural variation of mesophyll conductance (Carriquí et al., 2015, 2020; Faralli and Lawson, 2020). A summary of published studies reveals significant variation of mesophyll conductance in some of the world's most important crops, where the measured values vary by as much as a factor of ten within a species (Table 2). These studies indicate a useful source of unexploited genetic diversity that could be used for selective breeding to improve g_m and potentially the photosynthetic efficiency of these crops.

Synthetic biology has also been used successfully in agriculture to produce disease-resistant and nutrient-enriched crops (Kumar et al., 2020). Genetic engineering of plants via the introduction of one or multiple transgenes can be used to express genes from other plants or organisms that may facilitate enhancement of mesophyll conductance, whether it is through altered cell morphology, cell wall structure, membrane composition or other biochemical processes. This strategy can also be used to stack multiple favourable traits

Table 2 Variation of mesophyll conductance (g_m) in different crops

Reference	Crop	Number of cultivars/ genotypes	Observed g_m range ($mol\ m^{-2}\ s^{-1}$)
Barbour et al. (2010)	Barley	6	0.05-0.50
Muir et al. (2014)	Tomato	8	0.16-0.28
Jahan et al. (2014)	Wheat	10	*0.51-1.05
Tomás et al. (2014)	Grapevine	7	0.14-0.21
Ouyang et al. (2017)	Rice	7	0.05-0.12
Tomeo and Rosenthal (2017)	Soybean	12	*0.039-0.24
De Souza et al. (2020)	Cassava	12	*0.20-0.50

*Values were converted to units of $mol\ m^{-2}\ s^{-1}$ assuming pressure of 1 standard atmosphere.

together, for example, g_m traits with other photosynthetic traits. Additionally genome-editing technology such as CRISPR can be used to knockdown, knockout or upregulate target genes of interest within the plant genome. Photosynthetic improvements via genetic engineering are a promising avenue that is currently underutilised partially due to insufficient knowledge and understanding about several aspects of mesophyll conductance.

At present, there are relatively few studies that have used genetic engineering approaches to alter mesophyll conductance. Studies utilizing RNAi technology to knockdown carbonic anhydrase activity in tobacco have shown little effect on mesophyll conductance to CO_2 (Price et al., 1994; Williams et al., 1996). Increases in mesophyll conductance and net photosynthesis were observed in rice through combined ectopic expression of cyanobacteria ICTB, a possible bicarbonate transporter and FBP/Sbpase, which plays an important role in ribulose-1,5-bisphosphate (RuBP) regeneration (Gong et al., 2015). This study illustrates the potential utility of using multigene constructs to engineer improved mesophyll conductance. Most success with engineering enhanced g_m has been limited to changes in aquaporin expression. Transgenic overexpression of tobacco aquaporin 1 (NtAQP1) increased mesophyll conductance along with photosynthetic rates and CO_2 concentrations in the chloroplasts of tobacco plants (Flexas et al., 2006), while RNAi knockdown of NtAQP1 decreased photosynthesis and CO_2 conductance within tobacco leaves (Uehlein et al., 2008). Ectopic expression of the barley aquaporin PIP2;1 and overexpression of the rice aquaporin PIP1;2 both resulted in increased mesophyll conductance in rice, with the latter also increasing plant biomass and grain yield (Hanba et al., 2004; Xu et al., 2019). Together these studies highlight the importance of aquaporins as key targets for improving CO_2 conductance within the leaf. However, in some species, stacking aquaporin overexpression with other CO_2 diffusion traits will be necessary to see measurable differences in g_m.

12 Conclusion and future trends

Mesophyll conductance to CO_2 is influenced by a complex combination of anatomical and biochemical traits within leaves. Engineering improvements to mesophyll conductance presents an appealing target for boosting photosynthesis, especially in limited water environments where improved CO_2 supply to Rubisco would allow stomatal conductance (and therefore water loss) to be reduced without a photosynthetic penalty. The stacking of improved g_m traits with improvements to Rubisco or RUBP regeneration efficiency should also deliver further enhancements to photosynthesis as CO_2 concentration at the site of fixation becomes limiting. Manipulation of individual resistances to CO_2 within leaves and cells can also be stacked together for further improvements to the overall mesophyll conductance.

In order to deliver this, further comprehensive studies are needed to facilitate engineering enhanced mesophyll conductance and photosynthetic efficiency, such as large-scale mutant screens to identify suitable genes for additional study and manipulation. Specifically, improvements in the fundamental knowledge of genetic controls in the following areas are required:

- cell size and shape;
- cell packing and leaf thickness;
- cell wall structure and function;
- membrane channel regulation.

13 Where to look for further information

Further useful information about mesophyll conductance and efforts to improve it to benefit plant productivity can be found in the following resources:

- Cellular perspectives for improving mesophyll conductance. Lundgren and Fleming 2020. https://doi.org/10.1111%2Ftpj.14656.
- Recent developments in mesophyll conductance in C_3, C_4 and crassulacean acid metabolism plants. Cousins et al. 2020. https://doi.org/10.1111/tpj.14664.
- Mesophyll conductance: the leaf corridors for photosynthesis. Gago et al. 2020. https://doi.org/10.1042/BST20190312.
- Mesophyll conductance: walls, membrane and spatial complexity. John R. Evans 2021. https://doi.org/10.1111/nph.16968.

14 Acknowledgements

This work was supported by a sub-award from the University of Illinois as part of the Realizing Increased Photosynthetic Efficiency (RIPE) project, funded by the Bill & Melinda Gates Foundation to V. C. C. and C. E.-S. S.

15 Author contributions

C. E.-S. S., S. M. D. and V. C. C. all wrote and edited the manuscript and are equal co-authors.

16 References

Badger, M. R. and Price, G. D. 1994. The role of carbonic anhydrase in photosynthesis. *Annual Review of Plant Physiology and Plant Molecular Biology* 45(1), 369–392.

Bailey-Serres, J., Parker, J. E., Ainsworth, E. A., Oldroyd, G. E. D. and Schroeder, J. I. 2019. Genetic strategies for improving crop yields. *Nature* 575(7781), 109–118.

Barbour, M. M., McDowell, N. G., Tcherkez, G., Bickford, C. P. and Hanson, D. T. 2007. A new measurement technique reveals rapid post-illumination changes in the carbon isotope composition of leaf-respired CO_2. *Plant, Cell and Environment* 30(4), 469–482.

Barbour, M. M., Warren, C. R., Farquhar, G. D., Forrester, G. and Brown, H. 2010. Variability in mesophyll conductance between barley genotypes, and effects on transpiration efficiency and carbon isotope discrimination. *Plant, Cell and Environment* 33(7), 1176–1185.

Beebo, A., Mathai, J. C., Schoefs, B. and Spetea, C. 2013. Assessment of the requirement for aquaporins in the thylakoid membrane of plant chloroplasts to sustain photosynthetic water oxidation. *FEBS Letters* 587(14), 2083–2089.

Busch, F. A., Holloway-Phillips, M., Stuart-Williams, H. and Farquhar, G. D. 2020. Revisiting carbon isotope discrimination in C3 plants shows respiration rules when photosynthesis is low. *Nature Plants* 6(3), 245–258.

Carriquí, M., Cabrera, H. M., Conesa, M. À, Coopman, R. E., Douthe, C., Gago, J., Gallé, A., Galmés, J., Ribas-Carbo, M., Tomás, M. and Flexas, J. 2015. Diffusional limitations explain the lower photosynthetic capacity of ferns as compared with angiosperms in a common garden study. *Plant, Cell and Environment* 38(3), 448–460.

Carriquí, M., Douthe, C., Molins, A. and Flexas, J. 2019. Leaf anatomy does not explain apparent short-term responses of mesophyll conductance to light and CO_2 in tobacco. *Physiologia Plantarum* 165(3), 604–618.

Carriquí, M., Nadal, M., Clemente-Moreno, M. J., Gago, J., Miedes, E. and Flexas, J. 2020. Cell wall composition strongly influences mesophyll conductance in gymnosperms. *The Plant Journal: For Cell and Molecular Biology* 103(4), 1372–1385.

Carriquí, M., Nadal, M. and Flexas, J. 2021. Acclimation of mesophyll conductance and anatomy to light during leaf aging in Arabidopsis thaliana. *Physiologia Plantarum* 172(4), 1894–1907.

Clarke, V. C., Danila, F. R. and von Caemmerer, S. 2021. CO_2 diffusion in tobacco: a link between mesophyll conductance and leaf anatomy. *Interface Focus* 11(2), 20200040.

Clarke, V. C., Massey, B., George, A. M., Evans, J. R., von Caemmerer, S. and Groszmann, M. 2022. Mesophyll conductance is unaffected by expression of Arabidopsis PIP1 aquaporins in the plasmalemma of Nicotiana. *Journal of Experimental Botany.* 73(11), 3625–3636.

Clemente-Moreno, M. J., Gago, J., Díaz-Vivancos, P., Bernal, A., Miedes, E., Bresta, P., Liakopoulos, G., Fernie, A. R., Hernández, J. A. and Flexas, J. 2019. The apoplastic antioxidant system and altered cell wall dynamics influence mesophyll conductance and the rate of photosynthesis. *The Plant Journal: For Cell and Molecular Biology* 99(6), 1031–1046.

Cooper, G. J. and Boron, W. F. 1998. Effect of PCMBS on CO_2 permeability of Xenopus oocytes expressing aquaporin 1 or its C189S mutant. *American Journal of Physiology* 275(6), C1481–C1486.

Cosgrove, D. J. 2005. Growth of the plant cell wall. *Nature Reviews: Molecular Cell Biology* 6(11), 850–861.

Cosgrove, D. J. 2018. Diffuse growth of plant cell walls. *Plant Physiology* 176(1), 16–27.

Cosgrove, D. J. and Jarvis, M. C. 2012. Comparative structure and biomechanics of plant primary and secondary cell walls. *Frontiers in Plant Science* 3, 204.

Cousins, A. B., Mullendore, D. L. and Sonawane, B. V. 2020. Recent developments in mesophyll conductance in C3, C4, and crassulacean acid metabolism plants. *The Plant Journal: For Cell and Molecular Biology* 101(4), 816–830.

De Rosa, A., Watson-Lazowski, A., Evans, J. R. and Groszmann, M. 2020. Genome-wide identification and characterisation of aquaporins in Nicotiana tabacum and their relationships with other Solanaceae species. *BMC Plant Biology* 20(1), 266.

De Souza, A. P., Wang, Y., Orr, D. J., Carmo-Silva, E. and Long, S. P. 2020. Photosynthesis across African cassava germplasm is limited by RuBisCO and mesophyll conductance at steady state, but by stomatal conductance in fluctuating light. *New Phytologist* 225(6), 2498–2512.

Earles, J. M., Buckley, T. N., Brodersen, C. R., Busch, F. A., Cano, F. J., Choat, B., Evans, J. R., Farquhar, G. D., Harwood, R., Huynh, M., John, G. P., Miller, M. L., Rockwell, F. E., Sack, L., Scoffoni, C., Struik, P. C., Wu, A., Yin, X. and Barbour, M. M. 2019. Embracing 3D complexity in leaf carbon-water exchange. *Trends in Plant Science* 24(1), 15–24.

Ellsworth, P. V., Ellsworth, P. Z., Koteyeva, N. K. and Cousins, A. B. 2018. Cell wall properties in Oryza sativa influence mesophyll CO_2 conductance. *New Phytologist* 219(1), 66–76.

Endeward, V., Arias-Hidalgo, M., Al-Samir, S. and Gros, G. 2017. CO_2 permeability of biological membranes and role of CO_2 channels. *Membranes* 7(4), 61.

Endeward, V., Cartron, J. P., Ripoche, P. and Gros, G. 2008. RhAG protein of the rhesus complex is a CO_2 channel in the human red cell membrane. *FASEB Journal* 22(1), 64–73.

Ermakova, M., Osborn, H., Groszmann, M., Bala, S., Bowerman, A., McGaughey, S., Byrt, C., Alonso-cantabrana, H., Tyerman, S., Furbank, R. T., Sharwood, R. E. and von Caemmerer, S. 2021. Expression of a CO_2-permeable aquaporin enhances mesophyll conductance in the C4 species Setaria viridis. *eLife* 10, e70095.

Evans, J. R. 2021. Mesophyll conductance: walls, membranes and spatial complexity. *New Phytologist* 229(4), 1864–1876.

Evans, J. R. and Clarke, V. C. 2019. The nitrogen cost of photosynthesis. *Journal of Experimental Botany* 70(1), 7–15.

Evans, J. R., Kaldenhoff, R., Genty, B. and Terashima, I. 2009. Resistances along the CO_2 diffusion pathway inside leaves. *Journal of Experimental Botany* 60(8), 2235-2248.

Evans, J. R., Sharkey, T. D., Berry, J. A. and Farquhar, G. D. 1986. Carbon isotope discrimination measured concurrently with gas exchange to investigate CO_2 diffusion in leaves of higher plants. *Functional Plant Biology* 13(2), 281-292.

Evans, J. R., von Caemmerer, S., Setchell, B. A. and Hudson, G. S. 1994. The relationship between CO_2 transfer conductance and leaf anatomy in transgenic tobacco with a reduced content of RuBisCO. *Functional Plant Biology* 21(4), 475-495.

Fabre, N., Reiter, I. M., Becuwe-Linka, N., Genty, B. and Rumeau, D. 2007. Characterization and expression analysis of genes encoding alpha and beta carbonic anhydrases in Arabidopsis. *Plant, Cell and Environment* 30(5), 617–629.

Faralli, M. and Lawson, T. 2020. Natural genetic variation in photosynthesis: an untapped resource to increase crop yield potential? *The Plant Journal: For Cell and Molecular Biology* 101(3), 518-528.

Farquhar, G., O'Leary, M. H. O. and Berry, J. 1982. On the relationship between carbon isotope discrimination and the intercellular carbon dioxide concentration in leaves. *Functional Plant Biology* 9(2), 121–137.

Farquhar, G. D., von Caemmerer, S. and Berry, J. A. 1980. A biochemical model of photosynthetic CO_2 assimilation in leaves of C3 species. *Planta* 149(1), 78-90.

Fleischer, A., O'Neill, M. A. and Ehwald, R. 1999. The pore size of non-graminaceous plant cell walls is rapidly decreased by borate ester cross-linking of the pectic polysaccharide rhamnogalacturonan II1. *Plant Physiology* 121(3), 829–838.

Flexas, J., Ribas-Carbó, M., Hanson, D. T., Bota, J., Otto, B., Cifre, J., McDowell, N., Medrano, H. and Kaldenhoff, R. 2006. Tobacco aquaporin NtAQP1 is involved in mesophyll conductance to CO_2 in vivo. *The Plant Journal: For Cell and Molecular Biology* 48(3), 427–439.

Gago, J., Daloso, D. M., Carriquí, M., Nadal, M., Morales, M., Araújo, W. L., Nunes-Nesi, A. and Flexas, J. 2020. Mesophyll conductance: the leaf corridors for photosynthesis. *Biochemical Society Transactions* 48(2), 429–439.

Gillon, J. and Yakir, D. 2001. Influence of carbonic anhydrase activity in terrestrial vegetation on the ^{18}O Content of atmospheric CO_2. *Science* 291(5513), 2584–2587.

Gong, H. Y., Li, Y., Fang, G., Hu, D. H., Jin, W. B., Wang, Z. H. and Li, Y. S. 2015. Transgenic rice expressing Ictb and FBP/Sbpase derived from cyanobacteria exhibits enhanced photosynthesis and mesophyll conductance to CO_2. *PLoS ONE* 10(10), e0140928.

Groszmann, M., Osborn, H. L. and Evans, J. R. 2017. Carbon dioxide and water transport through plant aquaporins. *Plant, Cell and Environment* 40(6), 938–961.

Gutknecht, J., Bisson, M. A. and Tosteson, F. C. 1977. Diffusion of carbon dioxide through lipid bilayer membranes: effects of carbonic anhydrase, bicarbonate, and unstirred layers. *Journal of General Physiology* 69(6), 779–794.

Hanba, Y. T., Kogami, H. and Terashima, I. 2002. The effect of growth irradiance on leaf anatomy and photosynthesis in Acer species differing in light demand. *Plant, Cell and Environment* 25(8), 1021–1030.

Hanba, Y. T., Miyazawa, S. I., Kogami, H. and Terashima, I. 2001. Effects of leaf age on internal CO_2 transfer conductance and photosynthesis in tree species having different types of shoot phenology. *Functional Plant Biology* 28(11), 1075–1084.

Hanba, Y. T., Shibasaka, M., Hayashi, Y., Hayakawa, T., Kasamo, K., Terashima, I. and Katsuhara, M. 2004. Overexpression of the barley aquaporin HvPIP2;1 increases internal CO(2) conductance and CO(2) assimilation in the leaves of transgenic rice plants. *Plant and Cell Physiology* 45(5), 521–529.

Harley, P. C., Loreto, F., Di Marco, G. and Sharkey, T. D. 1992. Theoretical considerations when estimating the mesophyll conductance to CO_2 flux by analysis of the response of photosynthesis to CO_2. *Plant Physiology* 98(4), 1429–1436.

Heckwolf, M., Pater, D., Hanson, D. T. and Kaldenhoff, R. 2011. The Arabidopsis thaliana aquaporin AtPIP1;2 is a physiologically relevant CO_2 transport facilitator. *The Plant Journal: For Cell and Molecular Biology* 67(5), 795–804.

Hoshika, Y., Haworth, M., Watanabe, M. and Koike, T. 2020. Interactive effect of leaf age and ozone on mesophyll conductance in Siebold's beech. *Physiologia Plantarum* 170(2), 172–186.

Houston, K., Tucker, M. R., Chowdhury, J., Shirley, N. and Little, A. 2016. The plant cell wall: a complex and dynamic structure as revealed by the responses of genes under stress conditions. *Frontiers in Plant Science* 7, 984.

Huang, X., Wang, Z., Huang, J., Peng, S. and Xiong, D. 2021. Mesophyll conductance variability of rice aquaporin knockout lines at different growth-stages and growing environments. *The Plant Journal: For Cell and Molecular Biology* 107(5), 1503–1512.

Jahan, E., Amthor, J. S., Farquhar, G. D., Trethowan, R. and Barbour, M. M. 2014. Variation in mesophyll conductance among Australian wheat genotypes. *Functional Plant Biology* 41(6), 568-580.

Kelly, G., Sade, N., Attia, Z., Secchi, F., Zwieniecki, M., Holbrook, N. M., Levi, A., Alchanatis, V., Moshelion, M. and Granot, D. 2014. Relationship between hexokinase and the aquaporin PIP1 in the regulation of photosynthesis and plant growth. *PLoS ONE* 9(2), e87888.

Kromdijk, J., Głowacka, K. and Long, S. P. 2020. Photosynthetic efficiency and mesophyll conductance are unaffected in Arabidopsis thaliana aquaporin knock-out lines. *Journal of Experimental Botany* 71(1), 318-329.

Kumar, K., Gambhir, G., Dass, A., Tripathi, A. K., Singh, A., Jha, A. K., Yadava, P., Choudhary, M. and Rakshit, S. 2020. Genetically modified crops: current status and future prospects. *Planta* 251(4), 91.

Lehmeier, C., Pajor, R., Lundgren, M. R., Mathers, A., Sloan, J., Bauch, M., Mitchell, A., Bellasio, C., Green, A., Bouyer, D., Schnittger, A., Sturrock, C., Osborne, C. P., Rolfe, S., Mooney, S. and Fleming, A. J. 2017. Cell density and airspace patterning in the leaf can be manipulated to increase leaf photosynthetic capacity. *The Plant Journal: For Cell and Molecular Biology* 92(6), 981-994.

Liu, T., Barbour, M. M., Yu, D., Rao, S. and Song, X. 2022. Mesophyll conductance exerts a significant limitation on photosynthesis during light induction. *New Phytologist* 233(1), 360-372.

Long, S. P., Marshall-Colon, A. and Zhu, X. G. 2015. Meeting the global food demand of the future by engineering crop photosynthesis and yield potential. *Cell* 161(1), 56-66.

Loreto, F., Di Marco, G., Tricoli, D. and Sharkey, T. D. 1994. Measurements of mesophyll conductance, photosynthetic electron transport and alternative electron sinks of field grown wheat leaves. *Photosynthesis Research* 41(3), 397-403.

Lu, Z., Lu, J., Pan, Y., Lu, P., Li, X., Cong, R. and Ren, T. 2016. Anatomical variation of mesophyll conductance under potassium deficiency has a vital role in determining leaf photosynthesis. *Plant, Cell and Environment* 39(11), 2428-2439.

Lundgren, M. R. and Fleming, A. J. 2020. Cellular perspectives for improving mesophyll conductance. *The Plant Journal: For Cell and Molecular Biology* 101(4), 845-857.

Marchi, S., Tognetti, R., Minnocci, A., Borghi, M. and Sebastiani, L. 2008. Variation in mesophyll anatomy and photosynthetic capacity during leaf development in a deciduous mesophyte fruit tree (Prunus persica) and an evergreen sclerophyllous Mediterranean shrub (Olea europaea). *Trees* 22(4), 559-571.

Marini, A. M., Urrestarazu, A., Beauwens, R. and André, B. 1997. The Rh (rhesus) blood group polypeptides are related to NH_4+ transporters. *Trends in Biochemical Sciences* 22(12), 460-461.

Michenkova, M., Taki, S., Blosser, M. C., Hwang, H. J., Kowatz, T., Moss, F. J., Occhipinti, R., Qin, X., Sen, S., Shinn, E., Wang, D., Zeise, B. S., Zhao, P., Malmstadt, N., Vahedi-Faridi, A., Tajkhorshid, E. and Boron, W. F. 2021. Carbon dioxide transport across membranes. *Interface Focus* 11(2), 20200090.

Missner, A. and Pohl, P. 2009. 110 years of the Meyer-Overton rule: predicting membrane permeability of gases and other small compounds. *ChemPhysChem: A European Journal of Chemical Physics and Physical Chemistry* 10(9-10), 1405-1414.

Miyazawa, S.-I., Makino, A. and Terashima, I. 2003. Changes in mesophyll anatomy and sink-source relationships during leaf development in Quercus glauca, an evergreen tree showing delayed leaf greening. *Plant, Cell and Environment* 26(5), 745-755.

Miyazawa, S.-I. and Terashima, I. 2001. Slow development of leaf photosynthesis in an evergreen broad-leaved tree, Castanopsis sieboldii: relationships between leaf anatomical characteristics and photosynthetic rate. *Plant, Cell and Environment* 24(3), 279-291.

Momayyezi, M., McKown, A. D., Bell, S. C. S. and Guy, R. D. 2020. Emerging roles for carbonic anhydrase in mesophyll conductance and photosynthesis. *The Plant Journal: For Cell and Molecular Biology* 101(4), 831-844.

Mori, I. C., Rhee, J., Shibasaka, M., Sasano, S., Kaneko, T., Horie, T. and Katsuhara, M. 2014. CO_2 transport by PIP2 aquaporins of barley. *Plant and Cell Physiology* 55(2), 251-257.

Muir, C. D., Hangarter, R. P., Moyle, L. C. and Davis, P. A. 2014. Morphological and anatomical determinants of mesophyll conductance in wild relatives of tomato (Solanum sect. Lycopersicon, sect. Lycopersicoides; Solanaceae). *Plant, Cell and Environment* 37(6), 1415-1426.

Nakhoul, N. L., Davis, B. A., Romero, M. F. and Boron, W. F. 1998. Effect of expressing the water channel aquaporin-1 on the CO_2 permeability of Xenopus oocytes. *American Journal of Physiology* 274(2), C543-C548.

Niinemets, Ü., Garcia Plazaola, J. I. and Tosens, T. 2012. Photosynthesis during leaf development and ageing. In: Flexas, J. (Ed.) *Terrestrial Photosynthesis in a Changing Environment: A Molecular, Physiological, and Ecological Approach*. Cambridge: Cambridge University Press, 353-372.

Ogée, J., Wingate, L. and Genty, B. 2018. Estimating mesophyll conductance from measurements of C(18)OO photosynthetic discrimination and carbonic anhydrase activity. *Plant Physiology* 178(2), 728-752.

Oguchi, R., Hikosaka, K. and Hirose, T. 2003. Does the photosynthetic light-acclimation need change in leaf anatomy? *Plant, Cell and Environment* 26(4), 505-512.

Onoda, Y., Wright, I. J., Evans, J. R., Hikosaka, K., Kitajima, K., Niinemets, Ü., Poorter, H., Tosens, T. and Westoby, M. 2017. Physiological and structural tradeoffs underlying the leaf economics spectrum. *New Phytologist* 214(4), 1447-1463.

Ort, D. R., Merchant, S. S., Alric, J., Barkan, A., Blankenship, R. E., Bock, R., Croce, R., Hanson, M. R., Hibberd, J. M., Long, S. P., Moore, T. A., Moroney, J., Niyogi, K. K., Parry, M. A. J., Peralta-Yahya, P. P., Prince, R. C., Redding, K. E., Spalding, M. H., van Wijk, K. J., Vermaas, W. F. J., von Caemmerer, S., Weber, A. P. M., Yeates, T. O., Yuan, J. S. and Zhu, X. G. 2015. Redesigning photosynthesis to sustainably meet global food and bioenergy demand. *Proceedings of the National Academy of Sciences of the United States of America* 112(28), 8529-8536.

Otto, B., Uehlein, N., Sdorra, S., Fischer, M., Ayaz, M., Belastegui-Macadam, X., Heckwolf, M., Lachnit, M., Pede, N., Priem, N., Reinhard, A., Siegfart, S., Urban, M. and Kaldenhoff, R. 2010. Aquaporin tetramer composition modifies the function of tobacco aquaporins. *Journal of Biological Chemistry* 285(41), 31253-31260.

Ouyang, W., Struik, P. C., Yin, X. and Yang, J. 2017. Stomatal conductance, mesophyll conductance, and transpiration efficiency in relation to leaf anatomy in rice and wheat genotypes under drought. *Journal of Experimental Botany* 68(18), 5191-5205.

Published by Burleigh Dodds Science Publishing Limited, 2023.

Perez-Martin, A., Michelazzo, C., Torres-Ruiz, J. M., Flexas, J., Fernández, J. E., Sebastiani, L. and Diaz-Espejo, A. 2014. Regulation of photosynthesis and stomatal and mesophyll conductance under water stress and recovery in olive trees: correlation with gene expression of carbonic anhydrase and aquaporins. *Journal of Experimental Botany* 65(12), 3143-3156.

Pons, T. L., Flexas, J., von Caemmerer, S., Evans, J. R., Genty, B., Ribas-Carbo, M. and Brugnoli, E. 2009. Estimating mesophyll conductance to CO_2: methodology, potential errors, and recommendations. *Journal of Experimental Botany* 60(8), 2217-2234.

Popper, Z. A., Michel, G., Hervé, C., Domozych, D. S., Willats, W. G., Tuohy, M. G., Kloareg, B. and Stengel, D. B. 2011. Evolution and diversity of plant cell walls: from algae to flowering plants. *Annual Review of Plant Biology* 62, 567-590.

Poschenrieder, C., Fernández, J. A., Rubio, L., Pérez, L., Terés, J. and Barceló, J. 2018. Transport and use of bicarbonate in plants: current knowledge and challenges ahead. *International Journal of Molecular Sciences* 19(5), 1352.

Price, G. D., von Caemmerer, S., Evans, J. R., Yu, J.-W., Lloyd, J., Oja, V., Kell, P., Harrison, K., Gallagher, A. and Badger, M. R. 1994. Specific reduction of chloroplast carbonic anhydrase activity by antisense RNA in transgenic tobacco plants has a minor effect on photosynthetic CO_2 assimilation. *Planta* 193(3), 331-340.

Ray, D. K., Mueller, N. D., West, P. C. and Foley, J. A. 2013. Yield trends are insufficient to double global crop production by 2050. *PLoS ONE* 8(6), e66428.

Ren, T., Weraduwage, S. M. and Sharkey, T. D. 2019. Prospects for enhancing leaf photosynthetic capacity by manipulating mesophyll cell morphology. *Journal of Experimental Botany* 70(4), 1153-1165.

Reszczyńska, E. and Hanaka, A. 2020. Lipids composition in plant membranes. *Cell Biochemistry and Biophysics* 78(4), 401-414.

Ripoche, P., Goossens, D., Devuyst, O., Gane, P., Colin, Y., Verkman, A. S. and Cartron, J. P. 2006. Role of RhAG and AQP1 in NH_3 and CO_2 gas transport in red cell ghosts: a stopped-flow analysis. *Transfusion Clinique et Biologique: journal de la Societe Francaise de Transfusion Sanguine* 13(1-2), 117-122.

Roig-Oliver, M., Nadal, M., Clemente-Moreno, M. J., Bota, J. and Flexas, J. 2019. Cell wall components regulate photosynthesis and leaf water relations of Vitis vinifera cv. Grenache acclimated to contrasting environmental conditions. *Journal of Plant Physiology* 244, 153084.

Roig-Oliver, M., Rayon, C., Roulard, R., Fournet, F., Bota, J. and Flexas, J. 2021. Reduced photosynthesis in Arabidopsis thaliana atpme17.2 and atpae11.1 mutants is associated to altered cell wall composition. *Physiologia Plantarum* 172(3), 1439-1451.

Rondeau-Mouro, C., Defer, D., Leboeuf, E. and Lahaye, M. 2008. Assessment of cell wall porosity in Arabidopsis thaliana by NMR spectroscopy. *International Journal of Biological Macromolecules* 42(2), 83-92.

Sage, T. L. and Sage, R. F. 2009. The functional anatomy of rice leaves: implications for refixation of photorespiratory CO_2 and efforts to engineer C4 photosynthesis into rice. *Plant and Cell Physiology* 50(4), 756-772.

Sakoda, K., Yamori, W., Groszmann, M. and Evans, J. R. 2021. Stomatal, mesophyll conductance, and biochemical limitations to photosynthesis during induction. *Plant Physiology* 185(1), 146-160.

Sasaki, H., Hirose, T., Watanabe, Y. and Ohsugi, R. 1998. Carbonic anhydrase activity and CO_2-transfer resistance in Zn-deficient rice leaves. *Plant Physiology* 118(3), 929-934.

Sugiura, D., Terashima, I. and Evans, J. R. 2020. A decrease in mesophyll conductance by cell wall thickening contributes to photosynthetic down-regulation. *Plant Physiol*, pp.00328.02020.

Tenhaken, R. 2015. Cell wall remodeling under abiotic stress. *Frontiers in Plant Science* 5, 771.

Terashima, I., Fujita, T., Inoue, T., Chow, W. S. and Oguchi, R. 2009. Green light drives leaf photosynthesis more efficiently than red light in strong white light: revisiting the enigmatic question of why leaves are Green. *Plant and Cell Physiology* 50(4), 684-697.

Terashima, I., Hanba, Y. T., Tholen, D. and Niinemets, Ü. 2011. Leaf functional anatomy in relation to photosynthesis. *Plant Physiology* 155(1), 108-116.

Terashima, I. and Saeki, T. 1983. Light environment within a leaf I. Optical properties of paradermal sections of camellia leaves with special reference to differences in the optical properties of palisade and spongy tissues. *Plant and Cell Physiology* 24(8), 1493-1501.

Théroux-Rancourt, G. and Gilbert, M. E. 2017. The light response of mesophyll conductance is controlled by structure across leaf profiles. *Plant, Cell and Environment* 40(5), 726-740.

Tholen, D. and Zhu, X. G. 2011. The mechanistic basis of internal conductance: a theoretical analysis of mesophyll cell photosynthesis and CO_2 diffusion. *Plant Physiology* 156(1), 90-105.

Tomás, M., Flexas, J., Copolovici, L., Galmes, J., Hallik, L., Medrano, H., Ribas-Carbó, M., Tosens, T., Vislap, V. and Niinemets, Ü. 2013. Importance of leaf anatomy in determining mesophyll diffusion conductance to CO_2 across species: quantitative limitations and scaling up by models. *Journal of Experimental Botany* 64(8), 2269-2281.

Tomás, M., Medrano, H., Brugnoli, E., Escalona, J. M., Martorell, S., Pou, A., Ribas-Carbó, M. and Flexas, J. 2014. Variability of mesophyll conductance in grapevine cultivars under water stress conditions in relation to leaf anatomy and water use efficiency. *Australian Journal of Grape and Wine Research* 20(2), 272-280.

Tomeo, N. J. and Rosenthal, D. M. 2017. Variable mesophyll conductance among soybean cultivars sets a tradeoff between photosynthesis and water-use-efficiency. *Plant Physiology* 174(1), 241-257.

Tosens, T., Niinemets, U., Vislap, V., Eichelmann, H. and Diez, P. C. 2012. Developmental changes in mesophyll diffusion conductance and photosynthetic capacity under different light and water availabilities in Populus tremula: how structure constrains function. *Plant, Cell and Environment* 35(5), 839-856.

Uehlein, N., Lovisolo, C., Siefritz, F. and Kaldenhoff, R. 2003. The tobacco aquaporin NtAQP1 is a membrane CO_2 pore with physiological functions. *Nature* 425(6959), 734-737.

Uehlein, N., Otto, B., Hanson, D. T., Fischer, M., McDowell, N. and Kaldenhoff, R. 2008. Function of Nicotiana tabacum aquaporins as chloroplast gas pores challenges the concept of membrane CO_2 permeability. *The Plant Cell* 20(3), 648-657.

Uemura, M., Joseph, R. A. and Steponkus, P. L. 1995. Cold acclimation of Arabidopsis thaliana (effect on plasma membrane lipid composition and freeze-induced lesions). *Plant Physiology* 109(1), 15-30.

Vidal, D., Griera, E., Marin, P. and Sabido, J. 1990. Anatomical and physiological acclimation of Fatsia japonica leaves to irradiance. *American Journal of Botany* 77(9), 1149–1158.

Vogelmann, T. C. and Evans, J. R. 2002. Profiles of light absorption and chlorophyll within spinach leaves from chlorophyll fluorescence. *Plant, Cell and Environment* 25(10), 1313–1323.

Vogelmann, T. C. and Martin, G. 1993. The functional significance of palisade tissue: penetration of directional versus diffuse light. *Plant, Cell and Environment* 16(1), 65–72.

Wang, C., Hu, H., Qin, X., Zeise, B., Xu, D., Rappel, W. J., Boron, W. F. and Schroeder, J. I. 2016. Reconstitution of CO_2 regulation of SLAC1 anion channel and function of CO_2-permeable PIP2;1 aquaporin as CARBONIC ANHYDRASE4 interactor. *The Plant Cell* 28(2), 568–582.

Williams, T. G., Flanagan, L. B. and Coleman, J. R. 1996. Photosynthetic gas exchange and discrimination against $13CO_2$ and C18O16O in tobacco plants modified by an antisense construct to have low chloroplastic carbonic anhydrase. *Plant Physiology* 112(1), 319–326.

Wu, A., Hammer, G. L., Doherty, A., von Caemmerer, S. and Farquhar, G. D. 2019. Quantifying impacts of enhancing photosynthesis on crop yield. *Nature Plants* 5(4), 380–388.

Xiao, Y. and Zhu, X. G. 2017. Components of mesophyll resistance and their environmental responses: a theoretical modelling analysis. *Plant, Cell and Environment* 40(11), 2729–2742.

Xiong, D., Douthe, C. and Flexas, J. 2018. Differential coordination of stomatal conductance, mesophyll conductance, and leaf hydraulic conductance in response to changing light across species. *Plant, Cell and Environment* 41(2), 436–450.

Xu, F., Wang, K., Yuan, W., Xu, W., Shuang, L., Kronzucker, H. J., Chen, G., Miao, R., Zhang, M., Ding, M., Xiao, L., Kai, L., Zhang, J. and Zhu, Y. 2019. Overexpression of rice aquaporin OsPIP1;2 improves yield by enhancing mesophyll CO_2 conductance and phloem sucrose transport. *Journal of Experimental Botany* 70(2), 671–681.

Yin, X. and Struik, P. C. 2017. Can increased leaf photosynthesis be converted into higher crop mass production? A simulation study for rice using the crop model GECROS. *Journal of Experimental Botany* 68(9), 2345–2360, doi:10.1093/jxb/erx085.

Yin, X., van der Putten, P. E. L., Belay, D. and Struik, P. C. 2020. Using photorespiratory oxygen response to analyse leaf mesophyll resistance. *Photosynthesis Research* 144(1), 85–99.

Zhang, D., Li, Y. and Li, Y. 2021. The potential implications of a plasma membrane aquaporin in improving CO_2 transport capacity, photosynthetic potential and water use efficiency under contrasting CO_2 source in Solanum Lycopersicum (tomato). *Scientia Horticulturae* 283, 110122.

Chapter 7

Modifying canopy architecture to optimize photosynthesis in crops

Anthony Digrado and Elizabeth A. Ainsworth, Global Change and Photosynthesis Research Unit, USDA-ARS and University of Illinois at Urbana-Champaign, USA

1 Introduction

Crop yield potential (Y_p) per surface area is defined as the maximum achievable yield under optimum conditions, i.e. without nutrient or water limitation and with biotic and abiotic stresses controlled (Boyer, 1982; Evans and Fischer, 1999). Y_p has been defined by Monteith and Moss (1977) as $Y_p = 0.487 \cdot St \cdot \varepsilon_i \cdot \varepsilon_c \cdot \varepsilon_p$, where $0.487 \cdot St$ is the available photosynthetically active radiation across the growing season, ε_i is the radiation interception efficiency, ε_c is the conversion efficiency of intercepted radiation into biomass (equivalent to the radiation-use efficiency, RUE), and ε_p is the partitioning efficiency of biomass into harvestable product [also referred to as the harvest index (HI)]. Canopy architecture is strongly coupled to interception, scattering, and emission of radiation and therefore determines ε_i (Welles, 1990). It has been hypothesized that past breeding efforts have managed to raise ε_i close to its theoretical maximum, resulting in the interception of almost 90% of the available photosynthetically active radiation across the growing season in modern soybean cultivars (Dermody et al., 2008; Zhu et al., 2010; Koester et al., 2014) and close to 80% in grass species such as *Miscanthus* and switchgrass (Heaton et al., 2008). Modern

http://dx.doi.org/10.19103/AS.2022.0119.11

maize, rice, and wheat also intercept more than 90% of incident light at times during the growing season (Lindquist et al., 2005; Sheehy and Mitchell, 2013; Liu et al., 2019). ε_p has also been considered close to its theoretical maximum in modern crops (Austin et al., 1980). Typical values observed for rice and maize are ~50% (Tollenaar et al., 2004; Prasad et al., 2006; Sheehy and Mitchell, 2013; Liu et al., 2017). In modern soybean cultivars, ε_p approaches 60% (Morgan et al., 2005; Zhu et al., 2010; Koester et al., 2014; Slattery et al., 2017). Even though the theoretical maximum ε_p for other crops is uncertain, the value of 60% has been widely adopted as an upper target, given that some biomass in the stem needs to hold grain aboveground for harvest. In contrast with the two previous parameters of the Monteith equation, ε_c is considered to be well below the theoretical optimum and thus remains a target for improvement. The theoretical maximal values for $\varepsilon_{c,}$ when expressed as the ratio of the photosynthetically active radiation (400–700 nm), are estimated to be 9.4% and 12.3% for C_3 and C_4 crops at 30°C (Zhu et al., 2010). Estimates of ε_c for C_3 crops, including soybean, potato, rice, and other cereals, range from 1.9% to 4.1% (Slattery and Ort, 2015). Higher values are observed in C_4 grain crops, due to the efficiency of C_4 photosynthesis. Still, similar gaps exist in sorghum and maize, with average values of 4.6% and 4.9% (Slattery and Ort, 2015). In general, C_4 energy crops, including switchgrass, sugarcane, and *Miscanthus*, show higher ε_c values ranging between 4.8% and 7.5% (Heaton et al., 2008; Slattery and Ort, 2015) but still fall short of the theoretical maximum. ε_c results from the combined gross photosynthesis of all leaves within the canopy minus all plant respiratory losses (Sinclair and Muchow, 1999; Zhu et al., 2010). Canopy architecture, which strongly influences canopy photosynthesis, is a key determinant of ε_c. Leaf-level photosynthetic capacity can also impact ε_c, and strategies to improve leaf-level photosynthesis have been described in detail in previous reviews (Long et al., 2006; Parry et al., 2010; Zhu et al., 2010, 2020; Evans, 2013; Furbank et al., 2015, 2020; Foyer et al., 2017; Orr et al., 2017; Simkin et al., 2019). Canopy photosynthesis is determined by the cumulative photosynthetic activity of each individual leaf in the canopy. Leaf photosynthesis is determined by photochemical capacity and the microclimatic conditions the leaf experiences (also known as 'phylloclimate'). Temperature, humidity, and carbon dioxide concentration can all vary within a crop canopy, but in particular, the light environment within the canopy strongly determines the rate of photosynthesis. Therefore, this chapter focuses on how canopy architectural features affect the light environment within the canopy and hence canopy gas exchange.

2 Modeling light within crop canopies

Modeling the light environment throughout the canopy can identify key features of the canopy that impact light distribution. Light penetration in the canopy can be modeled as a function of the incoming radiation above the canopy (I_0) and canopy architectural traits (Goudriaan, 2016). The amount of light intercepted by the different layers of the canopy can be simply described by an exponential function like the Beer-Lambert equation (Fig. 1; Monsi and Saeki, 2005):

$$I = I_0 \cdot \exp(-k \cdot LAI) \quad (1)$$

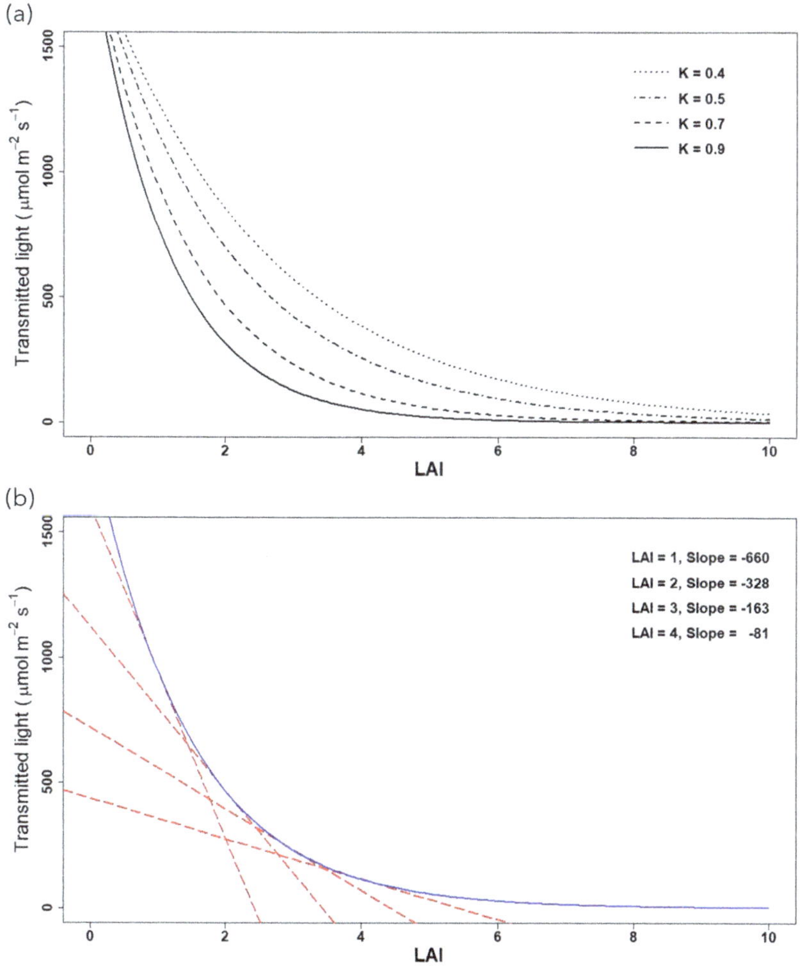

Figure 1 Light transmission as a function of the leaf area index (*LAI*). (a) Variation in light transmission in response to changes in the extinction coefficient *k*. Light transmission curves were produced using Equation 1. (b) Variation in absorption per leaf area at different depths. The blue line shows light transmission estimated by the Equation 1 with k = 0.7. Absorption at *LAI* = 1, 2, 3, and 4 is represented by the slope of the red dotted lines, obtained using Equation 2, with I = 2000 μmol m^{-2} s^{-1} and ρ_C= 0.05.

where I is the light intensity under the leaf area index (LAI) and k is the extinction coefficient. k can be calculated based on the leaf angle and the solar zenith angle assuming an ellipsoidal leaf angle distribution (Campbell, 1986) and represents the fraction of absorbed incoming downward radiation per leaf area at a given canopy depth. As a result, k equals 1 in canopies with horizontal leaves (planophile leaf angle distribution) and has lower values in canopies with vertical leaves (erectophile leaf angle distribution), where a greater proportion of light reaches the lower canopy layers (Fig. 1a). Because k is also affected by the solar zenith angle, it varies temporally with lower values when the sun is in zenith and higher values when the sun is closer to the horizon. Spatial variation in k can also be observed due to the spatial distribution of leaves, quantified by the clumping index Ω. Depending on how k is calculated, sun zenith angle and/or leaf clumping will be implicitly included in estimates of k (Zhang et al., 2014).

The absorption per leaf area can be obtained from the rate of light extinction in the canopy. This rate can be obtained for any depth i by taking the first derivative of Equation 1, giving $I_0 \cdot k\exp(-k \cdot LAI_i)$, where LAI_i is the leaf area index above the depth i (Fig. 1b). However, not all light that reaches the leaf surface is absorbed. Some light is reflected or transmitted (scattered). As a result of leaf scattering, the extinction coefficient k is lower than estimated for a black leaf absorbing all incoming radiation (k'). Radiation reaching the soil also can be scattered upward, though soil reflection plays a negligible role in dense canopies. Thus, the radiation at a given depth can be estimated by the net radiation between the flux of radiation downward (I_d) and upward radiation (I_u), $I_d - I_u$, which also equals $I_d - \rho_C I_d$, with ρ_C representing the fraction of light that escaped the canopy. The net absorbed radiation per leaf area at depth i (I_c) is thus written as:

$$I_c = I_d(1-\rho_C)\cdot k\exp(-k \cdot LAI_i) \tag{2}$$

To more accurately model radiation absorption within the canopy, it is also necessary to distinguish between direct beam radiation and diffuse radiation. The behavior of these different components of light can be modeled by making a few changes to Equation 2. The absorption for diffuse light per leaf area at depth i can be estimated by using the extinction coefficient for diffuse light, k_{diff}, and the canopy reflectance for diffuse light, $\rho_{C,diff}$:

$$I_{c,diff} = I_{diff} \cdot (1-\rho_{C,diff}) \cdot k_{diff}\exp(-k_{diff} \cdot LAI_i) \tag{3}$$

The amount of absorbed light per leaf area can be modeled as a function of LAI at different depths, and leaf photosynthetic activity can be modeled through the entire profile of the canopy using mathematical functions describing leaf photosynthesis as a response of light. By expressing leaf photosynthesis as a function of absorbed light, leaf photosynthesis in different layers of the canopy can be estimated, i.e. $\text{Leaf photosynthesis} = f(\text{Absorbed light}) = f(f(LAI))$.

Canopy photosynthesis can then be obtained by integration of the leaf photosynthesis-*LAI* response curve (Goudriaan, 2016).

Absorbed direct radiation can also be estimated by modifying Equation 2. The extinction coefficient for direct radiation can be approximated by the extinction coefficient for black nonscattering leaves, k'. The extinction coefficient k' can then be used in place of k to estimate the rate of light decrease in the canopy for a direct beam. Thus, the rate of light decrease for a direct beam in the canopy can be described as $k'_{dir} exp\left(-k'_{dir} \cdot LAI\right)$ where LAI represents the fraction of sunlit leaves. As sunlit leaves will scatter a fraction σ of intercepted light, only a fraction $(1-\sigma)$ of the direct light will be absorbed. Thus, the absorption of direct radiation by the sunlit leaves at depth i can be estimated as:

$$I_{sunlit,dir} = I_{dir} \cdot (1-\sigma) \cdot k'_{dir} exp\left(-k'_{dir} \cdot LAI_{sunlit,i}\right) \tag{4}$$

Sunlit leaves also absorb the scattered fraction of light. This fraction can be isolated by estimating the mean absorption of direct light, including light scattering, and subtracting from it the absorption rate of direct radiation that was estimated in Equation 4. The mean absorption rate of direct light is then obtained from an equation similar to Equation 2:

$$I_{mean,dir} = I_{dir} \cdot (1-\rho_C) \cdot k exp(-k \cdot LAI_i) \tag{5}$$

where ρ_C represents the fraction of light reflected out of the canopy. $I_{dir} \cdot (1-\rho_C)$ isolates the portion of direct light that is directly absorbed by leaves and scattered light absorbed by other layers in the canopy. This equation represents the absorption of direct radiation by sunlit leaves and the absorption of scattered radiation absorbed by both sunlit and shaded leaves. The absorbed scattered radiation, for both sunlit and shaded leaves, can then be estimated as:

$$I_{scatt} = I_{sunlit,dir} - I_{mean,dir} \tag{6}$$

Using the equations above, it is possible to model diffuse (Equation 3), direct (Equation 4), and scattered light (Equation 6) through the canopy profile and estimate the fraction of light absorbed by sunlit and shaded fraction of leaves. The photosynthetic activity through the profile of the canopy can then be derived by using photosynthetic light response equations and canopy photosynthesis can be obtained by integration.

Morphological features of the crop can have a profound impact on the light distribution within the canopy (see summary in Table 1). The total leaf area largely determines the fraction of intercepted light by the canopy. Leaf angle and leaf transmission affect the vertical light profile in the canopy. Leaf optical properties also alter the amount of light absorbed and scattered by the canopy. In the following sections, we discuss how past breeding efforts have improved

Table 1 Summary table of the parameters used in equations modeling the light environment in the canopy and affected by the canopy architecture

Parameters	Description	Canopy features affecting the parameter
LAI	Leaf area index: the total leaf area per surface unit of ground	Plant leaf area Plant density
k, k'	Extinction coefficient: characterize the efficiency by which a canopy intercepts light	Leaf angle Leaf clumping Leaf optical properties (for k only) *LAI*
ρ_c	Canopy reflectance: represents the fraction of light being reflected and escaping the canopy	Leaf optical properties *LAI*
Σ	Scattering coefficient: represents the fraction of light being scattered, i.e. reflected and transmitted	Leaf optical properties *LAI*

canopy architecture. We also discuss how optimizing canopy architectural traits can be used to improve canopy photosynthesis and potentially yield. The discussion highlights the variation in canopy architectural features in diverse crops and how different environments require different optimization strategies.

3 Impacts of breeding on modern crop canopy architecture

Breeding has profoundly altered canopy architectural traits in major grain and oilseed crops. This is perhaps most apparent in wheat and rice, which were the primary targets of the Green Revolution. A key success of the Green Revolution was selection of semidwarf, high-yielding varieties. Conventional varieties of wheat and rice grown before the Green Revolution had harvest indices of ~0.3. When these varieties were grown with high nitrogen fertilizer inputs of ~40 kg ha^{-1}, they grew very tall and lodged (Sakamoto and Matsuoka, 2004). In response, breeders selected semidwarf varieties with harvest indices of ~0.5, which made them less prone to lodging and able to take advantage of nitrogen inputs (Khush, 1999; Sakamoto and Matsuoka, 2004). Breeding since the Green Revolution has continued to alter wheat and rice canopy architecture. Light interception in winter wheat has been improved by increasing leaf area per tiller, contributing to *LAI* of up to 6.0 in modern cultivars (Tian et al., 2011; Xiao et al., 2012; Gupta et al., 2017). The ability to develop a high *LAI* for durum wheat, however, remained constant in some countries in Europe between 1945 and 2000 with an *LAI* of ~2.8 at anthesis at a density of 230 plants m^{-2} (Álvaro et al., 2008). For bread wheat, an increase in *LAI* at the milk grain stage was reported in Spain between 1920 and 2010 (Sanchez-Garcia et al., 2015). New cultivars also exhibit greater leaf area and photosynthetic capacity of the flag leaf (Reynolds et al., 1994; Tian et al., 2011; Gupta et al., 2017). There are also reports of increased single spike yield (Tian et al., 2011), grain number per square meter (Perry and D'Antuono, 1989; Sayre et al., 1997), and grain weight (Calderini et al., 1995; Sadras and Lawson, 2011) in modern winter wheat cultivars. Those changes may have improved yields by reducing sink limitation (Martintello et al., 1987; Foulkes et al., 2010). Modern cultivars were also reported to have earlier anthesis dates compared to older cultivars (Austin et al., 1980; Trethowan et al., 2007) or to have delayed leaf senescence (Sanchez-Garcia et al., 2015), increasing the length of the seed filling period.

Rice has seen similar changes in canopy architecture over decades of selection. Efforts to develop semidwarf varieties with better lodging resistance and greater HI have been ongoing since the 1950s, and the International Rice Research Institute released the first commercial, semidwarf high-yielding rice line, IR8, in 1966 (Khush, 1987; Peng et al., 2008). In 1996, China started the 'Super Rice Breeding Program' to develop specific plant ideotype models adapted to different agronomic practices and growing regions. Several morphological traits were targeted for optimization with the aim to produce rice capable of reaching a yield potential of 15 tons ha^{-1}. The following attributes were described in the ideotype: plant height > 100 cm; panicle at 60–70 cm height; three long, erect, and V-shaped leaves above the panicle; 250 panicles per square meter; grain weight of 7 g per panicle; HI > 5.5 (Yuan, 1998, 2003, 2017). Since the beginning of the Super Rice project, breeding programs have aimed to develop this ideal rice architecture. Breeding efforts have increased the panicle size, which has contributed to reduced sink limitation and higher HI (Peng et al., 2008; Zhu et al., 2016). The leaf rolling index (one-leaf edge distance/leaf width) of several Chinese modern cultivars has also been a target of selection and is close to the estimated optimal value to maximize RUE, ~11% (Cheng et al., 2007; Hu et al., 2009). In contrast, breeding in Brazil between 1983 and 2009 selected for shorter rice cultivars with a reduction in average height from 108 cm to 95 cm. Modern Brazilian cultivars have been selected against late flowering, with modern rice flowering on average 6 days sooner (Breseghello et al., 2011).

Maize canopy architecture has changed over generations of selection, resulting in greater light penetration within the canopy at high plant density. Between 1931 and 1991, plants with more erect leaves were selected (Duvick and Cassman, 1999; Wang et al., 2011; Ma et al., 2014), which improved the light environment within the canopy even under higher density planting (Ma et al., 2014; Assefa et al., 2018). In European cultivars, leaf area became more evenly distributed over the plant height, and a greater proportion of leaf area shifted to lower layers of the canopy, improving the light environment at ear height (Perez et al., 2019). The resulting enhancement in photosynthetic activity in leaves at ear height contributed to greater carbohydrate supply to the forming ear (Wardlaw, 1990; Hammer et al., 2009; Perez et al., 2019). A slight decrease in individual plant leaf area over the years has also been reported for some European hybrids (Perez et al., 2019). The opposite trend was observed between 1950 and 2000 for Chinese hybrids, with an increase in both plant leaf area and *LAI* (Ci et al., 2012; Ma et al., 2014). For US hybrids in the Corn Belt, leaf area per plant did not change between 1930 and 1980, and *LAI* was constant at a value close to 3.2 at a density of 47 000 plants ha^{-1} (Duvick, 1984, 2005). Since the mid-1980s, there has been a linear increase in US maize planting density of 700 plants ha^{-1} $year^{-1}$ (Assefa et al., 2018). Modern high-yielding hybrids now reach *LAI* up to 7.8 under optimal growth conditions at 94 000 plants ha^{-1} (Lindquist et al., 2005). Delayed senescence was reported for US maize hybrids between 1930 and 1950, extending the period of active photosynthesis during grain filling (Duvick, 1984; Duvick and Cassman, 1999). Modern cultivars also show an increased tolerance to high density planting, with an ability to produce the same amount of grain with less leaf area (Mansfield and Mumm, 2014). In Canada, while the leaf area per plant remained constant, *LAI* increased, thanks to greater plant density, with *LAI* increasing from 2.4 to 4.8 between the 1930s and the 2000s (Tollenaar, 1991; Duvick, 2005; Lee and Tollenaar, 2007). Yield increase has also been attributed to an increase in kernel number per ear and individual kernel weight between the 1950s and 2000s (Duvick, 2005; Tollenaar and Lee, 2006; Sun et al., 2017).

Wild soybean has a viny morphotype and a determinate growth habit. During the early 1900s, breeders selected an upright plant with either indeterminate or determinate growth (Tian et al., 2010; Anderson et al., 2019). The case of soybean in the USA is particular, as most of the cultivars grown come from a narrow genetic pool, with only 28 ancestors contributing to 95% of the genes in cultivars released between 1947 and 1988 (Gizlice et al., 1994). Between 1902 and 1977, soybean architecture changed with selection, and height was reduced from ~97 cm to ~80 cm, improving lodging tolerance (Specht and Williams, 1984; Poehlman, 1987; Ustun et al., 2001; Rincker et al., 2014). This especially contributed to greater yields in high-fertility regions prone to lodging. A reduction in plant height down to ~80 cm has also been reported in some regions of China (Wu et al., 2015; Wang et al., 2016a). Between the 1940s and the 2000s, the number of branches and nodes per plant have also been reduced in some Chinese modern cultivars (Wu et al., 2015; Wang et al., 2016a). In contrast, a moderate increase in the number of nodes has been reported for southern US cultivars, but it did not play a major role in the yield increase (Kahlon et al., 2011). Modern Chinese and US cultivars also showed delayed leaf senescence, extending the period with active vegetation during grain filling (Kumudini et al., 2001; Koester et al., 2014; Rowntree et al., 2014). A lower leaf area has also been reported in modern cultivars from Northeast China and Canada (Morrison et al., 1999; Jin et al., 2010). Soybean cultivars from the southern USA also showed an increase in seed size and pod number between 1953 and 1999 (Kahlon et al., 2011). Selection for greater yields has dramatically altered crop canopies, as described for wheat, rice, maize, and soybean; however, there may be opportunity for further improvement.

4 Potential targets for canopy improvement

4.1 Optimizing leaf area

4.1.1 Leaf area and light interception

Canopy total leaf area is a major determinant of crop yield, as leaves are the principal solar panels and sites of photosynthetic carbon fixation. The relationship between leaf area and yield was established in the late 1940s (Watson, 1947, 1952, 1958) when Watson (1947) first introduced the notion of *LAI* [m^2 ... m^{-2}] as a means to quantify the leaf surface area per soil surface area of a crop. Since the introduction of *LAI*, scientists have asked if it is optimal for yield. Several studies on soybean suggested that the canopy should intercept at least 95% of incident radiation before seed development to produce optimum yield (Shibles and Weber, 1966; Taylor et al., 1982; Board et al., 1997; Board, 2004). An *LAI* of 3.5–4.0 is sufficient to capture 95% of incident radiation (Taylor et al., 1982; Board and Harville, 1992; Malone et al., 2002; Müller et al., 2017), although *LAI* can reach 7 or 8 in high-yielding, optimal growth conditions (Setiyono et al., 2008). Similar observations were made in potato crops where significant losses in yield were observed once *LAI* dropped below 4.0 (Bastos et al., 2011; Bangemann et al., 2014), reducing light interception below 90% (Ziems et al., 2006; Jin et al., 2013). The optimal *LAI* values reported for monocot crops such as wheat, rice, and maize ranged between 3.0 and 6.0 (Davidson, 1965; Angus et al., 1972; Pepper et al., 1977; Shuting, 1994; Zheng et al., 2009; Liu et al., 2017; Xue et al., 2017; Li et al., 2018). Optimal canopies also have rapid development of leaves with high specific leaf area (leaf area/dry mass) before anthesis, which enables rapid canopy establishment and light interception along with high *LAI* during the reproductive growth period. For this reason, it has been suggested to breed for canopies with high specific leaf area during the vegetative period and high *LAI* at the reproductive stage (Dingkuhn et al., 1998).

4.1.2 Leaf area in excess

Once a canopy reaches a certain *LAI*, additional canopy layers do not increase light interception or A_C (Shuting, 1994). Excessive *LAI* increases self-shading, and leaves in the lower canopy can have a negative net carbon balance. This means that the cost for producing and maintaining excess leaf area (which also involves the production and maintenance cost of roots, stems, and branches needed to support the leaf area) can be greater than the carbon assimilated (Reich et al., 2009; Anten, 2016). This can eventually lead to a reduction in the canopy RUE, especially in situations of nitrogen deficiency (Bonelli and Andrade, 2020). The hypothesis that *LAI* may be in excess in major crops has led several studies to assess the impact of defoliation on yield. Results showed a yield decline in response to leaf removal in most cases (Fig. 2), yet a handful of studies have reported a consistent yield increase following defoliation (Liu et al., 2015, 2020; Srinivasan et al., 2016; Wei et al., 2018; Raza et al., 2019). This raises the question of what distinguishes these studies that have different results. In the case of maize, positive results were usually observed when *LAI* was >6.0 before defoliation treatments were initiated, and modest leaf removal at R1 (i.e. beginning of flowering) then stimulated yields (Raza et al., 2019). The same pattern was observed for soybean where a yield benefit was observed in very high *LAI* canopies when even after defoliation *LAI* was ~6.0 at R1 (Srinivasan et al., 2016). The interaction between the plant population density and plant architecture also affected the outcome of *LAI* reduction. Liu et al. (2020) showed that maize plants with a more compact canopy were less tolerant to leaf removal than plants with a more open canopy. Another common feature of successful soybean and maize defoliation studies was improved A_C as a result of greater light penetration in the canopy, which resulted in yield benefits of 4.6–13.5% (Liu et al., 2015, 2020; Srinivasan et al., 2016; Wei et al., 2018; Chen et al., 2020).

In many cases, moderate defoliation of soybean did not significantly affect yield even when *LAI* was initially >6.0 (Haile et al., 1998a; Malone et al., 2002). Soybean is very tolerant of defoliation, and most defoliation studies had no significant impact on yield (Fig. 2). Plants with indeterminate growth tolerate greater defoliation later in the season because of their ability to compensate for the loss of leaf area (Ziems et al., 2006). In contrast to other studies, Srinivasan et al. (2016) performed the defoliation treatment both during vegetative and reproductive growth. It is possible that other studies with a single early defoliation treatment failed to see any result because new leaves produced after defoliation reduced any benefit of early season leaf removal on the canopy light environment. It is also possible that defoliation negatively impacted the vertical nitrogen distribution in the canopy. Sunlit leaves at the top of the canopy have higher A_{max} and greater N content (Anten, 2016; D'Odorico et al., 2019). It is possible that removing sunlit leaves to enhance light in shaded leaves with lower A_{max} negatively impacted the N balance.

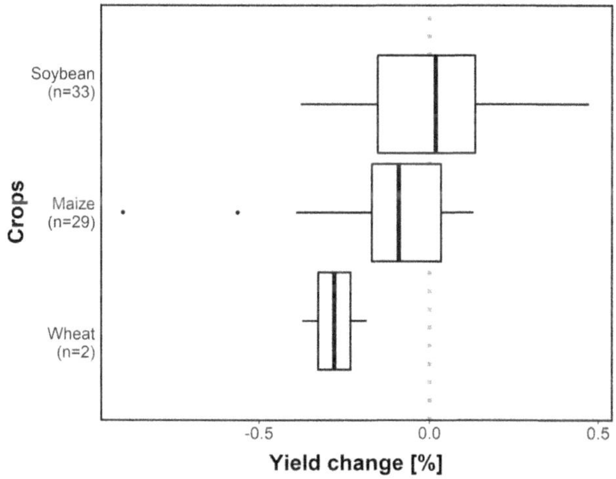

Figure 2 Percent change in yield following defoliation for different crops. For soybean, *LAI* was lowered from 5.17 ± 0.74 to 2.66 ± 0.97 (usually reported at the R2 stage) on average across studies. The median yield change was +1.9%. For maize, *LAI* was reduced from 6.16 ± 1.79 to 5.03 ± 1.50 (reported at the beginning of flowering) on average. The median yield change was -9.1%. For wheat, *LAI* was reduced from 6.15 to 3.57 ± 1.96 (reported at anthesis) on average. The median yield change was -27.9%. The numbers in parentheses represent the control-treatment pair comparisons used for the boxplots for each crop. Only studies on soybean, maize, and wheat for which defoliation was done before (or close to) anthesis and where control had an *LAI* > 4.0 (i.e. *LAI* close to or above optimum for light interception) were considered. The list of studies that were used in this figure are reported in Appendix.

When breeding for an optimal leaf area, the importance of considering leaves not just as photosynthetic organs but also as storage organs was emphasized in a study of Brazilian soybean cultivars (Tagliapietra et al., 2018). Using on-farm and research-station soybean data, the optimal *LAI* value at early flowering was ~4.0, which also maximized A_C. However, optimal *LAI* during pod filling was greater, between 6.0 and 6.5, which exceeds the optimum for A_C. Higher *LAI* allowed greater reserve storage as well as maintenance of sufficient photosynthetic activity during senescence. Based on this data, selecting for greater *LAI* and optimum biomass production by grain filling maximizes yields. This strategy may be especially important for crops with a high nitrogen demand and for which the nitrogen supply during senescence can limit yield (Borrás et al., 2004; Thomas and Ougham, 2014).

4.1.3 Leaf area and its distribution

The distribution of leaf area within the canopy and the proportion of leaf area that is direct vs. diffuse light can alter RUE and A_c. Early modeling work suggested that RUE was relatively insensitive to changes in *LAI* (Sinclair and

Horie, 1989) under non-nitrogen-limiting conditions. However, RUE increases with the proportion of *LAI* operating at nonsaturating light irradiance (Stöckle and Kemanian, 2009). This was demonstrated by a study of tomatoes which assessed the response of A_C to changes in the proportion of diffuse light relative to direct light, without changing the total light intensity (Li et al., 2014). The highest A_C was measured in the diffuse light treatment, which was attributed to more uniform light distribution, resulting in a greater leaf area exposed to light. The lower extinction coefficient in the canopy also increased the photosynthetic capacity of leaves in the middle layer of the canopy. Thus, it is likely that any change in canopy architecture that improves the distribution of light among the total canopy leaf area will benefit RUE. This was demonstrated in maize where canopies with vertical distribution of *LAI* had greater radiation interception efficiency (Perez et al., 2019). An erect distribution of leaves throughout the canopy may be an interesting target for breeding in other crops.

Leaf position in the canopy is affected by the petiole length, with longer petioles reducing self-shading (Takenaka, 1994) and hence increasing light-use efficiency. Longer internodes also benefit light interception efficiency (Takenaka, 1994; Pearcy et al., 2005). Similarly, crops with taller canopies showed higher A_C at saturating light, both under direct and diffuse conditions (Emmel et al., 2020), which may indicate greater light-use efficiency. Together, petiole length, internode length, and plant height are traits that affect leaf clumping, making them targets to improve the light extinction coefficient *K* and thus the light environment in the canopy.

In cereals such as wheat, rice, and maize, specific leaves contribute more to grain filling than others. Several studies have shown a yield increase in response to an improved light environment around the sink location, i.e. at the inflorescence height (Wardlaw, 1990; Li et al., 2008; Hammer et al., 2009; Perez et al., 2019). The flag leaf in wheat, for instance, has received a lot of attention as its photosynthetic capacity is positively correlated with grain yield (Simpson, 1968), and increases in its photosynthetic capacity, angle, light interception, and area have been suggested to improve yield. The light environment for those leaves thus deserves greater attention. There are also other organs that can contribute photosynthate to developing sinks. The leaf sheath was shown to account for up to 25% of A_c in barley (Sadok et al., 2020) and contributed to about 15% of grain yield (Archbold, 1942). Ear photosynthesis has also been shown to play an important role in the final grain yield, with a contribution estimated between 13% and 59% (Tambussi et al., 2007; Maydup et al., 2010; Wang et al., 2016b).

4.1.4 Leaf area through time

Maintaining actively photosynthetic leaves for a longer period of time dramatically improves annual interception efficiency (Dohleman and Long,

2009). The maintenance of green leaf area during grain filling, also referred to as the stay-green phenotype, has also been associated with higher yields (Duvick and Cassman, 1999; Blake et al., 2007; Koester et al., 2014). A functional stay-green phenotype can be achieved by delaying senescence, the degenerative process of catabolism of pigments and other components of photosynthetic proteins, organelles, and cellular components (Gregersen et al., 2013). Delayed leaf senescence maintains nitrogen in leaves, which in turn promotes photosynthetic activity. This extended period of carbon assimilation during the grain filling period can result in additional carbon remobilized to the grains. Early flowering has also been shown to extend the period of grain filling (Trethowan et al., 2007; de Felipe et al., 2016). On a landscape scale, replacing annuals with perennials can have a profound effect on seasonal C gain by extending the growing season, especially in temperate regions (Dohleman and Long, 2009). There is growing interest for developing perennial agriculture which could have other benefits such as preventing erosion, building soil organic matter, supporting diverse microbial communities, and providing other ecosystem services (Ciotir et al., 2019).

4.1.5 Leaf area and plant density

Theoretically, for any crop, there is a seeding density that maximizes yield (Deng et al., 2012) and balances resource availability and competition. Increasing the number of plants per unit area exacerbates competition and potentially abiotic stresses for individual plants. Historical selection of plants with improved tolerance to crowding density has enabled increased stand density and higher yields. But, selection at high densities has not been uniform across crops, and maize and soybean in the USA provide an interesting contrast. For decades, maize has been selected under very high plant density, which has increased yield on an area basis (Fig. 3a) and also resulted in very uniform row spacing in the USA (Fig. 3c) (Tollenaar and Wu, 1999; Mansfield and Mumm, 2014; Assefa et al., 2018). High planting densities lead to more rapid canopy closure, greater *LAI*, and increased light interception (Orlowski et al., 2012; Board and Harville, 1992). For US maize, there is no correlation between row spacing (which varies little) and yield (Fig. 3e). US soybean has also seen dramatic yield increases over the past few decades (Fig. 3b), but is still planted under a wide range of conditions, with plant density varying between 200 000 plants ha^{-1} and 600 000 plants ha^{-1} and row spacing varying between 18 cm and 102 cm (Fig. 3d; Heatherly and Welmore, 2004; Mourtzinis et al., 2018). The literature suggests that narrow row spacing (38 cm or less) offers on average a greater yield benefit over wider row spacing (Oplinger and Philbrook, 1992; De Bruin and Pedersen, 2008; Pedersen, 2008; Cox and Cherney, 2011; Orlowski et al., 2012), but soybeans' ability to compensate for a reduced population density

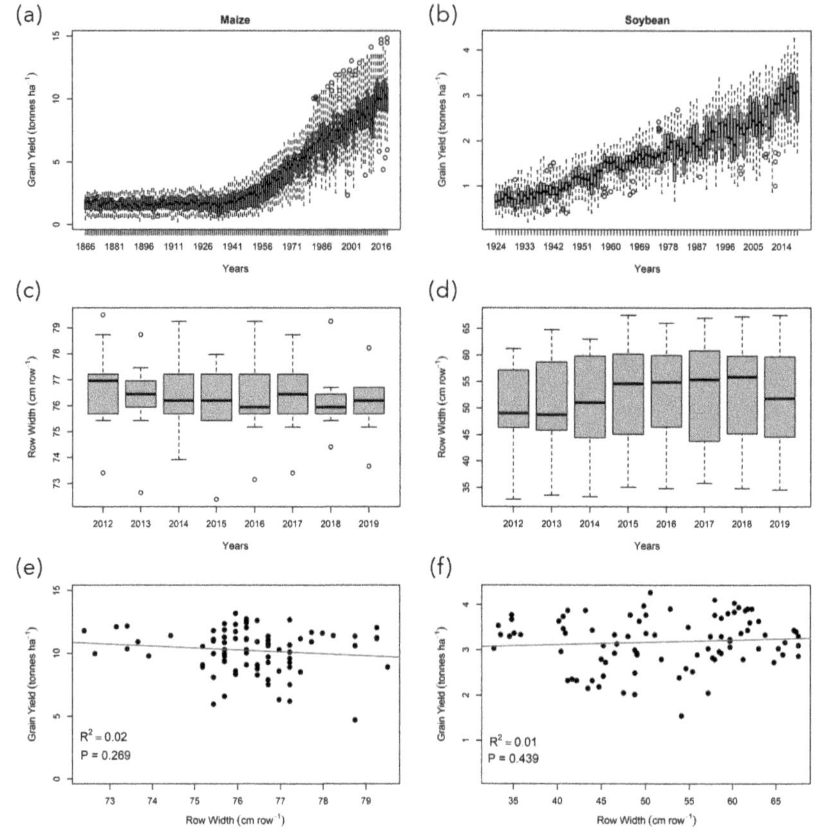

Figure 3 US county-level yield from 1866 and 2019 for maize (a) and from 1924 and 2019 for soybean (b). Row spacing between 2012 and 2019 in the USA for maize (c) and soybean (d). Linear regression relationship between grain yield and row width for maize (e) and soybean (f). Data were taken from the USDA National Agricultural Statistics Service, and each dot in E, F represents the average for a given year and state.

through branching results in high yields at lower populations (Board, 2000; De Bruin and Pedersen, 2008; Cox and Cherney, 2011). As a result of greater flexibility in branching, there is more variability in row spacing and population size in soybean, and no evidence that row spacing impacts yield (Fig. 3f). The advantage of narrow rows over wide rows is especially evident under short growing seasons, in situations of late sowing date and/or when early-maturity group cultivars are used. The benefit of narrow rows is less pronounced under other circumstances and can even be negligible in situations of early sowing in a full growing season (Lee, 2006; Andrade et al., 2019). Unsurprisingly, the transition to narrow row spacing appears to be slow, and the adoption of 19 cm row spacing remains marginal (Specht et al., 2014).

In contrast with the USA, narrow row spacing (38 cm or less) for soybean has been adopted in Northeast China and has contributed to more uniform distribution of leaf area and faster canopy closure in later maturing cultivars (Liu et al., 2008). The adoption of narrow row spacing systems has been accompanied by a reduction of leaf area in modern cultivars from Northeast China (Jin et al., 2010), which has contributed to improved soybean tolerance to high density planting by reducing self-shading. Comparable observations have been made for modern cultivars in Canada, which are characterized by a lower total leaf area and an improved tolerance to high planting density (Buttery and Buzzell, 1972; Morrison et al., 1999; Cober et al., 2005). On average, this has led to a reduction in *LAI* from about 5.0 to 4.0 in modern cultivars (Morrison et al., 1999; Jin et al., 2010).

Little is known about how much plant density can be increased and which density would be considered a theoretical maximum. Deng et al. (2012) presented a simple theoretical model showing that most of the variation in growth and yield of annual crops can be explained by the maximal size of the plant at maturity and the intrinsic growth rate. They suggested that the optimum planting density for all crops is intermediate, with lower yields resulting from incomplete use of resources at low planting densities and competition reducing yields at higher than optimal planting densities. While this simple allometric theory holds under optimum water and nutrient availability, those conditions are not always met in farmers' fields. Further investigation of trade-offs of changes in plant density under a range of environmental conditions will help determine optimal seeding rates.

4.2 Optimizing leaf traits

4.2.1 Leaf angle

At high *LAI*, self-shading progressively increases in the canopy, while upper leaves remain light saturated. As a result, the maintenance cost for leaves deep in the canopy can be greater than CO_2 assimilation, which reduces canopy RUE (Bonelli and Andrade, 2020). The advantage of erect leaves on yield and carbon assimilation manifests in canopies with $LAI > 4$ (Loomis and Williams, 1969; Pepper et al., 1977; Sinclair and Muchow, 1999) when self-shading starts to limit A_C. This was illustrated in a study of maize where a planophile variety showed reduced yields when planted at high plant density (maximum $LAI = 6$ at 7.5 plants m^{-2}), while an erectophile variety exhibited a yield increase when planted under the same conditions (Shuting et al., 1993). A modeling study showed similar results for rice where the optimum leaf angle was dependent on *LAI*. A_C was maximized at a low leaf angle when $LAI = 4.8$, but at $LAI = 7.7$, A_C was maximized at a high leaf angle (Song et al., 2013). More planophile leaves may be beneficial at low *LAI*, while at high *LAI* high leaf angle allows deeper light penetration within the canopy (Duncan, 1971). Hence, greater biomass can be achieved at higher leaf angles. In a study comparing ~1000 wheat lines with varied leaf angle, erectophile canopies exhibited on average 11% more aboveground biomass than planophile canopies (Richards et al., 2019). Greater biomass in erectophile canopies offered more resource storage in leaves and stems and was associated with higher yields. At an early developmental stage, horizontal leaves maximize light capture at low *LAI*, which benefits early growth rates and establishment of the canopy (Hammer et al., 2009). At higher *LAI*, however, higher leaf angles prevented photoinhibition of the top leaves by reducing the amount of light absorbed (Tollenaar and Dwyer, 1999; Werner et al., 1999; Pearcy et al., 2004) while raising the proportion of leaf area operating at light saturation. This means that the optimal leaf angle is dependent on *LAI* and needs to adapt with plant growth. Leaf angle is also not uniform in the canopy (Blad and Baker, 1972; Nobel et al., 1993), and optimizing the differential leaf angle distribution throughout the plant has been suggested to improve the light environment within the canopy (Mantilla-Perez and Salas Fernandez, 2017). Ideally, erect leaves at the top allow better light penetration, and more horizontal leaves at the bottom effectively capture all of the transmitted light (Duncan, 1971).

Leaves move and track the sun, adjusting light interception in a process called heliotropism. Such movements are observed in many crop plants, including cotton, soybean, cowpea, alfalfa, and sunflower. Two types of leaf movement are recognized within heliotropism: diaheliotropism and paraheliotropism. Diaheliotropic leaves orient themselves perpendicularly to direct solar rays, while paraheliotropic leaves orient their blades parallel to direct rays (Ehleringer and Forseth, 1980). Leaf movements can considerably affect the light environment in the canopy. Leaves at the top of the canopy with diaheliotropism behavior tend to reduce A_C by shading the lower leaves (Herbert, 1991). However, the impact of leaf movement on A_C is also dependent on *LAI*. By growing alfalfa exhibiting diaheliotropism on a rotating platform, Denison et al. (2010) showed an increase in A_C for $LAI > 5$ when diaheliotropism was disrupted by rotating the platform twice a day. This resulted from better light exposure of leaves previously shaded by the upper leaves. Paraheliotropic movement for the upper leaves seems to be more advantageous at high *LAI*. At low *LAI*, however, diaheliotropism can contribute to greater carbon gain by efficiently tracking sunlight and maximizing light interception (Rakocevic et al., 2010). This behavior thus offers an advantage at early developmental stages. This may be why soybean leaflets exhibit mainly diaheliotropic movements during the vegetative stages and more paraheliotropic movements during the reproductive stage (Rakocevic et al., 2010). Leaf movements usually remain consistent for all leaves (Levizou and Kyparissis, 2016). One exception to this uniform pattern has been observed for *Styrax camporum* Pohl., a native shrub from the Brazilian savanna. Observations have revealed that upper leaves exhibit a paraheliotropic behavior at noon, while lower leaves exhibit diaheliotropism (Habermann et al., 2008, 2011). Such asynchrony in leaf movement could be exploited to optimize leaf angle profiles in crop canopies and improve light-use efficiency.

4.2.2 Leaf shape

There is growing evidence that leaf shape can affect canopy RUE. Models have shown improved light capture efficiency in canopies with narrow leaves due to a reduction in self-shading (Takenaka, 1994; Falster and Westoby, 2003; Pearcy et al., 2004; Sarlikioti et al., 2011). This was further supported by a study showing higher *K* values observed in plant communities with larger leaves (Monsi and Saeki, 2005). Experiments have shown the advantage of narrow leaves on light interception efficiency in soybean (Bianchi et al., 2020). Improved light use was translated to a 28% yield increase at low planting density; however, no significant difference in yield was observed at high density. Similar observations have been made in rice where a mutant with narrow leaves showed greater penetration of light within the canopy (Hirotsu et al., 2017). This indicates that narrow leaves can constitute a strategy to improve the light environment in the canopy and offer an opportunity to improve canopy RUE. The advantage of narrow leaves was shown by models to be especially marked for plants with short internode length and distichous phyllotaxy (Takenaka, 1994). Thus, the benefit of a narrow leaf shape may depend on other aspects of plant architecture.

The leaf rolling index has also received considerable attention, especially in rice breeding programs. Moderate rolling contributes to leaf erectness in monocots such as rice, improving light penetration within the canopy. Leaf rolling also plays a photoprotective role. Because rolling alters incident light on leaves, it contributes to greater light saturation points of photosynthesis and prevents photoinhibition (Corlett et al., 1994; Hu et al., 2009; Feng et al., 2002; Kadioglu et al., 2012). By reducing photoinhibition under high light, moderately rolled leaves also contribute to maintaining high CO_2 assimilation and faster recovery from midday photoinhibition (Kromdijk et al., 2016). Studies have estimated that a leaf rolling index around 11% allows an optimum RUE in rice canopies (Cheng et al., 2007; Hu et al., 2009). The reduced leaf area exposed to light in moderately rolled leaves also contributes to reduce the amount of radiant energy absorbed by leaves which reduces leaf temperature (Heckathorn and DeLucia, 1991) and transpiration, thus improving water-use efficiency (Richards et al., 2002; Nar et al., 2009; Sariyeva et al., 2009; Sarieva et al., 2010). The reduced light absorption by rolled leaves in the upper layers of the canopy could also promote better light penetration to the lower layers of the canopy.

More studies are needed in other crops to understand the benefit of leaf rolling in canopy light-use efficiency and canopy gas exchange optimization.

Leaf shape and size also affects the thickness of the boundary layer, expressed as $\delta^{bl} = \alpha\sqrt{L/v}$, where α is a constant equal to 4.0 for a flat surface, v is the speed of the wind, and L is the length of the leaf in the direction of the wind (Nobel, 2009). The average boundary layer thickness tends to increase with leaf size. Also, the boundary layer is typically thinner at the edges, so leaves with irregular shapes have a thinner average boundary layer compared with circular leaves of equal area (Schuepp, 1993). This has consequences for the leaf thermal regulation, leading to increased resistance to sensible and latent heat exchange with thicker boundary layers. As a result, leaf size has been shown to be positively correlated with leaf temperature (Yates et al., 2010). This leads to a higher reliance on stomata for heat loss through transpiration (i.e. latent heat loss) to regulate leaf temperature (Leigh et al., 2017). Large leaves are commonly found in tropical environments such as rainforest, where the high humidity can reduce transpiration and latent heat loss (Meinzer et al., 1995). In contrast, smaller leaves have been hypothesized to be an adaptive mechanism to the Mediterranean-type climate. The thinner boundary layer enables a more efficient sensible heat loss in summer, reducing water loss through transpirational cooling, while still allowing a higher transpiration rate in winter which promotes nutrient acquisition (Yates et al., 2010). The impact of leaf size and shape on the boundary layer and temperature regulation can also be upscaled at the canopy level. When comparing two soybean isolines with only differences in leaf width, heat transfers were operating preferentially through sensible heat rather than latent heat in the canopy with narrow leaves (Baldocchi et al., 1985). This improved cooling through sensible heat transfer can contribute to improved water-use efficiency in narrow leaf canopies, especially in dry conditions with limited water for transpirational cooling (Baldocchi et al., 1985; Balota et al., 2008; Richard et al., 2019). Another study comparing three closely related winter wheat lines reported the coolest canopy temperature in line with the narrowest leaves (Balota et al., 2008). Higher wind speeds in canopies with narrow leaves also contribute to improved sensible heat exchanges by reducing the leaf boundary layer thickness. This was observed in a study comparing the vertical wind profile of four broad-leaved deciduous species, where wind speed attenuation was lower in species with narrower leaves (Barnard and Bauerle, 2016). The study also associated greater wind speed attenuation with higher *LAI* values. Air movements, by reducing the boundary layer thickness, can promote gas exchange and benefit A_C (Kitaya et al., 2000, 2003). Thus, altering leaf size and shape could be used as a strategy to reduce the share of transpirational cooling in canopies and improve A_C.

4.3 Optimizing leaf optical properties

Studies have suggested that reduction in canopy chlorophyll content could be used as a strategy to improve A_c (Zhu et al., 2010; Ort and Melis, 2011). This approach aims at decreasing the canopy light extinction coefficient by improving leaf transmittance. Field trials conducted with chlorophyll-deficient mutants have tested this hypothesis (Pettigrew et al., 1989; Slattery et al., 2017; Gu et al., 2017a, 2017b). While those studies showed mixed results regarding the benefit of chlorophyll reduction on A_c and yield, they all agreed that reductions in canopy chlorophyll content do not have a detrimental effect on yield. Walker et al. (2017) showed that canopy chlorophyll content could be reduced by at least 50% in soybean with no negative impact on yield, offering about 9% savings in leaf nitrogen. Similar conclusions have been drawn by studies that have observed an overaccumulation of nitrogen in wheat leaves (Grindlay, 1997; Pask et al., 2012). A threshold of 2.0 g N m^{-2} was reported over which additional nitrogen did not contribute to greater RUE. Different thresholds are reported for other crops (Sinclair and Horie, 1989). Those studies suggest that nitrogen savings could be reinvested in other processes limiting yield and photosynthesis.

The vertical nitrogen distribution within the canopy is also a target for optimization (Field, 1983; Hikosaka, 2016). For a canopy with a fixed amount of nitrogen, the distribution of nitrogen maximizes the daily A_C when the marginal gain of nitrogen investment on leaf photosynthesis is the same among leaves. This means that, for every leaf within the canopy, the rate of increase in daily carbon gain with changing nitrogen is the same. Another way to phrase it is that the reallocation of nitrogen across leaves depresses the daily carbon gain of the canopy when nitrogen is optimally distributed. This is represented by Field (1983) in this partial derivative:

$$\frac{\partial \int A dt}{\partial N_{area}} = \frac{\partial A_d}{\partial N_{area}} = \lambda \tag{7}$$

where A_d is the leaf daily CO_2 assimilation rate and λ is the Lagrange multiplier representing the marginal carbon gain per unit N_{area}. The nitrogen distribution is optimized and A_C is maximized when λ has the same value for all leaves. Farquhar (1989) further expressed the optimal nitrogen distribution as a distribution that would satisfy the following relation for a canopy with n layers:

$$I_1 : I_2 : \ldots : I_n = J_{max,1} : J_{max,2} : \ldots : J_{max,n} \tag{8}$$

where I is the absorbed irradiance and J_{max} is the maximum rate of electron transport at saturating light. This relation means that in canopies with optimally distributed nitrogen, the ratio between J_{max} and I is the same for each layer. Assuming a linear relationship between the light-saturated rate of photosynthesis (A_{max}), the respiration rate, and nitrogen content, it also implies that the ratio between A_{max} and *I* is equal for each layer. From those models, it can be derived that a higher proportion of nitrogen should be allocated to the upper part of the canopy where leaves are exposed to light and less nitrogen should be invested in shaded leaves (Hikosaka, 2014, 2016; Anten, 2016). A steep distribution of nitrogen content has been associated with higher yield in rice when cultivars with contrasting patterns of vertical nitrogen distribution were compared (Muryono et al., 2017). Hikosaka (2014) has estimated that a suboptimal nitrogen distribution could lead to a loss in carbon gain of about 20%. While the chlorophyll distribution theory aims at improving RUE by reducing the light extinction coefficient, the nitrogen distribution theory focuses on maximizing the carbon gain per nitrogen unit invested at each layer. Both suggest savings on nitrogen within the canopy by avoiding overinvestment in chlorophyll.

As the nitrogen distribution is dependent on the light availability, it is highly affected by any modifications of the canopy architecture with an impact on the light extinction coefficient. For instance, canopies characterized by planophile leaves require a steeper decline in the leaf nitrogen content compared to canopies with erect leaves (Hikosaka, 2014). As most of the light will be intercepted at the top of the canopy, more nitrogen will need to be invested in the upper part of the canopy. Alternatively, erect canopies can reduce the portion of nitrogen allocated to light harvesting components at the top of the canopy and more evenly distribute nitrogen through the canopy. It may also be that with improved light penetration in erectophile canopies, the benefits of reduced chlorophyll content may decline (Zhu et al., 2010).

Models of optimum nitrogen distribution have limitations, and several physiological and physical constraints are excluded from the analysis despite their profound impact on the nitrogen distribution, such as changes in leaf absorbance, light-use efficiency, hydraulic limitation, mesophyll conductance, and leaf structure (Buckley et al., 2013; Hikosaka, 2016). Nitrogen reallocation within plants from senescing leaves is also ignored from most models (Hikosaka, 2016; Hikosaka et al., 2016). However, dynamic models accounting for nitrogen loss in dropped leaves estimate a higher optimum *LAI* for maximizing A_C compared to static models (Anten, 2016). A possible explanation is that a high *LAI* allows more flexibility in shedding leaves if environmental conditions change, and the *LAI* becomes greater than optimal. The reallocation of nitrogen from old, shaded leaves also offers the opportunity to reinvest nitrogen in leaves that could more efficiently use it. Transgenic tobacco with delayed senescence had lower A_C compared to wild type because of greater respiration from shaded leaves while the wild type dropped leaves with a negative carbon balance (Boonman et al., 2006). There is also evidence that cultivars with lower nitrogen concentration per surface leaf area need to shed more leaves, and earlier, to meet the nitrogen demand of filling grains (Borrell and Hammer, 2000; Borrell et al., 2001; van Oosterom et al., 2010). However, improved uptake of soil nitrogen by the root system during grain filling could contribute to delayed leaf senescence (Rajcan and Tollenaar, 1999; Pommel et al., 2006). This indicates a possible compromise between 'staying green', nitrogen distribution, uptake, and reallocation.

Multiple criteria are needed to optimize the nitrogen distribution profile in the canopy. While in some canopies, it may be advantageous to invest nitrogen in light harvesting components in the top layers, in others reducing chlorophyll may be beneficial. Further nitrogen may be better invested in optimizing the quantum efficiency in the lower canopy layers where light availability is scarce and sun flecks are an important driver of carbon gain in those leaves. Nonetheless, optimization theory remains a useful tool in predictive modeling and provides a simple framework for guiding crop breeding.

5 Canopies under different environments

Most of the studies that have either modeled or experimentally determined the canopy architecture that would optimize yield have worked under ideal growth conditions. However, the environment can submit the plant to multiple, dynamic abiotic and biotic stresses. As a result, different phenotypes may be required to maximize yield under different environments.

The optimum *LAI*, for instance, is highly dependent on nutrient availability. Under low nitrogen conditions, increased *LAI* will reduce nitrogen content per leaf area, which will reduce leaf RUE (Niinemets, 1999; Walker et al., 2014). Cultivars with a lower *LAI* may therefore be preferred in regions with low nitrogen inputs as high *LAI* canopies require greater nitrogen availability to reach their maximum photosynthetic capacity and yield potential (Bonelli and Andrade, 2020). Because self-shading is reduced in low *LAI* canopies, the vertical distribution of nitrogen and *LAI* will be less critical for determining RUE (Li et al., 2014; Bonelli and Andrade, 2020). Early maturing cultivars may be preferred under conditions with poor soils and lower nutrient availability. Under those circumstances, high density planting may also enhance yields by promoting faster ground cover and better light interception (Carciochi et al., 2019), though the benefit of a higher plant density can be nullified under severe nitrogen shortage (Cooper and Jeffers, 1984).

There are also multiple ways to adapt the plant canopy in rain-fed production systems that are susceptible to drought events. Narrow row spacing enables canopies to reach critical *LAI* more quickly than wider row spacing. However, if narrow row spacing is accompanied by an increase in seed density, this can lead to greater competition for water. Experiments with maize tested the effects of three plant densities and different row spacings on crop performance. Plant density was the main factor determining root weight and length per plant in the upper soil layers, with lower values measured at higher plant densities (Gao et al., 2021). Less competition was found within the narrowest row spacing when there was a uniform planting layout which increased space and resource exploitation around individual roots (Gao et al., 2021). While the highest maize yields were recorded at the highest plant density in that well-watered study, different outcomes are expected in water-limited environments. In a sweet pearl millet experiment, increasing seeding rate decreased yields in a site with limited rainfall (Bouchard et al., 2011). Another benefit of narrow row spacing is faster ground cover early in the season, which can reduce water loss from soil evaporation by reducing the amount of radiation reaching the ground. However, higher biomass production early during the growth season can result in greater demand for water through more plant transpiration later in the season (Eberbach and Pala, 2005; Yang et al., 2021), which can be an issue if late-season drought events occur. The use of early maturing cultivars may also reduce the exposure to dehydration during reproductive stages (Richards, 1991; Semenov and Stratonovitch, 2013). Planophile canopies are also good candidates for dry environments, as they offer the advantage of greater light interception at lower *LAI* as well as faster ground cover (Richards et al., 2019).

With the projected increase in air temperature and CO_2 concentration, along with increased precipitation in some regions in the coming decades, greater *LAI* might be expected in crop canopies (Reza et al., 2020), leading to increased self-shading and water demand. The *LAI* value maximizing yield has been reported to be higher under elevated CO_2 (Srinivasan et al., 2016), likely due to the increase in RUE from reduced photorespiration (Slattery et al., 2013). The CO_2-induced increase in biomass can also cause an increase in growth and respiration as well as an increase in self-shading (Ryan, 1991; Leakey et al., 2009; Parry et al., 2010). In addition, rising air temperature will positively stimulate dark respiration (Li et al., 2021), which is suggested to reduce yields more than effects of temperature on photosynthesis (Li et al., 2021). It is unclear how canopies might be altered to minimize these effects of global change on leaf physiological processes, but perhaps light penetration deep into the canopy will be of increasing importance so that leaves maintain a positive carbon balance.

While increased air temperature can cause greater transpiration, elevated CO_2 has the opposite effect and was shown to lead to stomatal closure and reduced transpiration (Long et al., 2004; Bernacchi and VanLoocke, 2015). This suggests that optimum *LAI* for water-use efficiency in regions susceptible to drought with no access to irrigation systems will depend on the future climatic scenario. A reduction in leaf area and stomatal conductance at the top of the canopy has been suggested to reduce transpiration (Richards et al., 2002) and could be used to improve canopy water-use efficiency.

Elevated CO_2 can also alter leaf area duration. Accelerated leaf senescence in wheat (Zhu et al., 2009) and rice (Wang et al., 2020) has been reported, as the results of an increase in nitrogen sink demand (Sun et al., 2009). There are, however, contrasting reports showing delayed senescence in legumes (Dermody et al., 2006; Jin et al., 2017), which may be explained by enhanced nitrogen uptake from rhizobium symbioses. It is possible that early senescence in elevated CO_2 will be worse in a nitrogen-limited environment and in situations of increased nitrogen sink demands, shortening the duration of grain filling. Breeding for stay-green traits and improved nitrogen uptake could be important factors to maintain yield stability in the future.

6 Conclusion

As illustrated, the canopy light environment is a function of multiple canopy traits, whose interactions have profound effects on A_C and yield. Excess leaf area causes self-shading and increases respiration and water consumption, while suboptimal biomass production leads to insufficient light interception and reduced carbon and nitrogen storage. The optimum biomass will be a function of multiple leaf and canopy features as well as the environmental conditions and farming practices. For this reason, traits should be optimized with an integrative approach by considering the whole canopy, the environment, and management practices.

7 Future trends in research

Future research on crop canopy architecture is important for crop improvement (Murchie and Burgess, 2022). While traditionally it has been laborious and slow to measure canopy architecture, high throughput phenotyping of crop canopies using optical sensing-based platforms is increasingly used in research, breeding, and agriculture. Suites of sensors can rapidly identify morphological and physiological traits to maximize productivity (Sun et al., 2022), and improved sensors and more efficient processing of high throughput phenotyping data to identify features of crop canopies will undoubtedly be an important topic of research for the coming decades.

8 Acknowledegements

The authors thank Dr Jan Goudriaan and Dr Edward Lochocki for useful discussions and funding for the Realizing Increased Photosynthetic Efficiency (RIPE) project from the Bill & Melinda Gates Foundation, Foundation for Food and Agricultural Research (FFAR), and the UK Foreign Commonwealth & Development Office under grant no. OPP1172157.

9 Appendix

Yield and LAI relationship in defoliation studies used for the production of Fig. 2. LAI shows the LAI reported by the study, with the value in bracket indicating the percentage change in LAI relative to control after the defoliation treatment. Several rows for the same studies are different cultivars and/or treatment. Within a row, a comma separates measurements performed at different developmental stages. For convenience, some values reported in the table represent the average between replicated years when they show similar trends. Studies were identified from a search on Scopus with the following enquiry: *TITLE-ABS-KEY(LAI AND yield AND leaf AND area AND (defoliation OR removal OR excision)) on March 2020*. From 83 hits, only studies on soybean, maize, and wheat for which defoliation was done before (or close to) anthesis and where control had an LAI > 4.0 (i.e. LAI close to or above optimum for light interception) were considered.

Reference	LAI	Treatment	Yield (tons ha^{-1})
Soybean			
Haile et al. (1998b)	3.41 (−31.4) at R2, 5.22 (−11.1) at R4	Leaves removed by hand with a 2.5 LAI target at R2	4.9 (+6.5)
	3.52 (−33.4) at R2, 5.26 (−1.4) at R4		5.0 (+10.4)
	3.48 (−38.3) at R2, 5.62 (−7.3) at R4		3.5 (+25.4)
	3.05 (−26.6) at R2, 2.20 (−18.3) at R4		1.4 (−15.3)
	2.35 (−42.0) at R2, 1.68 (−28.4) at R4		0.7 (−37.9)
Haile et al. (1998b)	3.21 (−35.4) at R2, 5.39 (−8.2) at R4	Leaves removed by hand with a 1.5 LAI target at R2	4.3 (−6.3)
	2.65 (−49.9) at R2, 5.34 (−0) at R4		4.6 (+1.9)
	2.68 (−52.4) at R2, 6.24 (+3.0) at R4		3.1 (−11.1)
	2.32 (−44.2) at R2, 1.85 (−31.4) at R5		1.2 (−29.8)
	2.18 (−46.2) at R2, 1.63 (−30.6) at R5		0.7 (−35.7)
Haile et al. (1998a)	2.51 (−46.1) at R2, 5.02 (−6.8) at R4	Leaves removed by hand with a 2.5 LAI target at R2	2.9 (−0.7)
	3.51 (−37.9) at R2, 5.58 (−7.9) at R4		3.5 (+24.9)
	2.84 (−34.7) at R2, 5.36 (+7.4) at R4		3.4 (+5.5)
	3.12 (−36.9) at R2, 6.26 (+10.1) at R4		2.8 (+3.8)
	2.33 (−42.2) at R2, 1.65 (−29.8) at R5		0.7 (−37.9)
	2.31 (−42.4) at R2, 1.69 (−17.6) at R5		0.7 (−32.8)
	1.50 (−74.9) at R2, 3.39 (−34.5) at R5		2.5 (+11.8)
	1.97 (−68.1) at R2, 3.43 (−34.8) at R5		2.8 (+31.4)
	2.03 (−65.5) at R2, 3.51 (−28.3) at R5		2.8 (+25.9)
	2.14 (−64.2) at R2, 4.12 (−15.2) at R5		2.9 (+47.3)

Reference	LAI	Treatment	Yield (tons ha^{-1})
Haile et al. (1998a)	2.31 (−50.5) at R2, 5.33 (−1.2) at R4	Leaves removed by hand with a 1.5 LAI target at R2	2.9 (+1.7)
	2.72 (−51.8) at R2, 6.19 (+2.2) at R4		3.1 (+11.1)
	2.34 (−46.3) at R2, 4.49 (−10.1) at R4		3.1 (−5.2)
	2.82 (−43.0) at R2, 5.89 (+3.60) at R4		2.7 (+1.0)
	2.18 (−46.0) at R2, 1.63 (−30.6) at R5		0.7 (−35.7)
	2.05 (−49.0) at R2, 1.46 (−28.7) at R5		0.7 (−34.5)
	1.33 (−77.6) at R2, 3.35 (−35.2) at R5		2.6 (+13.7)
	1.30 (−78.9) at R2, 3.43 (−34.8) at R5		2.5 (+17.5)
	1.53 (−73.9) at R2, 3.21 (−34.3) at R5		2.5 (+14.9)
	1.37 (−77.1) at R2, 3.98 (−18.0) at R5		2.5 (+27.9)
Gustafson et al. (2006)	4.82 (−10.4) at R3	One leaflet per plant removed every 2 days	2.7 (−3.5)
Gustafson et al. (2006)	4.45 (−15.5) at R3	One leaflet per plant removed every day	2.3 (−15.9)
Srinivasan et al. (2016)	5.52 (−3.9) at R6	Emerging leaves	4.1 (+7.8)
Maize			
Singh and Nair (1975)	2.12 (−42.4) at R1, 1.84 (−31.6) at R2	Apical half of the leaf blade removed at R1	6.6 (−7.9)
	2.84 (−34.9) at R1, 1.97 (−41.2) at R2		7.0 (−9.1)
Subedi and Ma (2005)	3.48 (−26.9) at R1	All leaves below the earleaf removed at R1	9.6 (−9.2)
	1.83 (−61.7) at R1	All leaves above the earleaf removed at R1	1.1 (−88.9)
	4.32 (−9.4) at R1	Earleaf removed at R1	9.7 (−7.8)
	4.30 (−9.6) at R1	First leaf above the earleaf removed at R1	9.7 (−9.0)
	4.32 (−9.3) at R1	Second leaf above the earleaf removed at R1	9.3 (−11.8)
	4.37 (−8.2) at R1	Third leaf above the earleaf removed at R1	8.6 (−19.2)
	4.37 (−8.2) at R1	Fourth leaf above the earleaf removed at R1	9.2 (−11.2)
	4.42 (−7.1) at R1	Fifth leaf above the earleaf removed at R1	9.7 (−8.4)
	3.71 (−22.1) at R1	Sixth+ leaf above the earleaf removed at R1	6.8 (−35.2)
Liu et al. (2015)	6.70 (−5.4) at R1, 3.84 (+24.3) at R6	Two uppermost leaves removed at R1	17.1 (+13.5)
	5.96 (−15.9) at R1, 2.85 (−7.7) at R6	Four uppermost leaves removed at R1	11.1 (−26.7)
	4.71 (−33.5) at R1, 2.55 (−17.4) at R6	Six uppermost leaves removed at R1	6.6 (−56.1)

Reference	LAI	Treatment	Yield (tons ha⁻¹)
Wei et al. (2018)	5.87 (−5.8) at R1	Two uppermost leaves at R1	16.2 (+5.2)
	5.02 (−19.4) at R1	Four uppermost leaves at R1	13.7 (−1)
Raza et al. (2019)	6.02 (−6.0) at R2, 3.67 (+10.1) at R6	Two uppermost leaves removed at R1	8.4 (+11.2)
	5.40 (−15.7) at R2, 3.15 (−5.4) at R6	Four uppermost leaves removed at R1	6.8 (−10.3)
	4.52 (−29.4) at R2, 2.87 (−14.0) at R6	Six uppermost leaves removed at R1	6.1 (−18.6)
Liu et al. (2020)	8.28 (−5.8) at R1 7.41 (−7.4) at R1	Two uppermost leaves removed at R1	19.30 (+4.3) 19.6 (−0)
	7.44 (−15.4) at R1 6.11 (−23.6) at R1	Four uppermost leaves removed at R1	18.05 (−2.4) 18.6 (−5.1)
	5.61 (−36.2) at R1 6.37 (−20.4) at R1	Leaves below the third leaf below the ear removed at R1	19.15 (+3.5) 20.5 (+4.6)
	5.30 (−39.7) at R1 6.03 (−24.6) at R1	Two uppermost leaves and the leaves below the third leaf below the ear removed at R1	19.65 (+6.2) 20.35 (+3.8)
	4.23 (−51.9) at R1 4.70 (−41.2) at R1	Four uppermost leaves and the leaves below the third leaf below the ear removed at R1	17.05 (−7.8) 18.35 (−6.4)
Cereal			
Garcia del Moral (1992)	4.19 (−19.5) at anthesis 1.43 (−8.9) at maturity	Removing forage 2-3 cm above ground at Zadok 30	7.82 (−18.4)
	2.18 (−56.0) at anthesis, 0.62 (−56.6) at maturity	Removing forage 2-3 cm above ground at Zadok 30 and Zadok 31	4.89 (−37.5)

10 References

Álvaro, F., Isidro, J., Villegas, D., García Del Moral, L. F. and Royo, C. (2008). Breeding effects on grain filling, biomass partitioning, and demobilization in Mediterranean durum wheat, *Agron. J.* 100(2): 361-370.

Anderson, E. J., Ali, M. L., Beavis, W. D., Chen, P., Clemento, T. E., Diers, B. W., Graef, G. L., Grassini, P., Hyten, D. L., McHale, L. K., Nelson, R. L., Parrot, W. A., Patil, G. B., Stupar, R. M. and Tilmon, K. J. (2019). Soybean [*Glycine max* (L.) Merr.] breeding: history, improvement, production and future opportunities. In: Al-Khayri, J. M., Jain, S. and Johnson, D. V. (Eds), *Advances in Plant Breeding Strategies: Legumes*, 431-516. Springer International Publishing, Cham.

Andrade, J. F., Rattalino Edreira, J. I., Mourtzinis, S., Conley, S. P., Ciampitti, I. A., Dunphy, J. E., Gaska, J. M., Glewen, K., Holshouser, D. L., Kandel, H. J., Kyveryga, P., Lee, C. D., Licht, M. A., Lindsey, L. E., McClure, M. A., Naeve, S., Nafziger, E. D., Orlowski, J. M., Ross, J., Staton, M. J., Thompson, L., Specht, J. E. and Grassini, P. (2019). Assessing the influence of row spacing on soybean yield using experimental and producer survey data, *F. Crop. Res.* 230: 98–106.

Angus, J. F., Jones, R. and Wilson, J. H. (1972). A comparison of barley cultivars with different leaf inclinations, *Aust. J. Agric. Res.* 23(6): 945–957.

Anten, N. P. R. (2016). Optimization and game theory in canopy models. In: Hikosaka, K., Niinemets, Ü. and Anter, N. P. R. (Eds), *Canopy Photosynthesis: From Basics to Applications*, 355–377. Springer Netherlands, Dordrecht.

Archbold, H. K. (1942). Physiological studies in plant nutrition. XIII. Experiments with barley on defoliation and shading of the ear in relation to sugar metabolism, *Ann. Bot.* 4: 487–531.

Assefa, Y., Carter, P., Hinds, M., Bhalla, G., Schon, R., Jeschke, M., Paszkiewicz, S., Smith, S. and Ciampitti, I. A. (2018). Analysis of long term study indicates both agronomic optimal plant density and increase maize yield per plant contributed to yield gain, *Sci. Rep.* 8(1): 4937.

Austin, R. B., Bingham, J., Blackwell, R. D., Evans, L. T., Ford, M. A., Morgan, C. L. and Taylor, M. (1980). Genetic improvements in winter wheat yields since 1900 and associated physiological changes, *J. Agric. Sci.* 94(3): 675–689.

Baldocchi, D. D., Verma, S. B., Rosenberg, N. J., Blad, B. L. and Specht, J. E. (1985). Microclimate-plant architectural interactions: influence of leaf width on the mass and energy exchange of a soybean canopy, *Agric. Forest Meteorol.* 35(1–4): 1–20.

Balota, M., Payne, W. A., Evett, S. R. and Peters, T. R. (2008). Morphological and physiological traits associated with canopy temperature depression in three closely related wheat lines, *Crop Sci.* 48(5): 1897–1910.

Bangemann, L.-W., Sieling, K. and Kage, H. (2014). The effect of nitrogen and late blight on crop growth, solar radiation interception and yield of two potato cultivars, *F. Crop. Res.* 155: 56–66.

Barnard, D. M. and Bauerle, W. L. (2016). Seasonal variation in canopy aerodynamics and the sensitivity of transpiration estimates to wind velocity in broadleaved deciduous species, *J. Hydrol. Meteorol.* 17(12): 3029–3043.

Bastos, C., Whipple, S. D., Hoback, W. W. and Higley, L. G. (2011). Grasshopper injury to potato: consumption, effect on photosynthesis, and economic injury level, *Agron. J.* 103(6): 1655–1660.

Bernacchi, C. J. and VanLoocke, A. (2015). Terrestrial ecosystems in a changing environment: a dominant role for water, *Annu. Rev. Plant Biol.* 66(1): 599–622.

Bianchi, J. S., Quijano, A., Gosparini, C. O. and Morandi, E. N. (2020). Changes in leaflet shape and seeds per pod modify crop growth parameters, canopy light environment, and yield components in soybean, *Crop J.* 8(2): 351–364.

Blad, B. L. and Baker, D. G. (1972). Orientation and distribution of leaves within soybean canopies 1, *Agron. J.* 64(1): 26–29.

Blake, N. K., Lanning, S. P., Martin, J. M., Sherman, J. D. and Talbert, L. E. (2007). Relationship of flag leaf characteristics to economically important traits in two spring wheat crosses, *Crop Sci.* 47(2): 491–494.

Board, J. E. and Harville, B. G. (1992). Explanations for greater light interception in narrow- vs. wide-row, *Crop Sci.* 32(1): 198–202.

Board, J. E., Wier, A. T. and Boethel, D. J. (1997). Critical light interception during seed filling for insecticide application and optimum soybean grain yield, *Agron. J.* 89(3): 369-374.

Board, J. E. (2000). Light interception efficiency and light quality affect yield compensation of soybean at low plant populations, *Crop Sci.* 40(5): 1285-1294.

Board, J. E. (2004). Soybean cultivar differences on light interception and leaf area index during seed filling, *Agron. J.* 96(1): 305-310. doi:10.2134/agronj2004.3050.

Bonelli, L. E. and Andrade, F. H. (2020). Maize radiation use-efficiency response to optimally distributed foliar-nitrogen-content depends on canopy leaf-area index, *F. Crop. Res.* 247: 107557.

Boonman, A., Anten, N. P. R., Dueck, T. A., Jordi, W. J. R. M., van der Werf, A., Voesenek, L. A. C. J. and Pons, T. L. (2006). Functional significance of shade-induced leaf senescence in dense canopies: an experimental test using transgenic tobacco, *Am. Nat.* 168(5): 597-607.

Borrás, L., Slafer, G. A. and Otegui, M. E. (2004). Seed dry weight response to source-sink manipulations in wheat, maize and soybean: a quantitative reappraisal, *F. Crop. Res.* 86(2-3): 131-146.

Borrell, A. K. and Hammer, G. L. (2000). Nitrogen dynamics and the physiological basis of stay-green in sorghum, *Crop Sci.* 40(5): 1295-1307.

Borrell, A., Hammer, G. and Van Oostrerom, E. (2001). Stay-green: a consequence of the balance between supply and demand for nitrogen during grain filling?, *Ann. Appl. Biol.* 138: 91-95.

Bouchard, A., Vanasse, A., Seguin, P. and Bélanger, G. (2011). Yield and composition of sweet pearl millet as affected by row spacing and seeding rate, *Agron. J.* 103(4): 995-1001.

Boyer, J. S. (1982). Plant productivity and environment, *Science* 218(4571): 443-448.

Breseghello, F., de Morais, O. P., Pinheiro, P. V., Silva, A. C. S., da Maia de Castro, E., Guimarães, É. P., de Castro, A. P., Pereira, J. A., de Matos Lopes, A., Utumi, M. M. and de Oliveira, J. P. (2011). Results of 25 years of upland rice breeding in Brazil, *Crop Sci.* 51(3): 914-923.

Buckley, T. N., Cescatti, A. and Farquhar, G. D. (2013). What does optimization theory actually predict about crown profiles of photosynthetic capacity when models incorporate greater realism?, *Plant Cell Environ.* 36(8): 1547-1563.

Buttery, B. R. and Buzzell, R. I. (1972). Some differences between soybean cultivars observed by growth analysis, *Can. J. Plant Sci.* 52(1): 13-20.

Calderini, D. F., Dreccer, M. F. and Slafer, G. A. (1995). Genetic improvement in wheat yield and associated traits. A re-examination of previous results and the latest trends, *Plant Breed.* 114(2): 108-112.

Campbell, G. S. (1986). Extinction coefficients for radiation in plant canopies calculated using an ellipsoidal inclination angle distribution, *Agric. For. Meteorol.* 36(4): 317-321.

Carciochi, W. D., Schwalbert, R., Andrade, F. H., Corassa, G. M., Carter, P., Gaspar, A. P., Schmidt, J. and Ciampitti, I. A. (2019). Soybean seed yield response to plant density by yield environment in north America, *Agron. J.* 111(4): 1923-1932.

Chen, S., Liu, S., Yin, M., Zheng, X., Chu, G., Xu, C., Wang, D. and Zhang, X. (2020). Seasonal changes in crop growth and grain yield of different japonica rice cultivars in southeast China, *Agron. J.* 112(1): 215-227.

Cheng, S.-H., Cao, L.-Y., Zhuang, J.-Y., Chen, S.-G., Zhan, X.-D., Fan, Y.-Y., Zhu, D. and Min, S. (2007). Super hybrid rice breeding in China: achievements and prospects, *J. Integr. Plant Biol.* 49(6) : 805–810.

Ci, X., Li, M., Xu, J., Lu, Z., Bai, P., Ru, G., Liang, X., Zhang, D., Li, X., Bai, L., Xie, C., Hao, Z., Zhang, S. and Dong, S. (2012). Trends of grain yield and plant traits in Chinese maize cultivars from the 1950s to the 2000s, *Euphytica* 185(3): 395–406.

Ciotir, C., Applequist, W., Crews, T. E., Cristea, N., DeHaan, L. R., Frawley, E., Herron, S., Magill, R., Miller, J., Roskov, Y., Schlautman, B., Solomon, J., Townesmith, A., Van Tassel, D., Zarucchi, J. and Miller, A. J. (2019). Building a botanical foundation for perennial agriculture: global inventory of wild, perennial herbaceous Fabaceae species, *Plants People Planet* 1(4): 375–386.

Cober, E. R., Morrison, M. J., Ma, B. and Butler, G. (2005). Genetic improvement rates of short-season soybean increase with plant population, *Crop Sci.* 45(3): 1029–1034.

Cooper, R. L. and Jeffers, D. L. (1984). Use of nitrogen stress to demonstrate the effect of yield limiting factors on the yield response of soybean to narrow row systems 1, *Agron. J.* 76(2): 257–259.

Corlett, J. E., Jones, H. G., Massacci, A. and Masojidek, J. (1994). Water deficit, leaf rolling and susceptibility to photoinhibition in field grown sorghum, *Physiol. Plant.* 92: 423–430.

Cox, W. J. and Cherney, J. H. (2011). Growth and yield responses of soybean to row spacing and seeding rate, *Agron. J.* 103(1): 123–128.

D'Odorico, P., Emmel, C., Revill, A., Liebisch, F., Eugster, W. and Buchmann, N. (2019). Vertical patterns of photosynthesis and related leaf traits in two contrasting agricultural crops, *Funct. Plant Biol.* 46(3): 213–227.

Davidson, J. L. (1965). Some effects of leaf area control on the yield of wheat, *Aust. J. Agric. Res.* 16(5): 721–731.

De Bruin, J. L. and Pedersen, P. (2008). Effect of row spacing and seeding rate on soybean yield, *Agron. J.* 100(3): 704–710.

de Felipe, M., Gerde, J. A. and Rotundo, J. L. (2016). Soybean genetic gain in maturity groups III to V in Argentina from 1980 to 2015, *Crop Sci.* 56(6): 3066–3077.

Denison, R. F., Fedders, J. M. and Harter, B. L. (2010). Individual fitness versus whole-crop photosynthesis: solar tracking tradeoffs in alfalfa, *Evol. Appl.* 3(5–6): 466–472.

Deng, J., Ran, J., Wang, Z., Fan, Z., Wang, G., Ji, M., Liu, J., Wang, Y., Liu, J. and Brown, J. H. (2012). Models and tests of optimal density and maximal yield for crop plants, *Proc. Natl. Acad. Sci. U. S. A.* 109(39): 15823–15828.

Dermody, O., Long, S. P. and DeLucia, E. H. (2006). How does elevated CO_2 or ozone affect the leaf-area index of soybean when applied independently?, *New Phytol.* 169(1): 145–155.

Dermody, O., Long, S. P., McConnaughay, K. and DeLucia, E. H. (2008). How do elevated CO_2 and O_3 affect the interception and utilization of radiation by a soybean canopy?, *Glob. Change Biol.* 14(3): 556–564.

Dingkuhn, M., Jones, M. P., Johnson, D. E. and Sow, A. (1998). Growth and yield potential of *Oryza sativa* and *O. glaberrima* upland rice cultivars and their interspecific progenies, *F. Crop. Res.* 57: 57–69.

Dohleman, F. G. and Long, S. P. (2009). More productive than maize in the midwest: how does *Miscanthus* do it?, *Plant Physiol.* 150(4): 2104–2115.

Duncan, W. G. (1971). Leaf angles, leaf area, and canopy photosynthesis 1, *Crop Sci.* 11(4): 482–485.

Duvick, D. N. (1984). Genetic contributions to yield gains of U.S. Hybrid maize. In: Fehr, W. R. (Ed.). Genetic Contributions to Yield Gains of Five Major Crop Plants, 15-47. CSSA Special Publication No. 7. Crop Science Society of America, American Society of Agronomy, Madison, WI.

Duvick, D. N. and Cassman, K. G. (1999). Post-green revolution trends in yield potential of temperate maize in the north-central United States, *Crop Sci.* 39(6): 1622–1630.

Duvick, D. N. (2005). The contribution of breeding to yield advances in maize (*Zea mays* L.), *Adv. Agron.* 86: 83–145.

Eberbach, P. and Pala, M. (2005). Crop row spacing and its influence on the partitioning of evapotranspiration by winter-grown wheat in northern Syria, *Plant Soil* 268(1): 195-208.

Ehleringer, J. and Forseth, I. N. (1980). Solar tracking by plants, *Science* 210(4474): 1094–1098.

Emmel, C., D'Odorico, P., Revill, A., Hörtnagl, L., Ammann, C., Buchmann, N. and Eugster, W. (2020). Canopy photosynthesis of six major arable crops is enhanced under diffuse light due to canopy architecture, *Glob. Change Biol.* 26(9): 5164–5177.

Evans, L. T. and Fischer, R. A. (1999). Yield potential: it's definition, measurement, and significance, *Crop Sci.* 39(6): 1544–1551.

Evans, J. R. (2013). Improving photosynthesis, *Plant Physiol.* 162(4): 1780–1793.

Falster, D. S. and Westoby, M. (2003). Leaf size and angle vary widely across species: what consequences for light interception?, *New Phytol.* 158(3): 509–525.

Farquhar, G. D. (1989). Models of integrated photosynthesis of cells and leaves, *Phil. Trans. R. Soc. Lond.* 323: 357–367.

Feng, Y.-L., Cao, K.-F. and Feng, Z.-L. (2002). Thermal dissipation, leaf rolling and inactivation of PSII reaction centres in *Amomum villosum*, *J. Trop. Ecol.* 18(6): 865–876.

Field, C. (1983). Allocating leaf nitrogen for the maximization of carbon gain: leaf age as a control on the allocation program, *Oecologia* 56(2-3): 341–347.

Foulkes, M. J., Slafer, G. A., Davies, W. J., Berry, P. M., Sylvester-Bradley, R., Martre, P., Calderini, D. F., Griffiths, S. and Reynolds, M. P. (2010). Raising yield potential of wheat. III. Optimizing partitioning to grain while maintaining lodging resistance, *J. Exp. Bot.* 62(2): 469–486.

Foyer, C. H., Ruban, A. V. and Nixon, P. J. (2017). Photosynthesis solutions to enhance productivity, *Phil. Trans. R. Soc. B Biol. Sci.* 372(1730): 20160374.

Furbank, R. T., Quick, W. P. and Sirault, X. R. R. (2015). Improving photosynthesis and yield potential in cereal crops by targeted genetic manipulation: prospects, progress and challenges, *F. Crop. Res.* 182: 19–29.

Furbank, R. T., Sharwood, R., Estavillo, G. M., Silva-Perez, V., Condon, A. G. and Silva-Pérez, V. (2020). Photons to Food: genetic improvement of cereal crop photosynthesis, *J. Exp. Bot.* 71(7): 2226–2238.

Gao, J., Lei, M., Yang, L., Wang, P., Tao, H. and Huang, S. (2021). Reduced row spacing improved yield by optimizing root distribution in maize, *Eur. J. Agron.* 127: 126291.

García del Moral, L. F. (1992). Leaf area, grain yield and yield components following forage removal in Triticale, *J. Agron. Crop Sci.* 168(2): 100–107.

Gizlice, Z., Carter, T. E. and Burton, J. W. (1994). Genetic base for North American public soybean cultivars released between 1947 and 1988, *Crop Sci.* 34(5): 1143–1151.

Goudriaan, J. (2016). Light distribution. In: Hikosaka, K., Niinemets, Ü. and Anter, N. P. R. (Eds) *Canopy Photosynthesis: From Basics to Applications*, 3-22. Springer Netherlands, Dordrecht.

Gregersen, P. L., Culetic, A., Boschian, L. and Krupinska, K. (2013). Plant senescence and crop productivity, *Plant Mol. Biol.* 82(6): 603-622.

Grindlay, D. J. C. (1997). REVIEW Towards an explanation of crop nitrogen demand based on the optimization of leaf nitrogen per unit leaf area, *J. Agric. Sci.* 128(4): 377-396.

Gu, J., Zhou, Z., Li, Z., Chen, Y., Wang, Z., Zhang, H. and Yang, J. (2017a). Photosynthetic properties and potentials for improvement of photosynthesis in pale green leaf rice under high light conditions, *Front. Plant Sci.* 8: 1082.

Gu, J., Zhou, Z., Li, Z., Chen, Y., Wang, Z. and Zhang, H. (2017b). Rice (*Oryza sativa* L.) with reduced chlorophyll content exhibit higher photosynthetic rate and efficiency, improved canopy light distribution, and greater yields than normally pigmented plants, *F. Crop. Res.* 200: 58-70.

Gupta, S., Yadav, R., Gaikwad, K. B., Arora, A., Kumar, A., Kushwah, A. and Bainsla, N. K. (2017). Deciphering physiological basis of yield grain in Indian wheat cultivars, *Cereal Res. Commun.* 45(3): 512-524.

Gustafson, T. C., Knezevic, S. Z., Hunt, T. E. and Lindquist, J. L. (2006). Simulated insect defoliation and duration of weed interference affected soybean growth, *Weed Sci.* 54(4): 735-742.

Habermann, G., Machado, S. R., Guimarães, V. F. and Rodrigues, J. D. (2008). Leaf heliotropism in *Styrax camporum* Pohl from the Brazilian cerrado - distinct gas exchange and leaf structure, but similar leaf temperature and water relations, *Braz. J. Plant Physiol.* 20(1): 71-83.

Habermann, G., Ellsworth, P. F. V., Cazoto, J. L., Feistler, A. M., da Silva, L., Donatti, D. A. and Machado, S. R. (2011). Leaf paraheliotropism in *Styrax camporum* confers increased light use efficiency and advantageous photosynthetic responses rather than photoprotection, *Environ. Exp. Bot.* 71(1): 10-17.

Haile, F. J., Higley, L. G., Specht, J. E. and Spomer, S. M. (1998a). Soybean leaf morphology and defoliation tolerance, *Agron. J.* 90(3): 353-362.

Haile, F. J., Higley, L. G. and Specht, J. E. (1998b). Soybean cultivars and insect defoliation: yield loss and economic injury levels, *Agron. J.* 90(3): 344-352.

Hammer, G. L., Dong, Z., McLean, G., Doherty, A., Messina, C., Schussler, J., Zinselmeier, C., Paszkiewicz, S. and Cooper, M. (2009). Can changes in canopy and/or root system architecture explain historical maize yield trends in the U.S. corn belt?, *Crop Sci.* 49(1): 299-312.

Heatherly, L. G. and Welmore, R. W. (2004). Managing inputs for peak production. In: Boerma, H. R. and Specht, J. E. (Eds), *Soybeans: Improvement, Production, and Uses*, 451-536. CSSA, ASA-SSAJ. Madison Book Company, WI.

Heaton, E. A., Dohleman, F. G. and Long, S. P. (2008). Meeting US biofuel goals with less land: the potential of *Miscanthus*, *Glob. Change Biol.* 14(9): 2000-2014.

Heckathorn, S. A. and DeLucia, E. H. (1991). Effect of leaf rolling on gas exchange and leaf temperature of *Andropogon gerardii* and Spartina pectinate, *Bot. Gazette* 152(3): 263-268.

Herbert, T. J. (1991). Variation in interception of the direct solar beam by top canopy layers, *Ecology* 72(1): 17-22.

Hikosaka, K. (2014). Optimal nitrogen distribution within a leaf canopy under direct and diffuse light, *Plant Cell Environ.* 37(9): 2077-2085.

Hikosaka, K. (2016). Optimality of nitrogen distribution among leaves in plant canopies, *J. Plant Res.* 129(3): 299–311.

Hikosaka, K., Anten, N. P. R., Borjigidai, A., Kamiyama, C., Sakai, H., Hasegawa, T., Oikawa, S., Iio, A., Watanabe, M., Koike, T., Nishina, K. and Ito, A. (2016). A meta-analysis of leaf nitrogen distribution within plant canopies, *Ann. Bot.* 118(2): 239–247.

Hirotsu, N., Ujiie, K., Perera, I., Iri, A., Kashiwagi, T. and Ishimaru, K. (2017). Partial loss-of-function of NAL1 alters canopy photosynthesis by changing the contribution of upper and lower canopy leaves in rice, *Sci. Rep.* 7(1): 15958.

Hu, N., Lu, C., Yao, K. and Zou, J. (2009). Simulation on distribution of photosynthetically active radiation in canopy and optimum leaf rolling index in rice with rolling leaves, *Rice Sci.* 16(3): 217–225.

Jin, H., Liu, J., Song, B. and Xie, C. (2013). Impact of plant density on the formation of potato mimitubers derived from microtubers and tip-cuttings in plastic houses, *J. Integr. Agric.* 12(6): 1008–1017.

Jin, J., Liu, X., Wang, G., Mi, L., Shen, Z., Chen, X. and Herbert, S. J. (2010). Agronomic and physiological contributions to the yield improvement of soybean cultivars released from 1950 to 2006 in Northeast China, F. Crop. Res. 115(1): 116–123.

Jin, J., Li, Y., Liu, X., Wang, G., Tang, C., Yu, Z., Wang, X. and Herbert, S. J. (2017). Elevated CO_2 alters distribution of nodal leaf area and enhances nitrogen uptake contributing to yield increase of soybean cultivars grown in mollisols, *PLoS ONE* 12(5): e0176688.

Kadioglu, A., Terzi, R., Saruhan, N. and Saglam, A. (2012). Current advances in the investigation of leaf rolling caused by biotic and abiotic stress factors, *Plant Sci.* 182: 42–48.

Kahlon, C. S., Board, J. E. and Kang, M. S. (2011). An analysis of yield component changes for new vs. old soybean cultivars, *Agron. J.* 103(1): 13–22.

Khush, G. S. (1987). Rice breeding: past, present and future, *J. Genet.* 66(3): 195–216.

Khush, G. S. (1999). Green revolution: preparing for the 21st century, *Genome* 42(4): 646–655.

Kitaya, Y., Tsuruyama, J., Kawai, M., Shibuya, T. and Kiyota, M. (2000). Effects of air current on transpiration and net photosynthetic rates of plants in a closed plant production system. In: Kubota, C. and Chun, C. (Eds) *Transplant Production in the 21st Century*, 83–90. Kluwer Academic Publishers, Dordrecht.

Kitaya, Y., Tsuruyama, J., Shibuya, T., Yoshida, M. and Kiyota, M. (2003). Air current around single leaves and plant canopies and its effect on transpiration, photosynthesis, and plant organ temperatures. In: Kozai, T., Fujiwara, K. and Runkle, E. (Eds) *LED Lighting for Urban Agriculture*, 177–182. Springer, Singapore.

Koester, R. P., Skoneczka, J. A., Cary, T. R., Diers, B. W. and Ainsworth, E. A. (2014). Historical gains in soybean (*Glycine max* Merr.) seed yield are driven by linear increases in light interception, energy conversion, and partitioning efficiencies, *J. Exp. Bot.* 65(12): 3311–3321.

Kromdijk, J., Głowacka, K., Leonelli, L., Gabilly, S. T., Iwai, M., Niyogi, K. K. and Long, S. P. (2016). Improving photosynthesis and crop productivity by accelerating recovery from photoprotection, *Science* 354(6314): 857–861.

Kumudini, S., Hume, D. J. and Chu, G. (2001). Genetic improvement in short season soybeans. I. Dry matter accumulation, partitioning, and leaf area duration, *Crop Sci.* 41(2): 391–398.

Leakey, A. D. B., Ainsworth, E. A., Bernacchi, C. J., Rogers, A., Long, S. P. and Ort, D. R. (2009). Elevated CO_2 effects on plant carbon, nitrogen, and water relations: six important lessons from FACE, *J. Exp. Bot.* 60(10): 2859–2876.

Lee, C. D. (2006). Reducing row widths to increase yield: why it does not always work, *Crop Manag.* 5(1) doi:10.1094/CM-2006-0227-04-RV.

Lee, E. A. and Tollenaar, M. (2007). Physiological basis of successful breeding strategies for maize grain yield, *Crop Sci.* 47: S-202.

Leigh, A., Sevanto, S., Close, J. D. and Nicotra, A. B. (2017). The influence of leaf size and shape on leaf thermal dynamics: does theory hold up under natural conditions?, *Plant Cell Environ.* 40(2): 237–248.

Levizou, E. and Kyparissis, A. (2016). A novel pattern of leaf movement: the case of *Capparis spinosa* L., *Tree Physiol.* 36(9): 1117–1126.

Li, J., Xie, R. Z., Wang, K. R., Hou, P., Ming, B., Zhang, G. Q., Liu, G. Z., Wu, M., Yang, Z. S. and Li, S. K. (2018). Response of canopy structure, light interception and grain yield to plant density in maize, *J. Agric. Sci.* 156(6): 785–794.

Li, Q., Chen, Y., Liu, M., Zhou, X., Yu, S. and Dong, B. (2008). Effects of irrigation and planting patterns on radiation use efficiency and yield of winter wheat in North China, *Agric. Water Manag.* 95(4): 469–476.

Li, T., Heuvelink, E., Dueck, T. A., Janse, J., Gort, G. and Marcelis, L. F. M. (2014). Enhancement of crop photosynthesis by diffuse light: quantifying the contributing factors, *Ann. Bot.* 114(1): 145–156.

Li, G., Chen, T., Feng, B., Peng, S., Tao, L. and Fu, G. (2021). Respiration, rather than photosynthesis, determines rice yield loss under moderate high-temperature conditions, *Front. Plant Sci.* 12: 678653.

Lindquist, J. L., Arkebauer, T. J., Walters, D. T., Cassman, K. G. and Dobermann, A. (2005). Maize radiation use efficiency under optimal growth conditions, *Agron. J.* 97(1): 72–78.

Liu, G., Hou, P., Xie, R., Ming, B., Wang, K., Xu, W., et al. (2017). Canopy characteristics of high-yield maize with yield potential of 22.5 Mg ha^{-1}, *F. Crop. Res.* 213: 221–230.

Liu, G., Yang, Y., Liu, W., Guo, X., Xue, J., Xie, R., Ming, B., Wang, K., Hou, P. and Li, S. (2020). Leaf removal affects maize morphology and grain yield, *Agronomy* 10(2): 269.

Liu, J., Fan, Y., Ma, Y. and Li, Q. (2019). Response of photosynthetic active radiation interception, dry matter accumulation, and grain yield to tillage in two winter wheat genotypes, *Arch. Agron. Soil Sci.* 66(8): 1103–1114 doi:10.1080/03650340.2019.1657232.

Liu, T., Gu, L., Dong, S., Zhang, J., Liu, P. and Zhao, B. (2015). Optimum leaf removal increases canopy apparent photosynthesis, 13C-photosynthate distribution and grain yield of maize crops grown at high density, *F. Crop. Res.* 170: 32–39.

Liu, X., Jin, J., Wang, G. and Herbert, S. J. (2008). Soybean yield physiology and development of high-yielding practices in Northeast China, *F. Crop. Res.* 105: 157–171.

Long, S. P., Ainsworth, E. A., Rogers, A. and Ort, D. R. (2004). Rising atmospheric carbon dioxide: plants FACE the future, *Annu. Rev. Plant Biol.* 55: 591–628.

Long, S. P., Zhu, X. G., Naidu, S. L. and Ort, D. R. (2006). Can improvement in photosynthesis increase crop yields?, *Plant Cell Environ.* 29(3): 315–330.

Loomis, R. S. and Williams, W. A. (1969). Productivity and the morphology of crop stands: patterns with Leaves. In: Eastin, J. D., Haskins, F. A., Sullivan, C. Y. and Van Bavel, C. H. M. (Eds) *Physiological Aspects of Crop Yield*. 27–47. Am. Soc. Agron. Madison, WI.

Ma, D. L., Xie, R. Z., Niu, X. K., Li, S. K., Long, H. L. and Liu, Y. E. (2014). Changes in the morphological traits of maize genotypes in China between the 1950s and 2000s, *Eur. J. Agron.* 58: 1–10.

Malone, S., Herbert, D. A. Jr. and Holshouser, D. L. (2002). Relationship between leaf area index and yield in double-crop and full-season soybean systems, *J. Econ. Entomol.* 95(5): 945–951.

Mansfield, B. D. and Mumm, R. H. (2014). Survey of plant density tolerance in U.S. maize germplasm, *Crop Sci.* 54(1): 157–173.

Mantilla-Perez, M. B. and Salas Fernandez, M. G. (2017). Differential manipulation of leaf angle throughout the canopy: current status and prospects, *J. Exp. Bot.* 68(21–22): 5699–5717.

Martintello, P., Delocu, G., Boggini, G., Odoardi, M. and Stanca, A. M. (1987). Breeding progress in grain yield and selected agronomic characters of winter barley (Hordeum vulgare L.) over the Last Quarter of a Century, *Plant Breed.* 99(4): 289–294.

Maydup, M. L., Antonietta, M., Guiamet, J. J., Graciano, C., López, J. R. and Tambussi, E. A. (2010). The contribution of ear photosynthesis to grain filling in bread wheat (*Triticum aestivum* L.) , *F. Crop. Res.* 119: 48–58.

Meinzer, F. C., Goldstein, G., Jackson, P., Holbrook, N. M., Gutiérrez, M. V. and Cavelier, J. (1995). Environmental and physiological regulation of transpiration in tropical forest gap species: the influence of boundary layer and hydraulic properties, *Oecologia* 101(4): 514–522.

Monsi, M. and Saeki, T. (2005). On the factor light in plant communities and its importance for matter production. 1953, *Ann. Bot.* 95(3): 549–567.

Monteith, J. L. and Moss, C. J. (1977). Climate and the efficiency of crop production in Britain, *Phil. Trans. R. Soc. Lond. B* 281(980): 277–294.

Morgan, P. B., Bollero, G. A., Nelson, R. L., Dohleman, F. G. and Long, S. P. (2005). Smaller than predicted increase in aboveground net primary production and yield of field-grown soybean under fully open-air [CO_2] elevation, *Glob. Chang. Biol.* 11(10): 1856–1865.

Morrison, M. J., Voldeng, H. D. and Cober, E. R. (1999). Physiological changes from 58 years of genetic Improvement of short-season soybean cultivars in Canada, *Agron. J.* 91(4): 685–689.

Mourtzinis, S., Rattalino Edreira, J. I., Grassini, P., Roth, A. C., Casteel, S. N., Ciampitti, I. A., et al. (2018). Sifting and winnowing: analysis of farmer field data for soybean in the US North-Central region, *F. Crop. Res.* 221: 130–141.

Müller, M., Rakocevic, M., Caverzan, A. and Chavarria, G. (2017). Grain yield differences of soybean cultivars due to solar radiation interception, *Am. J. Plant Sci.* 08(11): 2795–2810.

Murchie, E.H., and Burgess, A.J. (2022) Casting light on the architecture of crop yield. *Crop and Environment* 1(1): 74–85.

Muryono, M., Chen, C. P., Sakai, H., Tokida, T., Hasegawa, T., Usui, Y., Nakamura, H. and Hikosaka, K. (2017). Nitrogen distribution in leaf canopies of high-yielding rice cultivar Takanari, *Crop Sci.* 57(4): 2080–2088.

Nar, H., Saglam, A., Terzi, R., Várkonyi, Z. and Kadioglu, A. (2009). Leaf rolling and photosystem II efficiency in *Ctenanthe setosa* exposed to drought stress, *Photosynthetica* 47(3): 429–436.

Niinemets, Ü. (1999), Research review. Components of leaf dry mass per area - thickness and density - alter leaf photosynthetic capacity in reverse directions in woody plants, *New Phytol.* 144(1): 35–47.

Nobel, P. S., Forseth, I. N. and Long, S. P. (1993). Canopy structure and light interception. In: Hall Scurlock, D. O., J. M. O., Bolhàr-Nordenkampf, H. R., Leegood, R. C. and Long, S. P. (Eds) *Photosynthesis and Production in a Changing Environment: A Field and Laboratory Manual*, 79–90. Springer Netherlands, Dordrecht.

Nobel, P. S. (2009). *Physicochemical and Environmental Plant Physiology. Physicochemical and Environmental Plant Physiology*. Academic Press, San Diego, CA.

Oplinger, E. S. and Philbrook, B. D. (1992). Soybean planting date, row width, and seeding rate response in three tillage systems, *J. Prod. Agric.* 5(1): 94–99.

Orlowski, J., Cox, W. J., Ditommaso, A. and Knoblauch, W. (2012). Planting soybean with a grain drill inconsistently increases yield and profit, *Agron. J.* 104(4): 1065-1073.

Orr, D. J., Pereira, A. M., da Fonseca Pereira, P., Pereira-Lima, I. A., Zsogon, A. and Araujo, W. L. (2017). *Engineering Photosynthesis: Progress and Perspectives, F1000Research 6(F1000 Faculty Rev)* (vol. 1891), doi: 10.12688/f1000research.12181.1. PMID: 29263782; PMCID: PMC5658708.

Ort, D. R. and Melis, A. (2011). Optimizing antenna size to maximize photosynthetic efficiency, *Plant Physiol.* 155(1): 79–85.

Parry, M. A. J., Reynolds, M., Salvucci, M. E., Raines, C., Andralojc, P. J., Zhu, X. G., Price, G. D., Condon, A. G. and Furbank, R. T. (2010). Raising yield potential of wheat. II. Increasing photosynthetic capacity and efficiency, *J. Exp. Bot.* 62(2): 453–467.

Pask, A. J. D., Sylvester-Bradley, R., Jamieson, P. D. and Foulkes, M. J. (2012). Quantifying how winter wheat crops accumulate and use nitrogen reserves during growth, *F. Crop. Res.* 126: 104–118.

Pearcy, R. W., Valladares, F., Wright, S. J. and de Paulis, E. L. (2004). A functional analysis of the crown architecture of tropical forest *Psychotria* species: do species vary in light capture efficiency and consequently in carbon gain and growth?, *Oecologia* 139(2): 163–177.

Pearcy, R. W., Muraoka, H. and Valladares, F. (2005). Crown architecture in sun and shade environments: assessing function and trade-offs with a three-dimensional simulation model, *New Phytol.* 166(3): 791–800.

Pedersen, P. (2008). *Row Spacing in Soybean*. Iowa State University, Extension.

Peng, S., Khush, G. S., Virk, P., Tang, Q. and Zou, Y. (2008). Progress in ideotype breeding to increase rice yield potential, *F. Crop. Res.* 108: 32–38.

Pepper, G. E., Pearce, R. B. and Mock, J. J. (1977). Leaf orientation and yield of maize 1, *Crop Sci.* 17(6): 883-886.

Perez, R. P. A., Fournier, C., Cabrera-Bosquet, L., Artzet, S., Pradal, C., Brichet, N., Chen, T. W., Chapuis, R., Welcker, C. and Tardieu, F. (2019). Changes in the vertical distribution of leaf area enhanced light interception efficiency in maize over generations of selection, *Plant Cell Environ.* 42(7): 2105–2119.

Perry, M. W. and D'Antuono, M. F. (1989). Yield improvement and associated characteristics of some Australian spring wheat cultivars introduced between 1860 and 1982, *Aust. J. Agric. Res.* 40(3): 457–472.

Pettigrew, W. T., Hesketh, J. D., Peters, D. B. and Woolley, J. T. (1989). Characterization of canopy photosynthesis of chlorophyll-deficient soybean isolines, *Crop Sci.* 29(4): 1025–1029.

Poehlman, J. M. (1987). Breeding soybeans. In: Poehlman, J. (Ed.) *Breeding Field Crops*, 421-450. Springer, Netherlands, Dordrecht.

Pommel, B., Gallais, A., Coque, M., Quilleré, I., Hirel, B., Prioul, J. L., Andrieu, B. and Floriot, M. (2006). Carbon and nitrogen allocation and grain filling in three maize hybrids differing in leaf senescence, *Eur. J. Agron.* 24(3): 203-211.

Prasad, P. V. V., Boote, K. J., Allen, L. H., Sheehy, J. E. and Thomas, J. M. G. (2006). Species, ecotype and cultivar differences in spikelet fertility and harvest index of rice in response to high temperature stress, *F. Crop. Res.* 95(2-3): 398-411.

Rajcan, I. and Tollenaar, M. (1999). Source:sink ratio and leaf senescence in maize: II. Nitrogen metabolism during grain filling, *F. Crop. Res.* 60: 255-265.

Rakocevic, M., Neumaier, N., Oliveira, G. Md de, Nepomuceno, A. L. and Farias, J. R. B. (2010). Heliotropic responses of soybean cultivars at three phenological stages and under two water regimes, *Pesq. Agropec. Bras.* 45(7): 661-670.

Raza, M. A., Feng, L. Y., Khalid, M. H. B., Iqbal, N., Meraj, T. A., Hassan, M. J., Ahmed, S., Chen, Y. K., Feng, Y. and Wenyu, Y. (2019). Optimum leaf excision increases the biomass accumulation and seed yield of maize plants under different planting patterns, *Ann. Appl. Biol.* 175(1): 54-68.

Reich, P. B., Falster, D. S., Ellsworth, D. S., Wright, I. J., Westoby, M., Oleksyn, J. and Lee, T. D. (2009). Controls on declining carbon balance with leaf age among 10 woody species in Australian woodland: do leaves have zero daily net carbon balances when they die?, *New Phytol.* 183(1): 153-166.

Reynolds, M. P., Acevedo, E., Sayre, K. D. and Fischer, R. A. (1994). Yield potential in modern wheat varieties: its association with a less competitive ideotype, *F. Crop. Res.* 37(3): 149-160.

Reza, M., Ali, R., Massah, R., Mostafa, B., Ali, J. and Stefan, B. (2020). Investigating the leaf area index changes in response to climate change (case study: Kasilian catchment, Iran), *SN Appl. Sci.* 2: 51.

Richard, B., Richter, G. M., Cerasuolo, M. and Shield, I. (2019). Optimizing the bioenergy water footprint by selecting SRC willow canopy phenotypes: regional scenario simulations, *Ann. Bot.* 124(4): 531-542.

Richards, R. A. (1991). Crop improvement for temperate Australia: future opportunities, *F. Crop. Res.* 26(2): 141-169.

Richards, R. A., Rebetzke, G. J., Condon, A. G. and Herwaarden, A. F. (2002). Breeding opportunities for increasing the efficiency of water use and crop yield in temperate cereals, *Crop Sci.* 42(1): 111-121.

Richards, R. A., Cavanagh, C. R. and Riffkin, P. (2019). Selection for erect canopy architecture can increase yield and biomass of spring wheat, *F. Crop. Res.* 244: 107649.

Rincker, K., Nelson, R., Specht, J., Sleper, D., Cary, T., Cianzio, S. R., Casteel, S., Conley, S., Chen, P., Davis, V., Fox, C., Graef, G., Godsey, C., Holshouser, D., Jiang, G.-L., Kantartzi, S. K., Kenworth, W., Lee, C., Mian, R., McHale, L., Naeve, S., Orf, J., Poysa, V., Schapaugh, W., Shannon, G., Uniatowski, R., Wang, D. and Diers, B. (2014). Genetic improvement of U.S. soybean in maturity groups II, III, and IV, *Crop Sci.* 0: 0.

Rowntree, S. C., Suhre, J. J., Weidenbenner, N. H., Wilson, E. W., Davis, V. M., Naeve, S. L., Casteel, S. N., Diers, B. W., Esker, P. D. and Conley, S. P. (2014). Physiological and phenological responses of historical soybean cultivar releases to earlier planting, *Crop Sci.* 54(2): 804-816.

Ryan, M. G. (1991). Effects of climate change on plant respiration, *Ecol. Appl.* 1(2): 157-167.

Sadok, W., Lopez, J. R., Zhang, Y., Tamang, B. G. and Muehlbauer, G. J. (2020). Sheathing the blade: significant contribution of sheaths to daytime and nighttime gas exchange in a grass crop, *Plant Cell Environ.* 43(8): 1844-1861.

Sadras, V. O. and Lawson, C. (2011). Genetic gain in yield and associated changes in phenotype, trait plasticity and competitive ability of South Australian wheat varieties released between 1958 and 2007, *Crop Pasture Sci.* 62(7): 533-549.

Sakamoto, T. and Matsuoka, M. (2004). Generating high-yielding varieties by genetic manipulation of plant architecture, *Curr. Opin. Biotechnol.* 15(2): 144-147.

Sanchez-Garcia, M., Álvaro, F., Peremarti, A., Trevaskis, B., Martín-Sánchez, J. A. and Royo, C. (2015). Breeding effects on dry matter accumulation and partitioning in Spanish bread wheat during the 20th century, *Euphytica* 203(2): 321-336.

Sariyeva, G., Kenjebaeva, S. and Lichtenthaler, H. (2009). Photosynthesis performance of non-rolling and flag leaf rolling wheat genotypes during temperature stress, *Acta Bot. Hung.* 51(1-2): 185-194.

Sarieva, G. E., Kenzhebaeva, S. S. and Lichtenthaler, H. K. (2010). Adaptation potential of photosynthesis in wheat cultivars with a capability of leaf rolling under high temperature conditions, *Russ. J. Plant Physiol.* 57(1): 28-36.

Sarlikioti, V., de Visser, P. H. B., Buck-Sorlin, G. H. and Marcelis, L. F. M. (2011). How plant architecture affects light absorption and photosynthesis in tomato: towards an ideotype for plant architecture using a functional-structural plant model, *Ann. Bot.* 108(6): 1065-1073.

Sayre, K. D., Rajaram, S. and Fischer, R. A. (1997). Yield potential progress in short bread wheats in Northwest Mexico, *Crop Sci.* 37(1): 36-42.

Schuepp, P. H. (1993). Tansley review no. 59 Leaf boundary layers, *New Phytol.* 125(3): 477-507.

Semenov, M. A. and Stratonovitch, P. (2013). Designing high-yielding wheat ideotypes for a changing climate, *Food Energy Secur.* 2(3): 185-196.

Setiyono, T. D., Weiss, A., Specht, J. E., Cassman, K. G. and Dobermann, A. (2008). Leaf area index simulation in soybean grown under near-optimal conditions, *Field Crops Res.* 108(1): 82-92.

Sheehy, J. E. and Mitchell., P. L. (2013). Designing rice for the 21st century: the three laws of maximum yield. In Discussion Paper Series 48. International Rice Research Institute, Los Nanos, p. 19.

Shibles, R. M. and Weber, C. R. (1966). Interception of solar radiation and dry matter production by various soybean planting patterns 1, *Crop Sci.* 6(1): 55-59.

Shuting, D., Changhao, H. and Rongqi, G. (1993). Rates of apparent photosynthesis, respiration and dry matter accumulation in maize canopies, *Biol. Plant.* 35(2): 273-277.

Shuting, D. (1994). Canopy apparent photosynthesis, respiration and yield in wheat, *J. Agric. Sci.* 122(1): 7-12.

Simpson, G. M. (1968). Association between grain yield per plant and photosynthetic area above the flag leaf node in wheat, *Can. J. Plant Sci.* 48(3): 253-260.

Simkin, A. J., López-Calcagno, P. E. and Raines, C. A. (2019). Feeding the world: improving photosynthetic efficiency for sustainable crop production, *J. Exp. Bot.* 70(4): 1119-1140.

Sinclair, T. R. and Horie, T. (1989). Leaf nitrogen, photosynthesis, and crop radiation use efficiency: a review, *Crop Sci.* 29(1): 90-98.

Sinclair, T. R. and Muchow, R. C. (1999). Radiation use efficiency. In: Sparks, D. L. (Ed.) *Advance in Agronomy*, 215-265. Academic Press.

Singh, R. P. and Nair, K. P. P. (1975). Defoliation studies in hybrid maize: II. Dry-matter accumulation, LAI, silking and yield components, *J. Agric. Sci.* 85(2): 247-254.

Slattery, R. A., Ainsworth, E. A. and Ort, D. R. (2013). A meta-analysis of responses of canopy photosynthetic conversion efficiency to environmental factors reveals major causes of yield gap, *J. Exp. Bot.* 64(12): 3723-3733.

Slattery, R. A. and Ort, D. R. (2015). Photosynthetic energy conversion efficiency: setting a baseline for gauging future improvements in important food and biofuel crops, *Plant Physiol.* 168(2): 383-392.

Slattery, R. A., VanLoocke, A., Bernacchi, C. J., Zhu, X. G. and Ort, D. R. (2017). Photosynthesis, light use efficiency, and yield of reduced-chlorophyll soybean mutants in field conditions, *Front. Plant Sci.* 8: 549.

Song, Q., Zhang, G. and Zhu, X. G. (2013). Optimal crop canopy architecture to maximise canopy photosynthetic CO_2 uptake under elevated CO_2 - a theoretical study using a mechanistic model of canopy photosynthesis, *Funct. Plant Biol.* 40(2): 108-124.

Specht, J. E. and Williams, J. H. (1984). Contribution of genetic technology to soybean productivity - retrospect and prospect. In: Fehr, W. R. (Ed.) *Genetic Contributions to Yield Gains of Five Major Crop Plants*, 49-74. Crop Science Society of America and American Society of Agronomy, Madison, WI.

Specht, J. E., Diers, B. W., Nelson, R. L., Francisco, J., de Toledo, F., Torrion, J. A., et al. (2014) Soybean, In Smith, S., Diers, B., Specht, J. and Carver, B. (Eds). *Yield Gains in Major U. S. Field Crops*, 311-356. American Society of Agronomy. Crop Science Society of America, and Soil Science Society of America, Madison, WI.

Srinivasan, V., Kumar, P. and Long, S. P. (2016). Decreasing, not increasing, leaf area will raise crop yields under global atmospheric change, *Glob. Chang. Biol.* 23(4): 1626-1635.

Stöckle, C. O. and Kemanian, A. R. (2009). Crop radiation capture and use efficiency: a framework for crop growth analysis. In: Sadras, V. and Calderini, D. (Eds) *Crop Physiology* (2nd edn.). *Applications for Genetic Improvement and Agronomy*, 145-170. Academic Press, San Diego.

Subedi, K. D. and Ma, B. L. (2005), Ear position, leaf area, and contribution of individual leaves to grain yield in conventional and leafy maize hybrids, *Crop Sci.* 45(6): 2246-2257.

Sun, D., Robbins, K., Morales, N., Shu, Q., and Cen, H. (2022) Advances in optical phenotyping of cereal crops. *Trends in Plant Science* 27(2): 191-208.

Sun, J., Yang, L., Wang, Y. and Ort, D. R. (2009). FACE-ing the global change: opportunities for improvement in photosynthetic radiation use efficiency and crop yield, *Plant Sci.* 177(6): 511-522.

Sun, Q., Liang, X., Zhang, D., Li, X., Hao, Z., Weng, J., Li, M. and Zhang, S. (2017). Trends in drought tolerance in Chinese maize cultivars from the 1950s to the 2000s, *F. Crop. Res.* 201: 175-183.

Tagliapietra, E. L., Streck, N. A., Rocha, T. S. M., Richter, G. L., Silva, M. R., Cera, J. C., Guedes, J. V. C. and Zanon, A. J. (2018). Optimum leaf area index to reach soybean yield potential in subtropical environment, *Agron. J.* 110(3): 932-938.

Takenaka, A. (1994). Effects of leaf blade narrowness and petiole length on the light capture efficiency of a shoot, *Ecol. Res.* 9(2): 109-114.

Tambussi, E. A., Bort, J., Guiamet, J. J., Nogués, S. and Araus, J. L. (2007). The photosynthetic role of ears in C3 cereals: metabolism, water use efficiency and contribution to grain yield, *CRC, Crit. Rev. Plant Sci.* 26(1): 1-16.

Taylor, H. M., Mason, W. K., Bennie, A. T. P. and Rowse, H. R. (1982). Responses of soybeans to two row spacings and two soil water levels. I. An analysis of biomass accumulation, canopy development, solar radiation interception and components of seed yield, *F. Crop. Res.* 5: 1-14.

Thomas, H. and Ougham, H. (2014). The stay-green trait, *J. Exp. Bot.* 65(14): 3889-3900.

Tian, Z., Jing, Q., Dai, T., Jiang, D. and Cao, W. (2011). Effects of genetic improvements on grain yield and agronomic traits of winter wheat in the Yangtze River basin of China, *F. Crop. Res.* 124(3): 417-425.

Tian, Z., Wang, X., Lee, R., Li, Y., Specht, J. E., Nelson, R. L., McClean, P. E., Qiu, L. and Ma, J. (2010). Artificial selection for determinate growth habit in soybean, *Proc. Natl. Acad. Sci. U. S. A.* 107(19): 8563-8568.

Tollenaar, M. (1991). Physiological basis of genetic improvement of maize hybrids in Ontario from 1959 to 1988, *Crop Sci.* 31(1): 119-124.

Tollenaar, M. and Dwyer, L. M. (1999). Physiology of maize. In: Smith, D. L. and Hamel, C. (Eds) *Crop Yield*, 169-204. Springer, Berlin, Heidelberg.

Tollenaar, M. and Wu, J. (1999). Yield improvement in temperate maize is attributable to greater stress tolerance, *Crop Sci.* 39(6): 1597-1604.

Tollenaar, M., Ahmadzadeh, A. and Lee, E. A. (2004). Physiological basis of heterosis for grain yield in maize, *Crop Sci.* 44(6): 2086-2094.

Tollenaar, M. and Lee, E. A. (2006). Dissection of physiological processes underlying grain yield in maize by examining genetic improvement and heterosis, *Maydica* 51: 399-408.

Trethowan, R. M., Reynolds, M. P., Ortiz-Monasterio, J. I. and Ortiz, R. (2007). The genetic basis of the green revolution in wheat production, *Plant Breed. Rev.* 28: 39 58.

Ustun, A., Allen, F. L. and English, B. C. (2001). Genetic progress in soybean of the U.S. Midsouth, *Crop Sci.* 41(4): 993-998.

van Oosterom, E. J., Chapman, S. C., Borrell, A. K., Broad, I. J. and Hammer, G. L. (2010). Functional dynamics of the nitrogen balance of sorghum. II. Grain filling period, *F. Crop. Res.* 115: 29-38.

Walker, A. P., Beckerman, A. P., Gu, L., Kattge, J., Cernusak, L. A., Domingues, T. F., Scales, J. C., Wohlfahrt, G., Wullschleger, S. D. and Woodward, F. I. (2014). The relationship of leaf photosynthetic traits - Vcmax and Jmax - to leaf nitrogen, leaf phosphorus, and specific leaf area: a meta-analysis and modeling study, *Ecol. Evol.* 4(16): 3218-3235.

Walker, B. J., Drewry, D. T., Slattery, R. A., VanLoocke, A., Cho, Y. B. and Ort, D. R. (2017). Chlorophyll can be reduced in crop canopies with little penalty to photosynthesis, *Plant Physiol.* 176(2): 1215-1232.

Wang, B., Cai, W., Li, J., Wan, Y., Li, Y., Guo, C., Wilkes, A., You, S.,Qin, X., Gao, Q. and Liu, K. (2020). Leaf photosynthesis and stomatal conductance acclimate to elevated [CO_2] and temperature thus increasing dry matter productivity in a double rice cropping system, *F. Crop. Res.* 248: 107735.

Wang, C., Wu, T., Sun, S., Xu, R., Ren, J., Wu, C., Jiang, B., Hou, W. and Han, T. (2016a). Seventy-five years of improvement of yield and agronomic traits of soybean cultivars released in the Yellow-Huai-Hai river valley, *Crop Sci.* 56(5): 2354-2364.

Wang, Y., Xi, W., Wang, Z., Wang, B., Xu, X., Han, M., Zhou, S. and Zhang, Y. (2016b). Contribution of ear photosynthesis to grain yield under rainfed and irrigation conditions for winter wheat cultivars released in the past 30 years in north China Plain, *J. Integr. Agric.* 15(10): 2247–2256.

Wang, T., Ma, X., Li, Y., Bai, D., Liu, C., Liu, Z., Tan, X., Shi, Y., Song, Y., Carlone, M., Bubeck, D., Bhardwaj, H., Jones, E., Wright, K. and Smith, S. (2011). Changes in yield and yield components of single-cross maize hybrids released in China between 1964 and 2001, *Crop Sci.* 51(2): 512–525.

Wardlaw, I. F. (1990). Tansley review no. 27 the control of carbon partitioning in plants, *New Phytol.* 116(3): 341–381.

Watson, D. J. (1947). Comparative physiological studies on the growth of field crops: I. Variation in net assimilation rate and leaf area between species and varieties, and within and between years, *Ann. Bot.* 11(1): 41–76.

Watson, D. J. (1952). The physiological basis of variation in yield. *Adv. Agron.* 4: 101–145.

Watson, D. J. (1958). The dependence of net assimilation rate on leaf-area index, *Ann. Bot.* 22(1): 37–54.

Wei, S., Wang, X., Jiang, D. and Dong, S. (2018). Physiological and proteome studies of maize (*Zea mays* L.) in response to leaf removal under high plant density, *BMC Plant Biol.* 18(1): 378.

Welles, J. M. (1990). Some indirect methods of estimating canopy structure, *Remote Sens. Rev.* 5(1): 31–43.

Werner, C., Correia, O. and Beyschlag, W. (1999). Two different strategies of Mediterranean macchia plants to avoid photoinhibitory damage by excessive radiation levels during summer drought, *Acta Oecol.* 20(1): 15–23.

Wu, T., Sun, S., Wang, C., Lu, W., Sun, B., Song, X., Han, X., Guo, T., Man, W., Cheng, Y., Niu, J., Fu, L., Song, W., Jiang, B., Hou, W., Wu, C. and Han, T. (2015). Characterizing changes from a century of genetic improvement of soybean cultivars in northeast China, *Crop Sci.* 55(5): 2056–2067.

Xiao, Y. G., Qian, Z. G., Wu, K., Liu, J. J., Xia, X. C., Ji, W. Q. and He, Z. H. (2012). Genetic gains in grain yield and physiological traits of winter wheat in Shandong Province, China, from 1969 to 2006, *Crop Sci.* 52(1): 44–56.

Xue, W., Lindner, S., Dubbert, M., Otieno, D., Ko, J., Muraoka, H., Werner, C. and Tenhunen, J. (2017). Supplement understanding of the relative importance of biophysical factors in determination of photosynthetic capacity and photosynthetic productivity in rice ecosystems, *Agric. For. Meteorol.* 232: 550–565.

Yang, W., Liu, W., Li, Y., Wang, S., Yin, L. and Deng, X. (2021). Increasing rainfed wheat yield by optimizing agronomic practices to consume more subsoil water in the loess plateau, *Crop J.* 9(6): 1418–1427.

Yates, M. J., Anthony Verboom, G., Rebelo, A. G. and Cramer, M. D. (2010). Ecophysiological significance of leaf size variation in Proteaceae from the Cape Floristic Region, *Funct. Ecol.* 24(3): 485–492.

Yuan, L. (1998). Hybrid rice breeding for super high yield. In: Denning, G. L. and Mew, T. W. (Eds) *China and IRRI: Improving China's Rice Productivity in the 21st Century* (vols. 10–12). International Rice Research Institute, Manila, Philippines.

Yuan, L. (2003). Recent progress in breeding super hybrid rice in China. In: Lu, Y. X. (Ed.), *Science Progress in China*, 231–236. Elsevier Science Ltd., Oxford.

Yuan, L. (2017). Progress in super-hybrid rice breeding, *Crop J.* 5(2): 100–102.

Zhang, L., Hu, Z., Fan, J., Zhou, D. and Tang, F. (2014). A meta-analysis of the canopy light extinction coefficient in terrestrial ecosystems, *Front. Earth Sci.* 8(4): 599-609.

Zheng, B., Ma, Y., Li, B., Guo, Y. and Deng, Q. (2009). Assessment of the effects of leaf angle combinations on potential photosynthesis capacity of rice with 3-D models using high performance computing. In *Plant Growth Modeling and Applications, International Symposium on*, 237-244. Beijing, China.

Zhu, C., Zhu, J., Zeng, Q., Liu, G., Xie, Z., Tang, H., Cao, J. and Zhao, X. (2009). Elevated CO_2 accelerates flag leaf senescence in wheat due to ear photosynthesis which causes greater ear nitrogen sink capacity and ear carbon sink limitation, *Funct. Plant Biol.* 36(4): 291-299.

Zhu, G., Peng, S., Huang, J., Cui, K., Nie, L. and Wang, F. (2016). Genetic improvements in rice yield and concomitant increases in radiation- and nitrogen-use efficiency in middle reaches of Yangtze River, *Sci. Rep.* 6: 21049.

Zhu, X. G., Long, S. P. and Ort, D. R. (2010). Improving photosynthetic efficiency for greater yield, *Annu. Rev. Plant Biol.* 61: 235-261.

Zhu, X. G., Ort, D. R., Parry, M. A. J. and von Caemmerer, S. (2020). A wish list for synthetic biology in photosynthesis research, *J. Exp. Bot.* 71(7): 2219-2225.

Ziems, J. R., Zechmann, B. J., Hoback, W. W., Wallace, J. C., Madsen, R. A., Hunt, T. E. and Higley, L. G. (2006). Yield response of indeterminate potato (*Solanum tuberosum* L.) to simulated insect defoliation, *Agron. J.* 98(6): 1435-1441.

Part 3

Improving photosynthesis: optimising chloroplast function/light conversion

Chapter 8

Modifying photorespiration to optimize crop performance

Xinyu Fu, Kaila Smith, Luke Gregory, Ludmila Roze and Berkley Walker, Michigan State University, USA

1 Introduction

Photorespiration is a major component of plant central metabolism in most crop plants and is responsible for large inefficiencies in photosynthesis. Photorespiration recycles the intermediates produced when rubisco, the initial enzyme of carbon fixation, catalyzes a reaction with oxygen instead of CO_2. Photorespiratory recycling is metabolically expensive, requiring 30–40% of plant energy under ambient light conditions and reducing net carbon assimilation by ~20% under typical growing conditions (Walker et al., 2016b, Sharkey, 1988). Photorespiration imposes these limitations in all crop plants that have not evolved mechanisms to concentrate CO_2 concentrations around rubisco to minimize oxygenation reactions using four-carbon intermediates (C_4 photosynthesis). Plants lacking the ability to concentrate CO_2 (and therefore with high rates of photorespiration) are classified as C_3 plants and include rice, wheat, potato, soybean, and most vegetables and fruits consumed around the world.

Our goals for this chapter are to introduce the core features of photorespiration and highlight the current state of the art in approaches to minimize the effects of this critical process in crop plants. There are many

http://dx.doi.org/10.19103/AS.2022.0119.12

excellent reviews highlighting various approaches to minimizing energy and carbon loss due to photorespiration, and here we extend these with some unique approaches to consider for 'next generation' improvements to this vital process (South et al., 2018, Betti et al., 2016, Kubis and Bar-Even, 2019). We also contend that future engineering efforts require a careful re-thinking of the root cause and design of successful photorespiratory bypasses and that untapped potential may lie in approaching photorespiration under non-steady-state conditions.

2 Photorespiration: the good, the bad and the inevitable

The basic biochemistry, physiology, and genes of photorespiration are well known and have been the subject of many excellent reviews (Fig. 1, Bauwe et al., 2010, Walker et al., 2016b, Busch, 2020, Ogren, 1984). The need for photorespiration was clear following the discovery that rubisco catalyzed a reaction between ribulose 1,5-bisphosphate (RuBP) and oxygen to produce the inhibitory molecule 2-phosphoglycolate (2-PG) in addition to the typical C_3 cycle intermediate 3-phosphoglycerate (3-PGA) (Ogren and Bowes, 1971). 2-PG is inhibitory to the C_3 cycle enzymes triosephosphate isomerase and sedoheptulose-1,7-bisphosphate phosphatase and represents a waste of previously fixed carbon, making it essential for the plant to recycle 2-PG (Anderson, 1971, Flügel et al., 2017). Resolving the biochemistry that recycles 2-PG back into intermediates of the C_3 cycle was one of the first major successes of forward genetics using *Arabidopsis thaliana* as a model organism (Somerville and Ogren, 1979, 1980a,b, 1981, Somerville, 2001). This core biochemistry is now well-resolved (save some interesting exceptions discussed below) but understanding the implications of photorespiration to plant productivity is less clear.

Often photorespiration is referred to in a negative sense, due to its high demands for energy and massive carbon loss, but viewed from a different angle photorespiratory metabolism is beneficial. While it is true that decreasing photorespiration by growing plants under elevated CO_2 often improves growth and yield (Ainsworth and Rogers, 2007, Ainsworth and Long, 2021), it is also true that disrupting flux through photorespiration by knocking out key genes is lethal or severely detrimental to growth under ambient CO_2 and oxygen concentrations (Foyer et al., 2009, Timm and Bauwe, 2013). Thus, photorespiration serves a metabolically expensive, yet vital, plant function stemming from rubisco oxygenation. In this sense, a clear distinction should be made between the source of the problem (rubisco oxygenation) and the particular solution that has evolved to handle the problem (photorespiration).

In addition to a vital role in detoxifying 2-PG, photorespiration is linked to beneficial processes like nitrate assimilation. Rates of nitrate assimilation

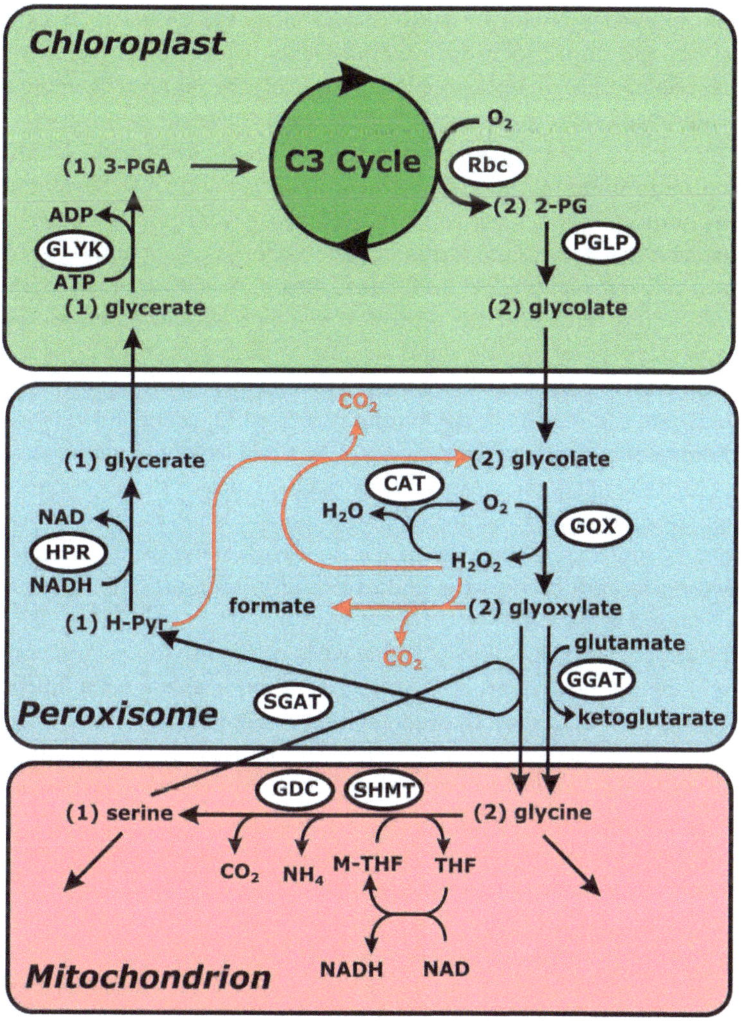

Figure 1 Simplified biochemical pathway of photorespiration illustrating the major metabolites and coordination between the chloroplast, peroxisome, and mitochondrion. Enzymatic reactions are shown in black, and putative non-enzymatic reactions are shown in red. The number in parenthesis shows the stoichiometry of each reaction, and the gene names are shown in the white ovals. Rbc, rubisco; PGLP, 2-phosphoglycolate phosphatase; GOX, glycolate oxidase; CAT, catalase; GGAT, glycine:glyoxylate aminotransferase; SHMT, serine hydroxymethyltransferase; GDC, glycine decarboxylase; SGAT, serine:glyoxylase aminotransferase; HPR, hydroxypyruvate reductase; GLYK, glycerate kinase.

decrease as rates of photorespiration relative to carbon fixation decrease in both wheat and *Arabidopsis* (Bloom et al., 2010, Bloom, 2006, Rachmilevitch et al., 2004). The mechanisms linking photorespiration to nitrate assimilation are not clear but could relate to increased cytosolic reductant concentrations available to reduce nitrate to nitrite due to increased shuttling of reducing equivalents, the surplus of reducing equivalents driven by the higher ATP to reductant demand of photorespiration, or possibly the removal of amino acids from photorespiration as discussed further below (Walker et al., 2020, Zhao et al., 2021). The link between photorespiration and nitrate assimilation may become more important in crop systems as increasing CO_2 concentrations reduce photorespiratory rates, which may decrease rates of nitrate assimilation and restrict crop growth. Future research should focus on confirming the exact nature of this mechanism and if this describes an initial response to altered photorespiratory rates or if this reduction in nitrate assimilation is long-term. This knowledge is vital to understand how plant nitrogen status may respond to future climates and possibly re-engineer this response to maintain or increase rates of nitrate assimilation under current or future CO_2 concentrations.

Photorespiration is also proposed to play a role in light energy dissipation when CO_2 is limiting to photosynthesis such as under drought stress and stomatal closure. Under this hypothesis, reducing equivalents produced in the light reactions are dissipated by the energy demand of photorespiration to prevent over-reduction of the photosynthetic apparatus (Pieruschka et al., 2006, Osmond et al., 1997).

The above discussion highlights that while photorespiration is responsible for high rates of energy and carbon loss, it performs a vital plant function. It also highlights the nuance between rubisco oxygenation and downstream photorespiration. For this chapter, we will ignore the important efforts to minimize rates of photorespiration discussed elsewhere in this volume (e.g. transferring C_4 traits into C_3 crops or improving rubisco specificity for CO_2) and treat rubisco oxygenation as inevitable. In this, we recognize that approaches to minimize rubisco oxygenation contain great promise but are outside of the purview of this chapter focused on the reactions downstream of rubisco oxygenation. We also do not consider the unique adaptation of C_2 photosynthesis and efforts to engineer this modification to photorespiration, where CO_2 release is localized into the bundle sheath for enhanced re-fixation (Lundgren, 2020, Khoshravesh et al., 2016). In the next sections, we will further break down the biochemistry of photorespiration to illustrate where the inefficiencies are and how they might be exploited to engineer higher rates of net photosynthesis, as well as discuss the future of photorespiration with climate change.

2.1 Where are carbon and energy lost during photorespiration?

Photorespiration requires massive amounts of previously reduced carbon and plant energy, but where do these losses specifically occur? Carbon loss is generally assumed to occur entirely during glycine decarboxylation, where two, two-carbon glycine molecules are converted to a single, three-carbon, serine with the loss of a single CO_2 and production of one NADH (Fig. 1). In this sense, photorespiration is 75% efficient at maintaining the carbon originally present in the initial 2-PG. The energy demand of photorespiration is more complex but revolves around the net energy demands of photorespiration and subsequent energy needed to regenerate RuBP back into the C_3 cycle (Edwards and Walker, 1983, von Caemmerer, 2000). The direct energy-requiring steps of photorespiration are the energy used to convert the 3-PGA from the oxygenation reaction to GAP, the ATP required to refix the released NH_4^+ into organic forms, and the ATP required to convert glycerate into 3-PGA and that 3-PGA into GAP. A reductant is required to refix the released NH_4^+ and reduce hydroxypyruvate, but this cost is assumed to be offset by the NADH generated from the glycine decarboxylase complex. When these energy demands are combined with the needs to process 3-PGA into RuBP, photorespiration requires net 3.5 ATP and 2 NADPH equivalents.

These energy and carbon requirements assume that carbon is cycled entirely from 2-PG and 3-PGA to RuBP. If the photorespiratory flux is partial, these losses are decreased. For example, there is clear evidence that serine and, to a lesser extent, glycine involved in photorespiration leave the cycle (Busch et al., 2018, Harley and Sharkey, 1991). The exit of serine and glycine decrease the energy needed to process 2-PG, and in the case of glycine, the amount of CO_2 lost per rubisco oxygenation. Indeed the most efficient detoxification of 2-PG might occur in a transient in mutants of serine hydroxymethyltransferase (a component of the glycine decarboxylase complex) (Somerville and Ogren, 1981). These mutants are unable to decarboxylate glycine and survive under photorespiratory conditions by accumulating high glycine pools. This strategy avoids glycine decarboxylation and the energy costs of NH_4^+ re-fixation and RuBP regeneration. This transient build-up of glycine is not sustainable since it requires continual de novo assimilation of NH_4^+ and results in continual production of glycine, but it illustrates the possible benefits of non-steady-state efficiency gains in photorespiration.

2.2 How will future climates affect photorespiration?

Rates of rubisco oxygenation directly respond to the two main effects of climate change, increasing CO_2 and temperature. On the one hand, elevated CO_2 decreases rates of rubisco oxygenation relative to carboxylation, decreasing necessary rates of photorespiration. This primary effect results generally in improved crop productivity under elevated CO_2, especially under conditions where CO_2 availability is limited (Ainsworth and Rogers, 2007, Ainsworth and Long, 2021). CO_2 availability is limited under lower-moisture conditions when stomata close to conserve water and restrict CO_2 diffusion into the intercellular airspace. On the other hand, increasing temperatures increase relative rates of photorespiration since the specificity of rubisco for CO_2 decreases as temperature increases and the solubility ratio of CO_2/O_2 decreases (Brooks and Farquhar, 1985). The net impact of climate change on photorespiration will depend then on the absolute degree of increases in temperature and CO_2. Under currently predicted climate change scenarios, there is a consensus that photorespiration will be reduced under future climates but still play a large role in plant metabolism and energetics. For example, when several combinations of CO_2 and temperature scenarios presented in the 2007 Intergovernmental Panel on Climate Change were used to parameterize an earth-systems model of photosynthesis, photorespiration universally decreased, but the magnitude of this decrease varied based on the specific scenario (IPCC, 2007, Walker et al., 2016b). In other words, the increased CO_2 concentrations reduced photorespiration more than the increases in temperature increased photorespiration, but we note that this depends on the specific physiological response of the plant to the modified atmosphere (Dusenge et al., 2019).

3 Recent efforts to improve photorespiration

Because photorespiration requires a large portion of energy and reduces net carbon assimilation, it has been identified as a clear target for crop improvement (Betti et al., 2016). The benefits are potentially massive; simulations estimate that ~150 trillion calories are lost annually to photorespiration and that a 5% reduction in photorespiratory loss would be worth ~$500 million in the US Midwest alone (Walker et al., 2016b). These benefits extend not only to food production but also to the sustainability of our food systems. For example, 30 times more energy is used by photorespiration as compared to the energy required to produce the fertilizer applied during the cultivation of potatoes (Roney and Walker, 2021). While improvements to photorespiration and photosynthesis, in general, do not directly offset carbon-producing human energy supplements, they provide an opportunity to grow more food using the same inputs, effectively lowering the carbon footprint per calorie. Recent efforts to improve photorespiration can be classified generally into efforts to speed flux through photorespiration or to create novel 'photorespiratory bypasses' that funnel 2-PG through alternative synthetic pathways (Kubis and Bar-Even, 2019, South et al., 2018).

3.1 Keeping traffic moving to prevent inhibition

Increasing flux capacity through photorespiration increases net photosynthetic rates in *Arabidopsis* and tobacco (Strategy shown in Fig. 2). Specifically, when glycine decarboxylase activity is increased, carbon fixation and growth rates

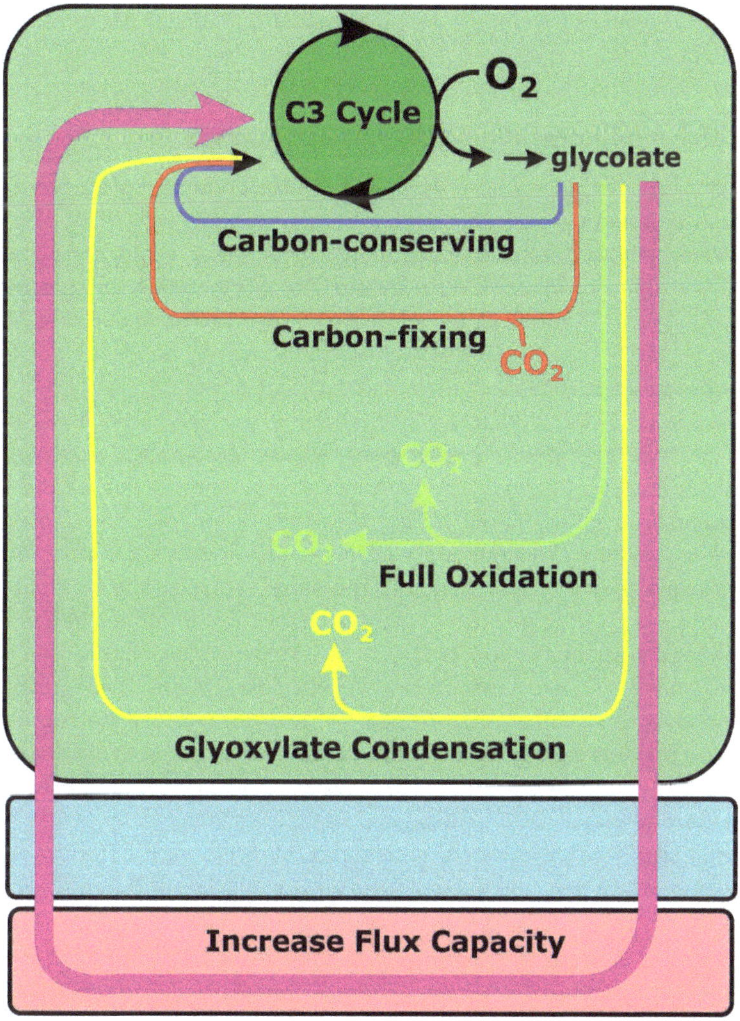

Figure 2 Conceptual approaches for increasing the efficiency of photorespiration. Featured are the five basic approaches to improving photorespiratory flux and efficiency, as discussed in more detail in the text with the pathway title as named in the main text. Organelles are unlabeled but follow the same pattern and color as in Figure 1.

increase substantially (Timm et al., 2012a, López-Calcagno et al., 2019). This increase in net photosynthesis likely occurs when inhibitory intermediates of photorespiration, like 2-PG, are processed more quickly and are therefore less inhibitory to the C_3 cycle (discussed in more detail in the next section). With less inhibition, the C_3 cycle and rubisco carboxylation is maintained at a higher rate resulting in higher rates of net CO_2 fixation and growth. Interestingly, minimizing the pool sizes of inhibitory intermediates of photorespiration may also help explain at least some of the apparent benefits of photorespiratory bypass lines.

3.2 Current photorespiratory bypass approaches

Many efforts to improve photorespiration involve engineering a novel set of reactions that can detoxify glycolate via alternative pathways (Fig. 2). These alternative pathways are referred to generally as 'photorespiratory bypasses' and have resulted in higher productivity and photosynthesis in a variety of model and crop species (Kebeish et al., 2007, Nölke et al., 2014, Maier et al., 2012, Dalal et al., 2015, South et al., 2019). Some pathways rely on the condensation and subsequent reduction of glyoxylate to form glycerate, which can enter the C_3 cycle following 2-PG to 3-PGA (Kebeish et al., 2007, Dalal et al., 2015). These 'glyoxylate condensation' bypasses release the same amount of carbon as native photorespiration but require less energy and do not require NH_4^+ release and re-assimilation. This carbon release also is moved to the chloroplast, which increases the chance that the CO_2 is re-assimilated and not lost to the intercellular airspace (Xin et al., 2014). Therefore, for glyoxylate condensation bypasses to work from a carbon balance perspective, photorespiration must natively leak substantial amounts of CO_2 into the intercellular airspace. It is not clear how much CO_2 is lost to the intercellular airspace when it is released in the mitochondria from native photorespiration, with some evidence suggesting high rates of release while others argue for low or modest rates of release and effective recapture (Busch et al., 2013, Walker and Ort, 2015, Walker et al., 2016a, Berghuijs et al., 2017, Tholen and Zhu, 2011, Tholen et al., 2012). These pathways also have various energy demands in the form of ATP or reducing equivalents which can make them more or less efficient from an energetics perspective (Xin et al., 2014).

In other classes of photorespiratory bypasses, carbon from glyoxylate is fully oxidized and decarboxylated in the chloroplast resulting in increased net CO_2 assimilation (South et al., 2019, Maier et al., 2012). The benefits of these 'full oxidation' bypasses are not clear from a carbon balance perspective since 100% of the carbon produced as 2-PG would be lost as CO_2 and require additional energy to be re-fixed, even if that re-fixation was more efficient when the carbon release occurs in the chloroplast. These schemes do not produce a simulated benefit from a carbon balance perspective and are actually predicted to reduce rates of net fixation when considered on a carbon balance basis (Xin et al., 2014).

One hypothesis is that glyoxylate condensation and/or full oxidation bypasses target the same mechanism of improvement as improvements to flux capacity outlined above. Perhaps these bypasses all serve to minimize pools of inhibitory intermediates of photorespiration and maintain higher rates of C_3

cycle activity. The higher rates of C_3 cycle activity could be large enough to offset the carbon-negative aspects of cycles exhibited in plants that appear to fully oxidize glyoxylate. Interestingly, partial expression of glyoxylate condensation pathways also increases net assimilation and growth, suggesting that other mechanisms (not simply increasing carbon recapture) explain the increases in net photosynthesis (Kebeish et al., 2007, Nölke et al., 2014). It is also argued that glyoxylate concentration might be beneficial to net carbon fixation and that pathways that increase glyoxylate concentration in the chloroplast may even suppress rates of rubisco oxygenation (Oliver and Zelitch, 1977, Oliver, 1980, Kubis and Bar-Even, 2019). Clearly, more research is needed to understand the exact mechanism by which these seemingly carbon-losing pathways result in higher rates of net fixation and growth, allowing us to target these mechanisms more fundamentally for crop improvement.

3.3 Can a photorespiratory bypass be carbon neutral or even carbon fixing?

'Carbon-conserving' photorespiratory bypasses schemes seek to minimize the issues of CO_2 release and recapture entirely by maintaining all carbon present originally in 2-PG. One proposed pathway involves the reduction of 2-PG and subsequent condensation with dihydroxyacetone phosphate to produce the C_3 cycle precursor for xylulose 5-phosphate (Ort et al., 2015). However, it has been noted that this pathway contains highly reversible reactions and the production of the inhibitory molecule xylulose bisphosphate (Kubis and Bar-Even, 2019, Yokota, 1991, Zhu and Jensen, 1991, Parry et al., 2008). Several carbon-conserving photorespiratory bypass schemes are predicted to be more favorable and involve the reduction of glycolate and ultimate aldol condensation to enter the C_3 cycle (Bar-Even, 2018, Trudeau et al., 2018). These schemes required the engineering of novel enzymes and resulted in the demonstration of in vitro photorespiratory schemes that do not release CO_2 (Trudeau et al., 2018).

Some proposed carbon-conserving bypass lines go a step further, actually fixing carbon as part of their activities. One implementation of a 'carbon-fixing' bypass showed no phenotype in cyanobacteria (Shih et al., 2014), while another resulted in elevated bicarbonate assimilation also in cyanobacteria (Yu et al., 2018). Rates of photorespiration are low in cyanobacteria, due to their functional carbon concentrating mechanisms. Therefore, it will be interesting to see how these bypasses perform in C_3 plants, which have orders of magnitude more flux through photorespiration.

4 How can photorespiration beat the heat?

Crop systems currently experience relatively high temperatures compared to laboratory conditions, and high-temperature episodes will only increase under future climate conditions. For example, growing temperatures in the US Midwest regularly reach 35°C, very different from the 'standard' 25°C often used to evaluate plant performance in the lab. The most recent Intergovernmental Panel on Climate Change estimates a global increase of 1.4–5.7°C by 2100, indicating that the ability of photorespiration to handle flux under elevated temperatures will become more important in the future climates (Masson-Delmotte et al., 2021). The next section will focus on specific reactions of photorespiration that are promising targets to improve the temperature response of net photosynthesis, an area of obvious interest to increasing crop productivity under present and future conditions.

4.1 2-phosphoglycolate-glycolate metabolism

Insufficient 2-PG phosphatase activities at elevated temperatures may restrict flux through photorespiration and result in the build-up of 2-PG and inhibition of the C_3 cycle and net CO_2 assimilation (Anderson, 1971, Kelly and Latzko, 1976). Additionally, 2-PG pools are high in phosphate, possibly decreasing available free phosphate from the sugar-phosphates of the C_3 cycle and also limiting the rate of net CO_2 assimilation (Timm et al., 2019, Sharkey, 1985, Harley and Sharkey, 1991, Yang et al., 2016). *Arabidopsis* lines overexpressing 2-PG phosphatase have higher rates of net photosynthesis at elevated temperatures compared to wild type lines, presumably because of the faster degradation of 2-phophoglycolate (Flügel et al., 2017, Timm et al., 2019). Therefore, maintaining high capacity for 2-PG cycling into downstream photorespiration may be important for minimizing inhibitory consequences at elevated temperatures and maintaining higher rates of net photosynthesis.

4.2 Glycolate oxidase

Elevated glycolate oxidase activities may also improve the photosynthetic efficiency of crops by maintaining carbon flux through photorespiration at elevated temperatures. If a higher level of peroxisomal glycolate oxidase is not available to degrade glycolate, large pools of glycolate can accumulate, which have been linked to the inhibition of rubisco (González-Moro et al., 1997, Wendler et al., 1992). Indeed, glycolate oxidase suppression lines show a photorespiratory phenotype, most likely from a severe reduction in assimilation rate (Xu et al., 2009, Cui et al., 2016, Lu et al., 2014). However, rice lines overexpressing glycolate oxidase maintain a higher photosynthetic rate under high temperatures compared to WT lines (Cui et al., 2016).

4.3 Exploiting the role of catalase in minimizing wasteful inefficiencies

It has long been hypothesized that excess hydrogen peroxide (H_2O_2) in the peroxisome can decrease the efficiency of photorespiration through non-enzymatic decarboxylation reactions. H_2O_2 is generated from the oxidation of glycolate and is broken down in the peroxisome by catalase. If catalase activity is insufficient, H_2O_2 can react with the photorespiratory intermediates glyoxylate and/or hydroxypyruvate and release excess CO_2, reducing the carbon recycling efficiency of photorespiration (Cousins et al., 2008, 2011, Keech et al., 2012, Grodzinski and Butt, 1976, Bao et al. 2021). Mutant analysis of photorespiratory genes indicates that under laboratory conditions, the enzymatic decarboxylation of photorespiratory intermediates predominates, but it is unclear if the efficiency of photorespiration is maintained under stress conditions. For example, combined measurements of gas exchange and rubisco biochemistry indicate that the stoichiometric amount of CO_2 released from photorespiration may increase under elevated temperatures in many model and crop species, consistent with increases in non-enzymatic decarboxylation reactions (Walker and Cousins, 2013, Walker et al., 2017). It is therefore possible that catalase activity is insufficient under some conditions to decrease rates of non-enzymatic decarboxylation, offering a potential route for increasing photorespiratory efficiency.

5 Photorespiration under non-steady-state conditions: could this improve carbon and nitrogen budgets?

The above approaches focus on improving photorespiration under steady-state conditions, but what approaches might target non-steady-state conditions? Environmental conditions are incredibly dynamic under actual cultivation conditions (Slattery et al., 2018), but the dynamics and efficiency of photorespiratory metabolism during non-steady-state conditions are unresolved. For example, photorespiratory metabolism is mostly studied under steady-state conditions, either after acclimation to various environmental conditions or genetic perturbations in the photorespiratory pathway (e.g. Wingler et al., 1999, Abadie et al., 2021, Timm et al., 2012b). Shifting plants from low- to high photorespiratory conditions often leads to a significant accumulation of photorespiratory intermediates (Abadie et al., 2016, Eisenhut et al., 2017). However, the dynamics of the changes in concentration of photorespiratory intermediates are rarely studied due to the challenges of fast sampling on a short time scale and a possible under-appreciation of how much time plants spend under a non-steady state. Therefore, to harness the potential of photorespiration to improve crop productivity, a potential avenue is to understand how changes in photorespiratory metabolism impact non-steady-state photosynthesis and how non-steady-state photorespiration will contribute to gross carbon gain and nitrogen assimilation under environmental fluctuations in nature.

Although it is technically challenging to capture the metabolic changes of photorespiratory intermediates under non-steady-state conditions, the CO_2 release from photorespiratory glycine directly impacts net CO_2 assimilation and can be non-invasively tracked within seconds using gas exchange analyzers (Walker et al., 2018). A sample gas exchange measurements captured the dynamic response of net CO_2 assimilation to changes in photorespiratory fluxes manipulated by O_2 mole fractions using a custom gas mixing system (Fig. 3). The delay in assimilation reaching a lower steady-state rate during the transient (shaded area) may be attributed to reduced CO_2 release from glycine decarboxylase. Specifically, when glycine accumulates under transient conditions, CO_2 release from the glycine decarboxylase reaction will be delayed until the glycine pool sizes stabilize and metabolism reaches a steady-state rate.

Alternatively, excess CO_2 release may occur as glycine pool sizes decline to a lower concentration. This excess release of CO_2 from photorespiration (as compared to steady-state rates of release) occurs during transients from high light to low light or darkness. This effect is often called the 'post-illumination burst' and results in more reduced rates of net assimilation (or more negative rates of

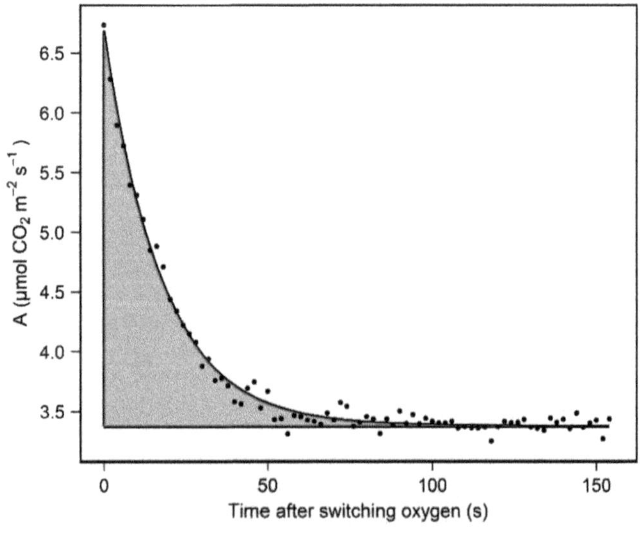

Figure 3 Dynamics of net CO_2 assimilation rate transitioning from 2% O_2 to 40% O_2. The delay in reaching the new steady-state assimilation rate is shown by the shaded region and is possibly due in part to accumulation in glycine pools. Measurements were performed on the youngest-fully expanded leaf of *Nicotiana tabacum* using an LI-Cor 6800 and a custom gas mixing system to control O_2 and N_2.

respiration in the dark) during these transitions (Bulley and Tregunna, 1971, Vines et al., 1983). The post-illumination burst most likely stems from the glycine pool that was previously accumulated before the light-dark transitions (Rawsthorne and Hylton, 1991). While the constant accumulation of glycine is not sustainable in the steady state, perhaps it is possible to relax glycine pools during non-steady-state conditions by re-routing carbon to other metabolic pathways that do not release CO_2 temporarily. Further development in transient approaches is needed to understand the efficiencies of photorespiration under non-steady-state conditions and how these efficiencies may be targeted for improvement.

The transient increase in the pools of photorespiratory intermediates could serve as a carbon and nitrogen reservoir supporting other metabolic processes that share intermediates with the photorespiratory pathway. A large portion of the photorespiratory intermediates (e.g. glycine, serine, and glycerate) exist as inactive pools, evidenced by the incomplete labeling of these pools at 1 h of feeding with $^{13}CO_2$ (Szecowka et al., 2013, Ma et al., 2014, Xu et al., 2021) and their presence in the vacuole (Szecowka et al., 2013). If these inactive pools of photorespiratory intermediates are remobilized during transient conditions to maintain high photorespiratory fluxes, there will be a substantial gain of carbon without compromising the photosynthetically fixed carbon. Besides fueling glycerate production and turnover of the Calvin-Benson cycle, the extra photorespiration-derived glycine and serine can be exported to sink tissues and used for protein synthesis in source leaves (Madore and Grodzinski, 1984, Cegelski and Schaefer, 2005).

6 Conclusion

Optimizing photosynthesis by modifying photorespiration is an exciting avenue for crop improvement, but much more work remains. We are encouraged by recent work placing these modifications in crop plants and look forward to additional validation under field conditions. Testing strategies under field conditions is especially important to ensure that improvements are robust and in practice contribute to increased crop yield. We are optimistic about the future of these general approaches and look forward to advances in the coming years.

7 Where to look for further information

1. For a comprehensive and updated picture of photorespiration and its connection to nitrogen metabolism (Busch, 2020).
2. International Congress on Photosynthesis Research, which is held every 4 years and often includes many talks on photorespiration. Hosted by the International Society for Photosynthesis Research.

3 CO_2 assimilation in Plants from Genome to Biome, a Gordon Research Conference, which is held every 3 years and includes intimate coverage of the latest advances in photorespiratory research.

8 References

Abadie, C., Boex-Fontvieille, E. R. A., Carroll, A. J. and Tcherkez, G. 2016. *In vivo* stoichiometry of photorespiratory metabolism. *Nature Plants* 2, 15220.

Abadie, C., Lalande, J., Limami, A. M. and Tcherkez, G. 2021. Non-targeted ^{13}C metabolite analysis demonstrates broad re-orchestration of leaf metabolism when gas exchange conditions vary. *Plant, Cell and Environment* 44(2), 445-457.

Ainsworth, E. A. and Long, S. P. 2021. 30 years of free-air carbon dioxide enrichment (FACE): what have we learned about future crop productivity and its potential for adaptation? *Global Change Biology* 27(1), 27-49.

Ainsworth, E. A. and Rogers, A. 2007. The response of photosynthesis and stomatal conductance to rising [CO_2]: mechanisms and environmental interactions. *Plant, Cell and Environment* 30(3), 258-270.

Anderson, L. E. 1971. Chloroplast and cytoplasmic enzymes II. Pea leaf triose phosphate isomerases. *Biochimica et Biophysica Acta* 235(1), 237-244.

Bao, H., Morency, M., Rianti, W., Saeheng, S., Roje, S., Weber, A. P. M. and Walker, B. J. 2021. Catalase protects against nonenzymatic decarboxylations during photorespiration. *Arabidopsis thaliana*. *Plant Direct* 5(12), e366. https://doi.org/10.1002/pld3.366.

Bar-Even, A. 2018. Daring metabolic designs for enhanced plant carbon fixation. *Plant Science* 273, 71-83.

Bauwe, H., Hagemann, M. and Fernie, A. R. 2010. Photorespiration: players, partners and origin. *Trends in Plant Science* 15(6), 330-336.

Berghuijs, H. N. C., Yin, X., Ho, Q. T., Retta, M. A., Verboven, P., Nicolaï, B. M. and Struik, P. C. 2017. Localization of (photo) respiration and CO_2 re-assimilation in tomato leaves investigated with a reaction-diffusion model. *PLoS ONE* 12(9), e0183746.

Betti, M., Bauwe, H., Busch, F. A., Fernie, A. R., Keech, O., Levey, M., Ort, D. R., Parry, M. A. J., Sage, R., Timm, S., Walker, B. and Weber, A. P. M. 2016. Manipulating photorespiration to increase plant productivity: recent advances and perspectives for crop improvement. *Journal of Experimental Botany* 67(10), 2977-2988.

Bloom, A. J. 2006. Rising carbon dioxide concentrations and the future of crop production. *Journal of the Science of Food and Agriculture* 86(9), 1289-1291.

Bloom, A. J., Burger, M., Asensio, J. S. R. and Cousins, A. B. 2010. Carbon dioxide enrichment inhibits nitrate assimilation in wheat and *Arabidopsis*. *Science* 328(5980), 899-903.

Brooks, A. and Farquhar, G. D. 1985. Effect of temperature on the CO_2/O_2 specificity of ribulose-1,5-bisphosphate carboxylase/oxygenase and the rate of respiration in the light. *Planta* 165(3), 397-406.

Bulley, N. R. and Tregunna, E. B. 1971. Photorespiration and the postillumination CO_2 burst. *Canadian Journal of Botany* 49(8), 1277-1284.

Busch, F. A. 2020. Photorespiration in the context of Rubisco biochemistry, CO_2 diffusion and metabolism. *The Plant Journal* 101(4), 919-939.

Busch, F. A., Sage, R. F. and Farquhar, G. D. 2018. Plants increase CO_2 uptake by assimilating nitrogen via the photorespiratory pathway. *Nature Plants* 4(1), 46-54.

Busch, F. A., Sage, T. L., Cousins, A. B. and Sage, R. F. 2013. C_3 plants enhance rates of photosynthesis by reassimilating photorespired and respired CO_2. *Plant, Cell and Environment* 36(1), 200–212.

Cegelski, L. and Schaefer, J. 2005. Glycine metabolism in intact leaves by in vivo ^{13}C and ^{15}N labeling. *Journal of Biological Chemistry* 280(47), 39238–39245.

Cousins, A. B., Pracharoenwattana, I., Zhou, W., Smith, S. M. and Badger, M. R. 2008. Peroxisomal malate dehydrogenase is not essential for photorespiration in *Arabidopsis* but its absence causes an increase in the stoichiometry of photorespiratory CO_2 release. *Plant Physiology* 148(2), 786–795.

Cousins, A. B., Walker, B. J., Pracharoenwattana, I., Smith, S. M. and Badger, M. R. 2011. Peroxisomal hydroxypyruvate reductase is not essential for photorespiration in *Arabidopsis* but its absence causes an increase in the stoichiometry of photorespiratory CO_2 release. *Photosynthesis Research* 108(2–3), 91–100.

Cui, L.-L., Lu, Y.-S., Li, Y., Yang, C. and Peng, X.-X. 2016. Overexpression of glycolate oxidase confers improved photosynthesis under high light and high temperature in rice. *Frontiers in Plant Science* 7, 1165.

Dalal, J., Lopez, H., Vasani, N. B., Hu, Z., Swift, J. E., Yalamanchili, R., Dvora, M., Lin, X., Xie, D., Qu, R. and Sederoff, H. W. 2015. A photorespiratory bypass increases plant growth and seed yield in biofuel crop *Camelina sativa*. *Biotechnology for Biofuels* 8, 175.

Dusenge, M. E., Duarte, A. G. and Way, D. A. 2019. Plant carbon metabolism and climate change: elevated CO_2 and temperature impacts on photosynthesis, photorespiration and respiration. *New Phytologist* 221(1), 32–49.

Edwards, G. E. and Walker, D. A. 1983. *C3, C4: Mechanisms, and Cellular and Environmental Regulation, of Photosynthesis*. London: Oxford.

Eisenhut, M., Bräutigam, A., Timm, S., Florian, A., Tohge, T., Fernie, A. R., Bauwe, H. and Weber, A. P. M. 2017. Photorespiration is crucial for dynamic response of photosynthetic metabolism and stomatal movement to altered CO_2 availability. *Molecular Plant* 10(1), 47–61.

Flügel, F., Timm, S., Arrivault, S., Florian, A., Stitt, M., Fernie, A. R. and Bauwe, H. 2017. The photorespiratory metabolite 2-phosphoglycolate regulates photosynthesis and starch accumulation in *Arabidopsis*. *The Plant Cell* 29(10), 2537–2551.

Foyer, C. H., Bloom, A. J., Queval, G. and Noctor, G. 2009. Photorespiratory metabolism: genes, mutants, energetics, and redox signaling. *Annual Review of Plant Biology* 60, 455–484.

González-Moro, B., Lacuesta, M., Becerril, J. M., Gonzalez-Murua, C. and Muñoz-Rueda, A. 1997. Glycolate accumulation causes a decrease of photosynthesis by inhibiting Rubisco activity in maize. *Journal of Plant Physiology* 150(4), 388–394.

Grodzinski, B. and Butt, V. S. 1976. Hydrogen peroxide production and the release of carbon dioxide during glycollate oxidation in leaf peroxisomes. *Planta* 128(3), 225–231.

Harley, P. C. and Sharkey, T. D. 1991. An improved model of C_3 photosynthesis at high CO_2: reversed O_2 sensitivity explained by lack of glycerate reentry into the chloroplast. *Photosynthesis Research* 27(3), 169–178.

IPCC 2007. *Climate Change 2007: The Physical Science Basis. Contribution of Working Group I to the Fourth Assessment Report of the Intergovernmental Panel on Climate Change*, Solomon, S., Qin, D., Manning, M., Chen, Z., Marquis, M., Averyt, K. B.,

Tignor, M. and Miller, H. L. (Eds.). Cambridge, UK and New York, NY, USA: Cambridge University Press.

Kebeish, R., Niessen, M., Thiruveedhi, K., Bari, R., Hirsch, H.-J., Rosenkranz, R., Stabler, N., Schonfeld, B., Kreuzaler, F. and Peterhansel, C. 2007. Chloroplastic photorespiratory bypass increases photosynthesis and biomass production in *Arabidopsis thaliana*. *Nature Biotechnology* 25(5), 593–599.

Keech, O., Zhou, W., Fenske, R., Colas-Des-Francs-Small, C., Bussell, J. D., Badger, M. R. and Smith, S. M. 2012. The genetic dissection of a short-term response to low CO_2 supports the possibility for peroxide-mediated decarboxylation of photorespiratory intermediates in the peroxisome. *Molecular Plant* 5(6), 1413–1416.

Kelly, G. J. and Latzko, E. 1976. Inhibition of spinach-leaf phosphofructokinase by 2-phosphoglycollate. *FEBS Letters* 68(1), 55–58.

Khoshravesh, R., Stinson, C. R., Stata, M., Busch, F. A., Sage, R. F., Ludwig, M. and Sage, T. L. 2016. C_3-C_4 intermediacy in grasses: organelle enrichment and distribution, glycine decarboxylase expression, and the rise of C_2 photosynthesis. *Journal of Experimental Botany* 67(10), 3065–3078.

Kubis, A. and Bar-Even, A. 2019. Synthetic biology approaches for improving photosynthesis. *Journal of Experimental Botany* 70(5), 1425–1433.

López-Calcagno, P. E., Fisk, S., Brown, K. L., Bull, S. E., South, P. F. and Raines, C. A. 2019. Overexpressing the H-protein of the glycine cleavage system increases biomass yield in glasshouse and field-grown transgenic tobacco plants. *Plant Biotechnology Journal* 17(1), 141–151.

Lu, Y., Li, Y., Yang, Q., Zhang, Z., Chen, Y., Zhang, S. and Peng, X.-X. 2014. Suppression of glycolate oxidase causes glyoxylate accumulation that inhibits photosynthesis through deactivating Rubisco in rice. *Physiologia Plantarum* 150(3), 463–476.

Lundgren, M. R. 2020. C_2 photosynthesis: a promising route towards crop improvement? *New Phytologist* 228(6), 1734–1740.

Ma, F., Jazmin, L. J., Young, J. D. and Allen, D. K. 2014. Isotopically nonstationary ^{13}C flux analysis of changes in Arabidopsis thaliana leaf metabolism due to high light acclimation. *Proceedings of the National Academy of Sciences of the United States of America* 111(47), 16967–16972.

Madore, M. and Grodzinski, B. 1984. Effect of oxygen concentration on ^{14}C-photoassimilate transport from leaves of *Salvia splendens* L. *Plant Physiology* 76(3), 782–786.

Maier, A., Fahnenstich, H., von Caemmerer, S., Engqvist, M. K., Weber, A. P., Flügge, U.-I. and Maurino, V. G. 2012. Transgenic introduction of a glycolate oxidative cycle into *A. thaliana* chloroplasts leads to growth improvement. *Frontiers in Plant Science* 3, 38.

Masson-Delmotte, V., Zhai, P., Pirani, A., Connors, S. L., Péan, C., Berger, S., Caud, N., Chen, Y., Goldfarb, L., Gomis, M. I., Huang, M., Leitzell, K., Lonnoy, E., Matthews, J. B. R., Maycock, T. K. and Waterfield, T., Yelekçi, O, Yu, R. and Zhou, B. (Eds), 2021. IPCC, 2021: summary for policymakers. In: *Climate Change 2021: The Physical Science Basis. Contribution of Working Group I to the Sixth Assessment Report of the Intergovernmental Panel on Climate Change*. Cambridge University Press.

Nölke, G., Houdelet, M., Kreuzaler, F., Peterhänsel, C. and Schillberg, S. 2014. The expression of a recombinant glycolate dehydrogenase polyprotein in potato (*Solanum tuberosum*) plastids strongly enhances photosynthesis and tuber yield. *Plant Biotechnology Journal* 12(6), 734–742.

Ogren, W. L. 1984. Photorespiration: pathways, regulation, and modification. *Annual Review of Plant Physiology* 35(1), 415-442.

Ogren, W. L. and Bowes, G. 1971. Ribulose diphosphate carboxylase regulates soybean photorespiration. *Nature: New Biology* 230(13), 159-160.

Oliver, D. J. 1980. The effect of glyoxylate on photosynthesis and photorespiration by isolated soybean mesophyll cells. *Plant Physiology* 65(5), 888-892.

Oliver, D. J. and Zelitch, I. 1977. Increasing photosynthesis by inhibiting photorespiration with glyoxylate. *Science* 196(4297), 1450-1451.

Ort, D. R., Merchant, S. S., Alric, J., Barkan, A., Blankenship, R. E., Bock, R., Croce, R., Hanson, M. R., Hibberd, J. M., Long, S. P., Moore, T. A., Moroney, J., Niyogi, K. K., Parry, M. A. J., Peralta-Yahya, P. P., Prince, R. C., Redding, K. E., Spalding, M. H., Van Wijk, K. J., Vermaas, W. F. J., von Caemmerer, S., Weber, A. P. M., Yeates, T. O., Yuan, J. S. and Zhu, X. G. 2015. Redesigning photosynthesis to sustainably meet global food and bioenergy demand. *Proceedings of the National Academy of Sciences of the United States of America* 112(28), 8529-8536.

Osmond, B., Badger, M., Maxwell, K., Björkman, O. and Leegood, R. 1997. Too many photons: photorespiration, photoinhibition and photooxidation. *Trends in Plant Science* 2(4), 119-121.

Parry, M. A. J., Keys, A. J., Madgwick, P. J., Carmo-Silva, A. E. and Andralojc, P. J. 2008. Rubisco regulation: a role for inhibitors. *Journal of Experimental Botany* 59(7), 1569-1580.

Pieruschka, R., Schurr, U., Jensen, M., Wolff, W. F. and Jahnke, S. 2006. Lateral diffusion of CO_2 from shaded to illuminated leaf parts affects photosynthesis inside homobaric leaves. *New Phytologist* 169(4), 779-788.

Rachmilevitch, S., Cousins, A. B. and Bloom, A. J. 2004. Nitrate assimilation in plant shoots depends on photorespiration. *Proceedings of the National Academy of Sciences of the United States of America* 101(31), 11506-11510.

Rawsthorne, S. and Hylton, C. M. 1991. The relationship between the post-illumination CO_2 burst and glycine metabolism in leaves of C_3 and C_3-C_4 intermediate species of *Moricandia*. *Planta* 186(1), 122-126.

Roney, H. E. and Walker, B. J. 2021. Photon to plate: A holistic view of photosynthetic and anthropogenic energy fluxes. *Food and Energy Security* 00, e348. https://doi.org/10.1002/fes3.348.

Sharkey, T. D. 1985. O_2-insensitive photosynthesis in C_3 plants: its occurrence and a possible explanation. *Plant Physiology* 78(1), 71-75.

Sharkey, T. D. 1988. Estimating the rate of photorespiration in leaves. *Physiologia Plantarum* 73(1), 147-152.

Shih, P. M., Zarzycki, J., Niyogi, K. K. and Kerfeld, C. A. 2014. Introduction of a synthetic CO_2-fixing photorespiratory bypass into a cyanobacterium. *Journal of Biological Chemistry* 289(14), 9493-9500.

Slattery, R. A., Walker, B. J., Weber, A. P. M. and Ort, D. R. 2018. The impacts of fluctuating light on crop performance. *Plant Physiology* 176(2), 990-1003.

Somerville, C. R. 2001. An early *Arabidopsis* demonstration. Resolving a few issues concerning photorespiration. *Plant Physiology* 125(1), 20-24.

Somerville, C. R. and Ogren, W. L. 1979. A phosphoglycolate phosphatase-deficient mutant of *Arabidopsis*. *Nature* 280(5725), 833-836.

Somerville, C. R. and Ogren, W. L. 1980a. Inhibition of photosynthesis in *Arabidopsis* mutants lacking leaf glutamate synthase activity. *Nature* 286(5770), 257-259.

Somerville, C. R. and Ogren, W. L. 1980b. Photorespiration mutants of *Arabidopsis thaliana* deficient in serine-glyoxylate aminotransferase activity. *Proceedings of the National Academy of Sciences of the United States of America* 77(5), 2684–2687.

Somerville, C. R. and Ogren, W. L. 1981. Photorespiration-deficient mutants of *Arabidopsis thaliana* lacking mitochondrial serine transhydroxymethylase activity. *Plant Physiology* 67(4), 666–671.

South, P. F., Cavanagh, A. P., Liu, H. W. and Ort, D. R. 2019. Synthetic glycolate metabolism pathways stimulate crop growth and productivity in the field. *Science* 363(6422), 609–616.

South, P. F., Cavanagh, A. P., Lopez-Calcagno, P. E., Raines, C. A. and Ort, D. R. 2018. Optimizing photorespiration for improved crop productivity. *Journal of Integrative Plant Biology* 60(12), 1217–1230.

Szecowka, M., Heise, R., Tohge, T., Nunes-Nesi, A., Vosloh, D., Huege, J., Feil, R., Lunn, J., Nikoloski, Z., Stitt, M., Fernie, A. R. and Arrivault, S. 2013. Metabolic fluxes in an illuminated *Arabidopsis* rosette. *The Plant Cell* 25(2), 694–714.

Tholen, D., Ethier, G., Genty, B., Pepin, S. and Zhu, X.-G. 2012. Variable mesophyll conductance revisited: theoretical background and experimental implications. *Plant, Cell and Environment* 35(12), 2087–2103.

Tholen, D. and Zhu, X. G. 2011. The mechanistic basis of internal conductance: a theoretical analysis of mesophyll cell photosynthesis and CO_2 diffusion. *Plant Physiology* 156(1), 90–105.

Timm, S. and Bauwe, H. 2013. The variety of photorespiratory phenotypes – employing the current status for future research directions on photorespiration. *Plant Biology* 15(4), 737–747.

Timm, S., Florian, A., Arrivault, S., Stitt, M., Fernie, A. R. and Bauwe, H. 2012a. Glycine decarboxylase controls photosynthesis and plant growth. *FEBS Letters* 586(20), 3692–3697.

Timm, S., Mielewczik, M., Florian, A., Frankenbach, S., Dreissen, A., Hocken, N., Fernie, A. R., Walter, A. and Bauwe, H. 2012b. High-to-low CO_2 acclimation reveals plasticity of the photorespiratory pathway and indicates regulatory links to cellular metabolism of Arabidopsis. *PLoS ONE* 7(8), e42809.

Timm, S., Woitschach, F., Heise, C., Hagemann, M. and Bauwe, H. 2019. Faster removal of 2-phosphoglycolate through photorespiration improves abiotic stress tolerance of Arabidopsis. *Plants* 8(12), 563.

Trudeau, D. L., Edlich-Muth, C., Zarzycki, J., Scheffen, M., Goldsmith, M., Khersonsky, O., Avizemer, Z., Fleishman, S. J., Cotton, C. A. R., Erb, T. J., Tawfik, D. S. and Bar-Even, A. 2018. Design and *in vitro* realization of carbon-conserving photorespiration. *Proceedings of the National Academy of Sciences of the United States of America* 115(49), E11455–E11464.

Vines, H. M., Tu, Z. P., Armitage, A. M., Chen, S. S. and Black, C. C. Jr. 1983. Environmental responses of the post-lower illumination CO_2 burst as related to leaf photorespiration 1. *Plant Physiology* 73(1), 25–30.

von Caemmerer, S. 2000. *Biochemical Models of Leaf Photosynthesis*. Collingwood: CSIRO.

Walker, B. J., Busch, F. A., Driever, S. M., Kromdijk, J. and Lawson, T. 2018. Survey of tools for measuring in vivo photosynthesis. In: Covshoff, S. (Ed.) *Photosynthesis: Methods and Protocols*, 3–24. New York, NY: Springer New York.

Walker, B. J. and Cousins, A. B. 2013. Influence of temperature on measurements of the CO_2 compensation point: differences between the Laisk and O_2-exchange methods. *Journal of Experimental Botany* 64(7), 1893–1905.

Walker, B. J., Kramer, D. M., Fisher, N. and Fu, X. 2020. Flexibility in the energy balancing network of photosynthesis enables safe operation under changing environmental conditions. *Plants* 9(3), 301.

Walker, B. J., Orr, D. J., Carmo-Silva, E., Parry, M. A. J., Bernacchi, C. J. and Ort, D. R. 2017. Uncertainty in measurements of the photorespiratory CO_2 compensation point and its impact on models of leaf photosynthesis. *Photosynthesis Research* 132(3), 245–255.

Walker, B. J. and Ort, D. R. 2015. Improved method for measuring the apparent CO_2 photocompensation point resolves the impact of multiple internal conductances to CO_2 to net gas exchange. *Plant, Cell and Environment* 38(11), 2462–2474.

Walker, B. J., Skabelund, D. C., Busch, F. A. and Ort, D. R. 2016a. An improved approach for measuring the impact of multiple CO_2 conductances on the apparent photorespiratory CO_2 compensation point through slope-intercept regression. *Plant, Cell and Environment* 39(6), 1198–1203.

Walker, B. J., Vanloocke, A., Bernacchi, C. J. and Ort, D. R. 2016b. The costs of photorespiration to food production now and in the future. *Annual Review of Plant Biology* 67, 107–129.

Wendler, C., Putzer, A. and Wild, A. 1992. Effect of glufosinate (phosphinothricin) and inhibitors of photorespiration on photosynthesis and ribulose-1,5-bisphosphate carboxylase activity. *Journal of Plant Physiology* 139(6), 666–671.

Wingler, A., Lea, P. J. and Leegood, R. C. 1999. Photorespiratory metabolism of glyoxylate and formate in glycine-accumulating mutants of barley and *Amaranthus edulis*. *Planta* 207(4), 518–526.

Xin, C., Tholen, D. and Zhu, X.-G. 2014. The benefits of photorespiratory bypasses: how can they work? *Plant Physiology* 167(2), 574–585.

Xu, H., Zhang, J., Zeng, J., Jiang, L., Liu, E., Peng, C., He, Z. and Peng, X. 2009. Inducible antisense suppression of glycolate oxidase reveals its strong regulation over photosynthesis in rice. *Journal of Experimental Botany* 60(6), 1799–1809.

Xu, Y., Fu, X., Sharkey, T. D., Shachar-Hill, Y. and Walker, A. B. J. 2021. The metabolic origins of non-photorespiratory CO_2 release during photosynthesis: a metabolic flux analysis. *Plant Physiology* 186(1), 297–314.

Yang, J. T., Preiser, A. L., Li, Z., Weise, S. E. and Sharkey, T. D. 2016. Triose phosphate use limitation of photosynthesis: short-term and long-term effects. *Planta* 243(3), 687–698.

Yokota, A. 1991. Carboxylation and detoxification of xylulose bisphosphate by spinach ribulose bisphosphate carboxylase/oxygenase. *Plant and Cell Physiology* 32, 755–762.

Yu, H., Li, X., Duchoud, F., Chuang, D. S. and Liao, J. C. 2018. Augmenting the Calvin-Benson–Bassham cycle by a synthetic malyl-CoA-glycerate carbon fixation pathway. *Nature Communications* 9(1), 2008.

Zhao, H.-L., Chang, T.-G., Xiao, Y. and Zhu, X.-G. 2021. Potential metabolic mechanisms for inhibited chloroplast nitrogen assimilation under high CO_2. *Plant Physiology* 187(3), 1812–1833.

Zhu, G. and Jensen, R. G. 1991. Xylulose 1,5-bisphosphate synthesized by ribulose 1,5-bisphosphate carboxylase/oxygenase during catalysis binds to decarbamylated enzyme 1. *Plant Physiology* 97(4), 1348–1353.

Chapter 9

Maximizing the efficiency of ribulose bisphosphate (RuBP) regeneration to optimize photosynthesis in crops

Thomas D. Sharkey, MSU-DOE Plant Research Laboratory, Michigan State University, USA

1 Introduction

Photosynthesis is the source of carbon and energy needed for the growth and maintenance of plants and it is widely believed that improving photosynthesis will improve crop yields (Faralli and Lawson, 2020, Raines, 2011, Simkin et al., 2019, Zhu et al., 2010). However, crop growth and yield are complex. In some cases, things other than photosynthetic rate per unit leaf area can control plant growth and crop yield (Araus et al., 2021; Körner, 2015; Kubis and Bar-Even, 2019; Poorter et al., 2013; Sinclair et al., 2019; Weraduwage et al., 2015). Nevertheless, increases in crop yield in Free Air CO_2 Enrichment (FACE) sites as well as growth analyses done in elevated CO_2 in greenhouses and growth chambers, make it clear that increasing photosynthesis increases crop growth and yield (Farquhar and Sharkey, 1994) even if not as much as anticipated (Long et al., 2006).

There are many methods that researchers are employing to increase the efficiency of photosynthesis. One possible outcome will be increased crop yields, but another useful outcome is a reduction in the resources required to achieve good yields. Many attempts focus on one or another process of photosynthesis and in this chapter the process of ribulose bisphosphate (RuBP)

http://dx.doi.org/10.19103/AS.2022.0119.13

regeneration will be explored. The phrase RuBP regeneration is used in two different ways depending on context. In one case it refers only to reactions from triose phosphates to RuBP in the Calvin-Benson cycle, but in another context it includes most of photosynthetic metabolism except rubisco capacity. In this chapter the more inclusive definition that arises from photosynthesis models will be emphasized. In this definition, all processes needed to make RuBP from phosphoglyceric acid (PGA) are included. Thus, both the reduction phase and regeneration phase of the Calvin-Benson cycle are included in the expanded definition of RuBP regeneration.

2 Component processes of photosynthesis

To examine the potential for increasing crop yield by increasing photosynthesis, it is important to consider that photosynthesis comprises several distinct processes. Generally, photosynthesis is defined as the conversion of CO_2 to sugar with the aid of light energy and with some involvement of water. The amount of water consumed in photosynthetic reactions is <1% of the amount of water involved in transpiration through stomata (discussed elsewhere in this volume) and can be ignored when talking about the biochemistry of photosynthesis. This leaves the two major inputs, CO_2 and light, as critical factors in photosynthesis. In many textbooks, these are split into light reactions and dark reactions, but this division does not reflect the mechanisms that control photosynthetic efficiency. In particular, the label of 'dark reactions' is misleading and should be replaced by 'photosynthetic carbon metabolism' (Buchanan, 2016, Sharkey, 2019).

The core carbon metabolism reactions are known as the Calvin-Benson cycle (sometimes just Calvin cycle, sometimes Benson-Calvin cycle, and sometimes Calvin-Benson-Bassham cycle, all of which are justified). A way to avoid issues of who contributed to the discovery is to use a descriptive name like Photosynthetic Carbon Reduction (PCR) cycle (Johnson and Berry, 2021, von Caemmerer, 2020) although that might be confused with Polymerase Chain Reaction (PCR). I will use Calvin-Benson cycle here.

The Calvin-Benson cycle is often presented in three phases: (1) the carboxylation phase catalyzed by rubisco, (2) the reduction phase where nearly all the light energy is used, and (3) the regeneration phase where 5/6 of the carbon from the reduction phase is remodeled to become the molecule that rubisco uses to fix CO_2, RuBP, so that the cycle can continue. The reaction converting triose phosphate to RuBP is one way to define RuBP regeneration, but a broader definition is used here.

A highly influential model of photosynthesis, often called the Farquhar, von Caemmerer, Berry (FvCB) model, was introduced in 1980 (Farquhar et al., 1980). That model resulted in characterizing photosynthetic behavior, when measured by CO_2 uptake by leaves of C_3 plants, as controlled by one of two processes, (1) rubisco, the enzyme that fixes CO_2 by carboxylation of RuBP operating with saturating levels of RuBP, and (2) RuBP regeneration, including all reactions from PGA to RuBP. In this context 'RuBP regeneration' includes the carbon metabolism reactions described above plus the reduction phase of the Calvin-Benson cycle and all the light-dependent electron- and proton-transport reactions. The predictions of this model were tested by gas exchange, enzyme activity assays, and what today would be called metabolite profiling (Badger et al., 1984, von Caemmerer and Farquhar, 1981). The model held up well.

2.1 Enzyme kinetics important to understanding photosynthesis controls

Biologists tend to think of enzymatic reactions as gradually saturating with increasing substrate as described by the Michalis-Menten equation (Fig. 1 black line). As changes in metabolite concentrations work their way through the photosynthetic system, it is expected that control of photosynthetic rate will be shared among many processes because the product of one reaction will be the precursor for the next reaction (Kacser and Burns, 1973). However, the

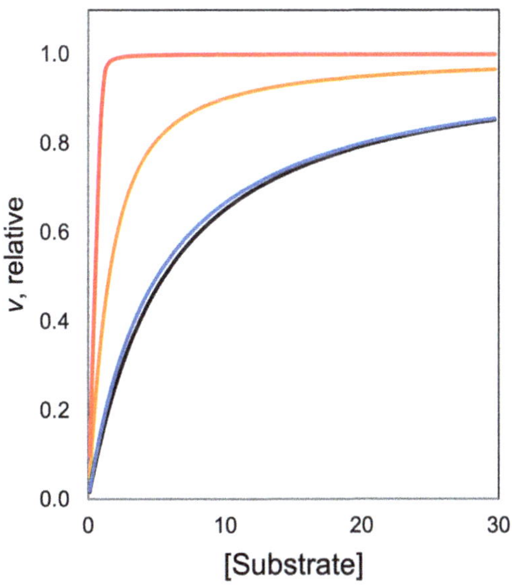

Figure 1 Enzyme kinetics relevant to rubisco. When the Michaelis-Menten constant (K_m) of an enzyme for a substrate is five times greater than the concentration of enzyme active sites, the relationship between the rate, *v*, and free substrate concentration ([*S*]) will show gradual saturation when calculated using either the Michalis-Menten equation (black line, Equation1) or the enzyme kinetic quadratic equation (blue line, Equation 2). The gold line shows the response when the Km and enzyme concentration are equal.

rubisco reaction has two substrates, one (RuBP) provided at a specific rate by metabolism, and one supplied as a gas (either CO or O_2). The reaction is ordered, with RuBP binding first (Pierce et al., 1986). The concentration of RuBP will be determined by the rate of production and rate of consumption, while the gases have an infinite supply but at a given concentration (or better, partial pressure), subject to effects of diffusion resistances. Farquhar (1979) explored the kinetics that can be expected given that rubisco violates important assumptions of the Michalis-Menten equation. Specifically, the concentration of rubisco active sites is very high relative to rubisco's affinity for RuBP. This condition is recognized in enzyme theory. When this occurs, a quadratic equation is required to describe enzyme kinetics (Fig. 1 red line) (Morrison, 1969). In extreme cases (such as rubisco) with very high concentrations of enzyme relative to the binding affinity of the substrate, every RuBP will be bound to rubisco for as long as there are unbound rubisco sites (nearly vertical red line in Fig. 1). When rubisco can go faster than RuBP can be made, RuBP regeneration will determine the rate of photosynthesis (Farquhar, 1979). This is called RuBP-regeneration-limited photosynthesis. Photosynthetic behavior in response to light, CO_2, and temperature will reflect the behavior of the processes that make up RuBP regeneration in RuBP-regeneration-limited photosynthesis.

$$v = \frac{[E_T] \cdot k_{cat} \cdot [S]}{[S] + K_m} \tag{1}$$

$$v = \frac{b - \sqrt{b^2 - 4 \cdot [E_T] \cdot [S_T]}}{2 \cdot [E_T]} \tag{2}$$

where $[E_T]$ is the total amount of enzyme present (bound to substrate plus unbound) and k_{cat} is the turnover number of the enzyme. Often, $[E_T] \cdot k_{cat}$ is called $V_{m.}$ $b = [E_T] \cdot [S_T] \cdot K_m$, and $[S_T]$ is the total amount of substrate. If the K_m and enzyme concentration are equal, the quadratic equation is less gradual (gold line). When the K_m is just 1% of the enzyme concentration, the quadratic equation appears as 2 nearly straight lines (red line). Rubisco has this kind of kinetics with respect to RuBP, though not CO_2. When rubisco is slow, RuBP supply exceeds the enzyme concentration and photosynthesis depends on how much rubisco is present (horizontal part of the red curve). But, when the rate of RuBP consumption by rubisco exceeds the capacity for RuBP regeneration, the nearly vertical part of the red curve will be relevant. This defines RuBP regeneration limitation and tells why transitions between behaviors is abrupt.

But, once the last rubisco site is bound with RuBP, there will be no increase in rate with increasing RuBP (horizontal red line in Fig. 1). With all sites bound with RuBP, the rate of photosynthesis will be determined by the amount and kinetics of rubisco and not by RuBP regeneration. When RuBP supply exceeds its potential use, photosynthetic behavior in response to light, CO_2, and temperature will reflect the behavior of rubisco. For example, light will have no effect on the photosynthetic rate. In this case photosynthesis is called rubisco limited. Because the affinity for RuBP is large relative to the rubisco enzyme concentration, the transition between the RuBP-regeneration-limited and rubisco-limited conditions is abrupt. For this reason, RuBP effects are 'either/or', *either* rubisco limited (RuBP saturated), *or* RuBP-regeneration limited. The gases are supplied at a specific partial pressure but variable rate and so in the RuBP-saturated condition, carboxylation and oxygenation reactions can be modeled with Michalis-Menten kinetics.

2.2 Photosynthetic responses to light and CO_2

If we consider two-dimensional space with CO_2 on the X-axis and light on the Y-axis, different zones can be predicted that will differ in how photosynthesis responds to CO_2 or light (Fig. 2). These zones represent rubisco-limited photosynthesis and RuBP-regeneration-limited photosynthesis. The triose

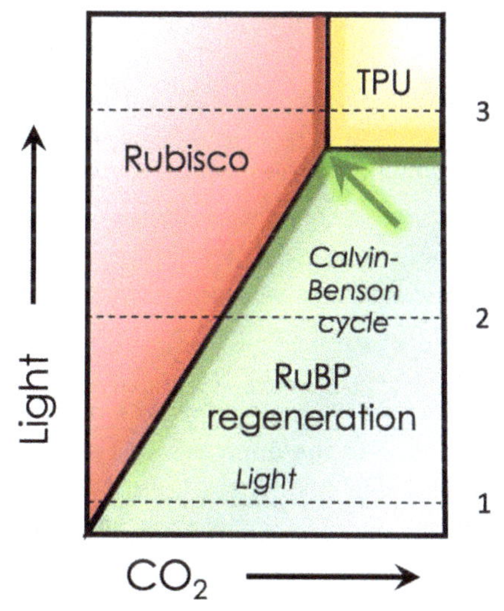

Figure 2 Light and CO_2 determine which component process will control the behavior of photosynthesis. Within RuBP regeneration light will be the primary determinant of photosynthetic responses but as light increases, the Calvin-Benson cycle may begin to affect RuBP regeneration and hence photosynthesis. CO_2 response curves at dotted lines 1, 2, and 3 are described in the text and Fig. 3. Photosynthesis is most efficient when it occurs at the juncture of all three potentially limiting factors. The RuBP regeneration section (green) includes both light limitation and PCR cycle reactions other than rubisco, possibly exerting simultaneous control. The green arrow indicates the optimum operating point where rubisco and RuBP regeneration are operating at their maximum capacity.

phosphate use (TPU) limitation shown in Fig. 2 will be described further. In each zone, photosynthetic responses are modeled using specific equations. For simplicity the equations will not be given here, instead the properties of photosynthetic responses will be described. An extensive review of the equations involved can be found in Yin et al. (2021).

2.2.1 When RuBP regeneration limits

The balance between RuBP regeneration and its use will determine what proportion of rubisco enzyme sites have RuBP and so are active. When RuBP regeneration limits - when the plant is operating at light and CO_2 levels to the right of the diagonal line in Fig. 2 - the potential for rubisco activity is greater than the potential for regeneration of RuBP. The concentration of RuBP, and of RuBP-bound active sites of rubisco, falls until the rate of consumption matches the rate of production. Upon a rapid decrease in light, which will slow the regeneration of RuBP (Mott et al., 1984), or increase in CO_2, which will increase consumption of RuBP (Sharkey et al., 1986b), the predicted decline in RuBP concentration can be observed. However, regulatory processes quickly take over, and rubisco is deactivated by losing an essential carbamate and metal ion leading to an increase in RuBP concentration, even though RuBP production is limiting the overall rate of photosynthesis. The specific processes that regulate decarbamylation, and thus deactivation of rubisco, are not known.

A major component of RuBP regeneration limitation is light availability. Under low light levels, regulatory processes are engaged to maximize the total absorption of light and to optimize the distribution of the absorbed light between Photosystem II (PS II) and Photosystem I (PS I), such that electron transport proceeds at the rate permitted by the reduced plastoquinone (PQH_2) supply and the maximum activity of the cytochrome b_6f complex (Cyt b_6f) (Johnson and Berry, 2021). Under these conditions, feedback control of the excitation distribution between PS II and PS I can adjust the relative rates of linear and cyclic electron flow in relation to the energetic demands of the Calvin-Benson cycle, photorespiration, and respiration in the light, R_L. However, the rate of RuBP regeneration still exerts kinetic control over the overall rate of photosynthesis. The behavior in the RuBP regeneration zone includes strong sensitivity to light, but also moderate sensitivity to CO_2 because CO_2 will compete with oxygen because of photorespiration (covered elsewhere in this volume). As CO_2 increases, oxygen interference decreases, and this results in a higher rate of photosynthesis.

2.2 2 When rubisco limits

As light increases, the increasing supply of reduced PQH_2 permits faster production of reductant and ATP, driving a higher rate of RuBP regeneration and permitting more rubisco to be maintained in an activated state. Once fully activated, and with excess RuBP, the rate of photosynthesis will be determined by the kinetics of RuBP-saturated rubisco. As a result, the rubisco-limited zone in Fig. 2 gets larger with increasing light. As light increases, the amount of rubisco needed to maximize photosynthesis increases. At high light, rubisco is typically fully activated (Sage, 1990). In this zone, light absorption is deactivated. At a biochemical level, this can result from processes called non-photochemical quenching (*NPQ*) and specifically, that part of NPQ called energy-dependent quenching (q_E) (Tietz et al., 2017). Another way that electron transport is regulated to match rubisco capacity is by reducing the activity of Cyt b_6f in the electron transport chain (Johnson and Berry, 2021, Kiirats et al., 2009). In leaves, light absorption can be modulated to match rubisco capacity by chloroplast movements (Wada and Kong, 2018) and leaf movements (Nilsen and Forseth, 2018) also contribute. Canopy architecture, covered elsewhere in this volume, also will affect light received by individual leaves.

When rubisco capacity is setting the rate of photosynthesis (red zone in Fig. 2), photosynthesis does not respond to light but does respond very strongly to CO_2. The strong response to CO_2 results from CO_2 being a substrate for the carboxylation reaction, in addition to the suppression of photorespiration by CO_2.

2.2.3 The triose phosphate use limitation

A third limitation sometimes occurs. When the products of photosynthesis (primarily three carbon sugars, triose phosphates) are not used as fast as they can be produced, ATP production can become limited by phosphate, since phosphate is not being liberated from triose phosphates (and other end products) as fast as required. This is called the TPU limitation. In this case, photosynthesis no longer responds to increases in light or CO_2. The transition to TPU limitation also appears to be abrupt but the reason for this abruptness is not known and could even be illusory since it is not known what a co-limitation between rubisco and TPU, or RuBP regeneration and TPU, would look like. Strictly speaking, TPU limitation is a form of RuBP regeneration limitation, but it results in unusual and diagnostic gas exchange characteristics, in particular lack of, or even reverse sensitivity to CO_2 (and O_2). For this reason, TPU is treated as being distinct from RuBP regeneration although at high rates of photosynthesis, limited TPU capacity could reduce RuBP regeneration without becoming the dominant limitation.

When TPU limits, both rubisco capacity and RuBP regeneration capacity are more than what can be used. Rubisco deactivation (Sharkey et al., 1986a) and NPQ (Sharkey et al., 1988, Takizawa et al., 2008) have been observed in TPU-limited conditions. The deactivation of rubisco could result from reduced rubisco activase activity because of a limited ATP supply (Sharkey et al., 1986b). Reduced rubisco activation state and NPQ are significant inefficiencies for photosynthesis. The convergence of the three interior lines in Fig. 2 (green arrow) is the optimum situation, where each process is being used to its maximum capacity.

The TPU limitation is frequently encountered when light, CO_2, and temperature are just slightly outside of the range a leaf normally experiences, but generally not seen under conditions the leaf has experienced the past few days (Kumarathunge et al., 2019, Sage et al., 1989). This suggests that TPU is highly regulated. It is likely that excessive capacity for starch and sucrose synthesis is not optimal, so TPU capacity is regulated to be just slightly more than what is likely to be needed. This may improve the functioning of regulatory interactions that keep rubisco capacity and RuBP regeneration capacity matched.

TPU is a form of feedback from sugar accumulation onto photosynthesis. In the short term, this feedback is believed to work through release of phosphate from carbon that leaves the Calvin-Benson cycle. In the longer term, the mechanism by which sugar accumulation reduces photosynthesis often involves complex sugar sensing pathways and regulation of gene expression for both rubisco and RuBP regeneration reactions (Griffiths et al., 2016, Lastdrager et al., 2014, Ono and Watanabe, 1997). Sucrose synthesis is one component of TPU and it is possible to alter the occurrence of TPU by overexpressing a gene for sucrose-phosphate synthase (SPS) (Laporte et al., 2001). There are also recent reports of genes that control the rate of starch accumulation in response to CO_2, providing a mechanism for increasing TPU when rubisco activity and RuBP regeneration capacity are increased by growth at high CO_2 (Fukayama et al., 2021). Manipulating source/sink ratios of rice plants was shown to affect the occurrence of TPU (Fabre et al., 2019).

2.3 Optimal use of resources

Photosynthesis will be most efficient if it operates at the point where the three component processes in Fig. 2 meet. However, because light and temperature (though for the most part not CO_2) vary throughout the day, and the zones of Fig. 2 do not vary rapidly, leaves will operate for parts of the day in one or another zone and either rubisco will be deactivated, or photosynthetic electron transport capacity will be limited by Cyt b_6f activity or processes associated with NPQ, or both.

Returning now to Fig. 2, the diagonal line starting in the lower left corner represents the condition where both rubisco and RuBP regeneration are fully engaged and so resources are being used optimally. As light increases, the leaf moves up the line. There is more potential for rubisco to limit at high light and more potential for RuBP regeneration to limit at low light. The further from the line, the more one process or the other will be in excess and so deactivated. In this case resources such as nitrogen will be wasted. These zones can be mapped using measurements of instantaneous rates of photosynthesis as a function of CO_2. Normally, the CO_2 partial pressure (or concentration, which is partial pressure divided by total atmospheric pressure and often reported as parts per million) is reported as that estimated to be present in the airspaces inside of leaves (C_i). Typically, A/C_i curves are reported. At low light, only at the lowest CO_2 partial pressure will rubisco limit photosynthesis. Over most of the range, RuBP regeneration, specifically light availability, determines the behavioral characteristics of photosynthesis, e.g. CO_2 and temperature sensitivity. Photosynthetic electron transport and RuBP usage will be independent of CO_2 partial pressure, but photosynthesis will not. This is because as CO_2 increases, it suppresses oxygenation, reducing the use of light-derived energy in photorespiration and increasing energy used for carboxylation. This condition is represented by line 1 in Fig. 2.

At higher, but not saturated, light (typically 1/8 to 1/4 of sunlight), photosynthesis is limited by rubisco at low CO_2 partial pressure but by RuBP regeneration at high CO_2 partial pressure (Fig. 2, line 2). When rubisco limits, RuBP use increases with CO_2 and so does photosynthetic electron transport, unlike during RuBP regeneration limitation, during which electron transport does not increase with CO_2. Therefore, in conditions represented by line 2, electron transport should increase with CO_2 up to a point and then abruptly change to being independent of the changes in CO_2. Line 3 shows a case where photosynthesis is rubisco limited at low to moderate CO_2 but then crosses into the TPU-limited zone without passing through the RuBP regeneration zone. In this case the electron transport rate should increase in the rubisco-limited zone and then begin to decline without any section where the electron transport rate is independent of CO_2.

The predictions for the three dotted lines in Fig. 2 can be tested by measuring the assimilation of CO_2 both by gas exchange and electron transport by chlorophyll fluorescence analysis. Data measured on a bean leaf is shown in Fig. 3. The lowest light level corresponds to dotted line 1 of Fig. 2. In this case, RuBP regeneration limits at all but the very lowest CO_2 partial pressures. The intermediate light level of Fig. 3 corresponds to dotted line 2 of Fig. 2. While in the rubisco limitation zone, RuBP usage, and therefore electron transport, increases with increasing CO_2, but then the leaf enters the RuBP regeneration zone and electron transport no longer changes with CO_2 even though photosynthesis does. The highest light level in Fig. 3 corresponds to dotted line 3 of Fig. 2. At low CO_2 in the rubisco-limited zone, electron transport increases with CO_2 and at the low CO_2 partial pressures electron transport is the same at the high and intermediate light levels. Unlike in the

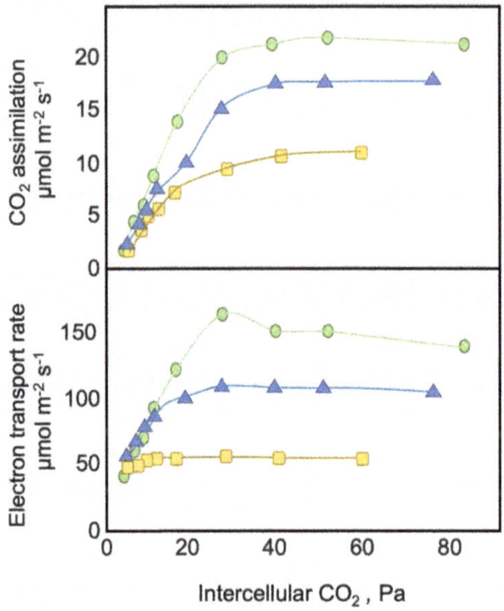

Figure 3 CO_2 responses of photosynthesis measured by gas exchange and electron transport calculated from fluorescence parameters, for a leaf of *Phaseolus vulgaris*. Photon flux densities were 950 (green circles), 380 (blue diamonds), and 175 (gold squares) μmol photons m^{-2} s^{-1}. Leaf temperature was 25°C.

intermediate light level data, at high light electron transport changes from increasing with CO_2 to decreasing with CO_2, an indication that the leaf has crossed from the rubisco-limited zone to the TPU-limited zone. The specific CO_2 partial pressures and light levels where these transitions occur vary from leaf to leaf. In addition, the TPU zone is often not observed, especially when measurements are made at a leaf temperature above the temperature at which the plants were grown because TPU increases with temperature more than rubisco or RuBP regeneration (Yang et al., 2016).

Considering shading, leaf flutter, light flecks, partly cloudy skies etc. it can be seen from Fig. 3 that leaves could experience conditions of rubisco and/or TPU limitation and no RuBP regeneration limitation (high light) and then RuBP regeneration only (low light) many times each day. This complicates the understanding of how to optimize photosynthesis.

At any given instant, photosynthesis appears to be under the control of one of three syndromes. But, over the course of a day or longer, conditions will vary and optimal investment into rubisco or the components of RuBP regeneration will depend on net carbon gain over the course of the day and the cost of resources, such as enzymes, needed to achieve that gain. The specific tradeoff of carbon gain versus costs of the photosynthetic apparatus made by any given plant will be affected by the environmental niche occupied by the plant.

3 Optimizing ribulose bisphosphate regeneration

Behavior in the rubisco-limited zone is dominated by the properties of rubisco but the RuBP-regeneration zone is more complex and includes both light-driven electron and proton fluxes needed to make NADPH and ATP, and the Calvin-Benson cycle, which consumes the NADPH and ATP. RuBP regeneration capacity is determined by light availability and electron transport at very low light, but Calvin-Benson cycle capacity may become important as light increases. However, the border between control by light availability and control by Calvin-Benson cycle activity is not well understood. It appears there is not an abrupt border like the border between rubisco and RuBP regeneration.

3.1 Resource investment

Photosynthesis depends on fixed nitrogen (Field and Mooney, 1986). Although the atmosphere is mostly N_2, N_2 is not generally available to plants. Instead, an energetically expensive process is required to make the nitrogen available in its reduced form. Rubisco accounts for as much as 20% of the total amount of nitrogen (Eckardt et al., 1997) and 40% of soluble protein (Evans and Seemann, 1984) in photosynthesizing leaves. However, photosynthetic electron transport is also very expensive in terms of nitrogen investment. It can account for 15% to as much as 60% of leaf nitrogen (Evans, 1989). Typically light-harvesting complexes, the two photosystems, the cyt b_6f complex, and ATP synthase account for 24% of leaf nitrogen (Evans and Clarke, 2018). On the other hand, enzymes of the Calvin-Benson cycle other than rubisco account for a relatively small percentage of leaf protein, perhaps just 4% (Evans and Clarke, 2018). The balance between nitrogen invested in rubisco versus that invested in light capture and conversion will determine, in part, how well matched rubisco and RuBP regeneration are, so that the leaf can, on average, operate near the transition between rubisco limited and RuBP-regeneration limited, which is the optimal condition (von Caemmerer and Farquhar, 1981).

Nitrogen investment in rubisco and RuBP regeneration increases approximately linearly with increasing availability of nitrogen (Evans, 1989). However, growth light intensity has a very large effect on the investment of nitrogen in RuBP regeneration. Low-light-grown plants invest more nitrogen in light-harvesting protein complexes. As a result, they have much more chlorophyll but a lower capacity for RuBP regeneration. The variation of nitrogen investment in the various photosynthetic processes is a way of optimizing RuBP regeneration.

A less obvious way to optimize resource investment between RuBP regeneration and rubisco is by altering specific leaf area (leaf area per leaf mass, SLA). A leaf with very high SLA will have more area per amount of leaf mass and so intercept more light per unit of leaf mass than a leaf with a low SLA (because it has less area per leaf mass). But the high SLA leaf will have less rubisco per unit leaf area and so less rubisco capacity. At low light this will not be a disadvantage since RuBP regeneration will be limiting, not rubisco. As SLA decreases (more mass per area of leaf) electron transport capacity and rubisco will increase, but light absorptivity will increase much less because even a relatively thin leaf absorbs most of the light that falls on it. Thus, light absorption of leaves does not increase with leaf mass per area as much as rubisco or Calvin-Benson cycle enzymes (Evans and Poorter, 2001). Optimizing canopy architecture will also improve RuBP regeneration and is covered elsewhere in this volume.

3.2 Light use efficiency

3.2.1 Wavelength dependence

It is generally thought that blue and red light are the most effective for photosynthesis. 'Photosynthetically active radiation' is a measure defined as all the photons between 400 nm and 700 nm. It is also generally thought that green light is not well absorbed reducing its effectiveness. However, light use by leaves is more complicated. One issue is that green light is absorbed surprisingly well, just a little less than red or blue. Green light can be scattered within a leaf more than red or blue and so there is more opportunity for the green light to be absorbed. At high light intensity, red and blue light will saturate the upper layer of leaves and the chloroplasts there will start to dissipate red and blue photons. Green photons will get deeper in the leaf, past the saturated layer. Therefore, at high light, green light may be absorbed and cause photosynthesis throughout the depth of the leaf while many blue and red photons are dissipated in the upper, saturated layer (Terashima et al., 2009). Just the opposite of conventional wisdom, green light can be the best color of light for photosynthesis.

It is generally assumed that light absorption is the same as light use by a leaf. However, careful measurements made years ago showed that absorbed blue light is less effective for photosynthesis than absorbed red light (McCree, 1972). The effect can be substantial so that even at low light, green light is as effective or more effective than blue light (McClain and Sharkey, 2020, McCree, 1972).

The cutoff for measuring the parameter 'photosynthetically active radiation' is 400 nm but light of wavelengths less than 400 nm, down to 375 to 350 depending on the species (McCree, 1972), will drive photosynthesis. At the other end of the active spectrum, red and far-red light, the situation is more complex. The two photosystems of photosynthetic electron transport differ in the wavelengths of red light they can use. When supplied as single wavelength light, photosynthesis has a strong cutoff at 700 nm and longer wavelengths are not effective. However, when longer wavelength light is supplied in addition to broad spectrum light, it can be very effective. This is known as the Emmerson Enhancement effect and recently it has been shown that crop growth can respond even if up to 70% of the photons have a wavelength greater than 700 nm (Zhen and Bugbee, 2020).

A large portion of the photons in sunlight have wavelengths longer than 700 nm and some bacteria have chlorophyll that can capture these longer wavelength photons. What is the opportunity to make use of these to improve RuBP regeneration? A recent report showed that incorporating chlorophyll *d* from a cyanobacterium to the chlorophylls *a* and *b* normally found in plants can extend the ability to use red light out by more than 25 nm (Elias et al., 2021). This should enhance RuBP regeneration. Other aspects of improving light harvesting for photosynthesis are covered in other chapters of this volume.

3.2.2 Efficiency of photon use

Photons reaching PS II reaction centers are, at best, used with 83% efficiency (Björkman and Demmig, 1987). The rest of the energy is lost as fluorescence or heat (non-radiative dissipation). From first principles, the ideal behavior would be that PS II efficiency would be high and stay high as light increased until light saturation was reached, at which point the efficiency would decrease to keep the rate of electron transport at the same rate as whatever process was setting the rate of photosynthesis. However, this is not what happens. This apparent efficiency of light capture declines with increasing light, even at quite low light (Fig. 4) (Johnson and Berry, 2021, Yin et al., 2009).

To understand this behavior (and determine if it can be modified to optimize RuBP regeneration) requires understanding the relationships between the Calvin-Benson cycle and photosynthetic electron transport. The fundamental relationship is that electron (and proton) transport makes NADPH and ATP, which are consumed in the Calvin-Benson cycle, but the regulatory interactions to keep all of the processes coordinated are complex.

3.3 Interactions between light use and carbon metabolism in ribulose bisphosphate regeneration

The RuBP-regeneration limitation encompasses both light availability and the Calvin-Benson cycle enzymes. Some models focus on Calvin-Benson cycle reactions to find which Calvin-Benson cycle enzyme could be increased to increase the rate of photosynthesis (Zhu et al., 2007){Raines, 2011 #7427} while other models treat RuBP regeneration as primarily reflecting light availability

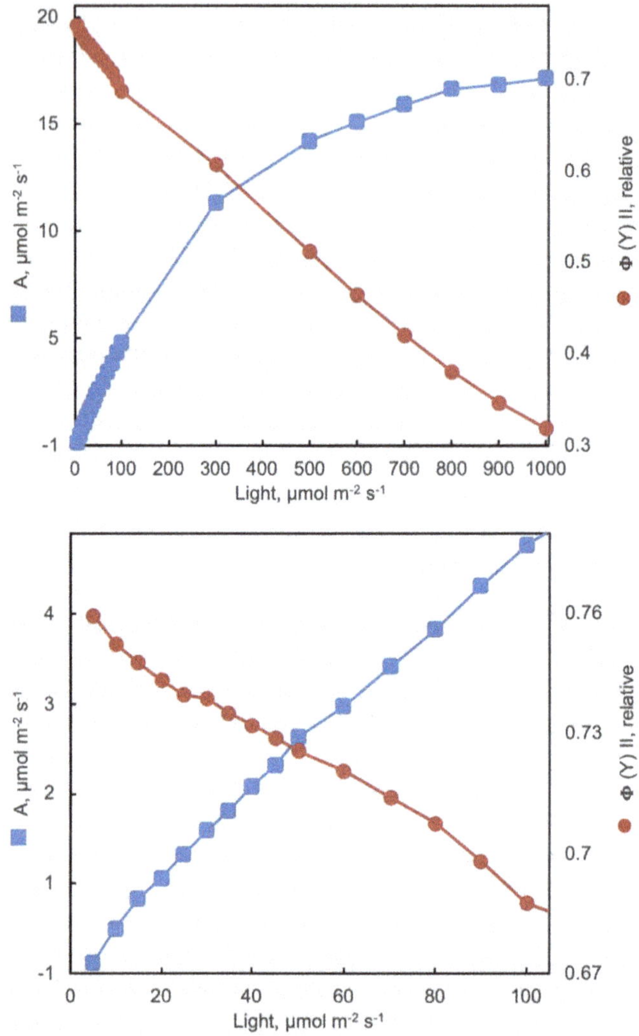

Figure 4 CO_2 assimilation and ϕ II (also described as Y II) measured by chlorophyll fluorescence analysis. The bottom panel is a closer look at the first 10% of light intensities shown in the top panel. Unpublished data of Stephanie C. Schmiege using a poplar leaf and LI-COR 6800.

(Johnson and Berry, 2021). How carbon metabolism and light capture and conversion to ATP and NADPH interact is an area ripe for study.

There are several processes, especially in photosynthetic electron transport, that are relevant to the interaction between carbon metabolism and electron transport. Regulation of photosynthetic electron transport has been studied extensively and parts of the story are covered in detail elsewhere in this volume. Especially important is understanding the balance between photosynthetic efficiency and protection from damage caused when light is absorbed but cannot be used (Alboresi et al., 2019). However, the capacity for carbon metabolism will affect how fast NADPH and ATP can be used and thus will interact with the regulation of photosynthetic electron transport. When PGA is high and triose phosphates are low, the ratios of NADPH/NADP and ATP/ADP can be modest and still allow high rates of photosynthesis. However, if triose phosphates are high or PGA is low then NADPH/NADP and ATP/ADP ratios must be higher to compensate for the unfavorable ratios of Calvin-Benson cycle intermediates. The requirement for higher ATP/ADP or NADPH/NADP ratios will in turn affect electron transport capacity. Therefore, there will be linkages in carbon metabolism regulation and electron transport regulation. A full understanding of these interactions is a goal of current research (Johnson and Berry, 2021).

Some of the important electron transport processes that may be affected by carbon metabolism are briefly described in the following sections.

3.3.1 Non-photochemical quenching

Non-photochemical quenching is primarily the result of energy-dependent quenching (q_E). This is a specific process that depends on the pH inside the chloroplast thylakoid lumen. Low pH in the lumen results in formation of zeaxanthin from violaxanthin and in protonation of a PS II protein called PsbS. When the thylakoid lumen pH is low, energy absorbed in the antenna complexes never reaches PS II and the excess energy is harmlessly dissipated. When RuBP use is limiting, it will feedback on ATP production. This will result in increased proton motive force, and one component of the proton motive force is low luminal pH. Low luminal pH leads to q_E and reduced efficiency of photosynthetic electron transport. Low luminal pH causes protonation of the PS II protein PsbS (Niyogi et al., 2005) and stimulation of zeaxanthin accumulation (Takizawa et al., 2007). Since ATP and NADPH usage depend on carbon metabolism, q_E is one way that carbon metabolism and electron transport are linked.

3.3.2 Photosynthetic control

Photosynthetic control is a similar process in which low luminal pH restricts electron flow (Colombo et al., 2016, Nishio and Whitmarsh, 1993). However, in this case the restriction is at the cytochrome b_6f complex situated between PS II and PS I. Limited flow at the cytochrome b_6f complex protects PS I from over reduction. PS I is easily damaged (Lima-Melo et al., 2019). Photosynthetic control regulation of electron flow can lead to PS II damage. Fortunately, PS II has an efficient repair cycle (Aro et al., 1993) so this damage can be easily repaired. Very recent evidence indicates that photosynthetic control, specifically regulation of the activation of the Cyt b_6f complex, plays a primary role in regulating photosynthetic electron flow (Johnson and Berry, 2021).

3.3.3 Cyclic electron flow

The ratio of ATP to NADPH required by the Calvin-Benson cycle is 3:2 but photosynthetic linear electron flow is believed to provide ATP:NADPH at 2.57:2 leaving an ATP deficit (Kramer and Evans, 2011). Other carbon metabolism can increase or decrease this deficit (Sharkey, 2021). In photosynthesis there are two mechanisms to increase ATP production independent of NADPH production (Shikanai and Yamamoto, 2017, Storti et al., 2019). The first of these is cyclic electron flow, in which electrons from PS I cycle back to the cytochrome b_6f complex and result in proton transport with no net electron production or consumption (Yamori and Shikanai, 2016). There are two pathways that have been described, one mediated by the NAD(P)H dehydrogenase-like complex (NDH) and the other is the ferredoxin: plastoquinone oxidoreductase (FQR)-dependent flow. The NDH pathway is particularly efficient in converting photons to ATP (Strand et al., 2017). The extent of cyclic electron flow in C_3 plants is debated but it is known to be activated by hydrogen peroxide (Strand et al., 2015), and by low ATP concentration (Fisher et al., 2019). High temperature also stimulates cyclic electron flow (Bukhov et al., 1999, Essemine et al., 2017), possibly by consuming ATP in a plastidial G6P shunt (Sharkey et al., 2020). In C_4 plants cyclic electron flow is essential and may rely mostly on the NDH pathway (Takabayashi et al., 2005).

3.3.4 Pseudo-cyclic electron flow

When electrons move from PS I back to oxygen to make water, the protons that were translocated during the electron transport can drive ATP production. This allows ATP production without net oxygen evolution and no NADPH production, mimicking cyclic electron flow, and thus is called pseudo-cyclic electron flow. If PS I is over-reduced, it can donate electrons directly to oxygen to make superoxide radicals, a form of reactive oxygen species (ROS) that can damage cells. Superoxide dismutase can convert the superoxide to hydrogen peroxide, a less dangerous ROS and an important signaling molecule. Hydrogen peroxide can be converted to water by catalase. In some photosynthetic organisms, flavodiiron proteins catalyze electron donation to oxygen. However, while these proteins are present in cyanobacteria, early diverging plant lineages, and gymnosperms, they have been mostly lost in angiosperms (Allahverdiyeva et al., 2015, Shirao et al., 2013).

3.3.5 *Alternative oxidase mitochondrial electron flow*

Plant mitochondria have a mechanism that allows electron transport to proceed without making ATP. This is accomplished primarily by an alternative terminal oxidase in mitochondria that passes electrons to oxygen without transporting protons, thus not making ATP and dissipating energy as heat. Sometimes this metabolism is used to cause plant parts, especially flowers, to heat up. This can be useful in volatilizing some pollinator attractants. In addition, there is evidence that in some situations, alternative oxidase electron flow can protect the photosynthetic apparatus from excess electron damage by providing a harmless mechanism for disposing of excess electrons (Dahal et al., 2015, Vanlerberghe et al., 2016).

4 Examples of improving ribulose bisphosphate regeneration through engineering

The regeneration RuBP is more likely to be limiting at elevated CO_2. The rapid change in atmospheric CO_2 provides a challenge and opportunity for improving crop growth (Dingkuhn et al., 2020) and increasing the capacity for RuBP regeneration is an important target for improvement. Two examples of improving RuBP regeneration to improve plant performance will be discussed here. The first is an improvement of the Calvin-Benson cycle through increased expression of sedoheptulose bisphosphatase and the second is an improvement of TPU through the expression of a corn gene for sucrose-phosphate synthase in tomatoes.

4.1 *Sedoheptulose bisphosphatase*

Sedoheptulose bisphosphatase (SBPase) is a critical step in the Calvin-Benson cycle. It distinguishes the pentose phosphate pathway reactions of the Calvin-Benson cycle from the more widely known non-oxidative branch of the pentose phosphate pathway (Sharkey, 2021). In the Calvin-Benson cycle, SBPase replaces transaldolase. As a result, this section of the Calvin-Benson cycle is irreversible. The dephosphorylation by SBPase provides a thermodynamic boost to pentose phosphate synthesis. Instead of equilibrium between hexose and triose sugars on one hand and pentose sugars on the other, replacing transaldolase with SBPase makes the metabolism unidirectional from hexose/triose to pentoses (Sharkey, 2021).

Given its central role in the Calvin-Benson cycle, it is not surprising that SBPase is a highly regulated enzyme (Raines et al., 2000; Woodrow and Walker, 1983). Both fructose bisphosphatase and SBPase are light-activated enzymes, meaning that they become active when reduced by thioredoxin. The effectiveness of reduction depends on the concentration of their substrate in both cases (Leegood et al., 1982, Woodrow and Walker, 1983). This gives rise to a very strong dependence on substrate since it acts as an activator as well as a substrate.

Cyanobacteria have an enzyme that has both fructose bisphosphatase and SBPase activity, but in plants the activities are provided by different enzymes, which can be regulated separately. SBPase activity is found in other organisms. In yeast, the enzyme coded by *Shb17* is an SBPase that is believed to direct carbon toward pentoses when needed for RNA and DNA synthesis (Clasquin et al., 2011).

A model of carbon metabolism of photosynthesis indicated that photosynthetic rate is most sensitive to changes in rubisco and SBPase and that SBPase might need to be increased by 2.5 fold for optimal distribution of nitrogen resources in photosynthesis (Zhu et al., 2007). In another study involving both modeling and experimental evidence, SBPase again was identified as the most critical enzyme controlling the rate of the Calvin-Benson cycle after rubisco (Poolman et al., 2001).

A series of reports have consistently found that reducing *SBPase* gene expression reduces photosynthesis and plant growth while increasing expression by various methods increases photosynthesis and plant growth in greenhouses and in the field (Driever et al., 2017; Harrison et al., 2001; Lawson et al., 2006; Lefebvre et al., 2005; Ogawa et al., 2015). Increasing SBPase along with other proteins needed for RuBP regeneration, including electron transport components such as the Cyt b_6f complex can have a synergistic effect on plant growth (Simkin et al., 2017). This may reflect that the rate-setting process varies over the course of a day, or it could reflect that RuBP regeneration capacity is determined by both carbon metabolism (other than rubisco) and electron transport simultaneously.

However, increased SBPase gene expression generally has a modest effect on the photosynthetic rate. When plants were grown in high CO_2 by free air CO_2 enrichment (FACE), the SBPase overexpressing plant had increased photosynthesis only at high CO_2 and then only at some times of the day (Rosenthal et al., 2011). It may be that light utilization is the more important component of RuBP regeneration and the Calvin-Benson cycle reactions are regulated to optimize light use.

4.2 Sucrose-phosphate synthase

While TPU is considered separately from RuBP regeneration, an example of increasing yield by manipulating TPU is presented here. Starch and sucrose synthesis are the primary uses of TPU exported from the Calvin-Benson cycle. Therefore, increasing sucrose synthesis by overexpression of SPS might be expected to increase TPU, increasing the concentration of phosphate available for ATP synthesis and increasing photosynthetic capacity. This was tested by expressing a corn gene for SPS in tomatoes (Worrell et al., 1991). It was shown that this modification increased the rate of photosynthesis that was possible before TPU began to limit (Laporte et al., 1997). In that study, yield was increased by 70–80% when grown in open-top chambers at ambient or elevated CO_2.

5 Conclusion

The division between rubisco-limited photosynthesis conditions and RuBP regeneration-limited conditions is abrupt. At any given instant, photosynthesis will respond to variations in the environment with characteristics of one or the other process but not both. Within RuBP regeneration the division between photosynthetic electron transport limitation and Calvin-Benson cycle limitation is decidedly mixed. Photosynthesis can respond to increases in light intensity at the same time as it responds to increases in Calvin-Benson cycle capacity (e.g. SBPase). Also, the efficiency of light use begins to decline at very low light suggesting shared control with Calvin-Benson cycle reactions and other metabolism. Genetic modifications have been made to increase RuBP regeneration and many of these show some increase in rate but not as much as might be hoped, likely reflecting the shared control of the capacity for RuBP regeneration. Increasing the capacity for photosynthesis by increasing the capacity for RuBP regeneration is very promising, especially in a high CO_2 world that appears inevitable, but progress will require understanding the interaction of (1) photosynthetic light capture and conversion to ATP and NADPH with (2) carbon metabolism in the photosynthesizing leaf.

6 Where to look for further information

An active group studying ways to improve RuBP regeneration capacity is in the international program "**Realizing Increased Photosynthetic Efficiency** for sustainable increases in crop yield". Their website is as follows:

- https://ripe.illinois.edu/objectives/rubp-regeneration.

Other websites include the following:

- https://www.life.illinois.edu/govindjee/honorsfrom.html.
- https://en.wikipedia.org/wiki/Calvin_cycle.
- https://www.capitalise.eu/calvin-benson-cycle/.
- https://blog.aspb.org/calvin-cycle-calvin-benson-cycle-or-other/.

7 Acknowledgements

My photosynthesis research is supported by the Division of Chemical Sciences, Geosciences and Biosciences, Office of Basic Energy Sciences of the United States Department of Energy (Grant DE-FG02-91ER20021) and I get partial salary support from Michigan AgBioResearch. I thank Dr. Jennifer Robinson and Dr. Stephanie Schmiege for very helpful comments on the manuscript.

8 References

Alboresi, A., Storti, M. and Morosinotto, T. 2019. Balancing protection and efficiency in the regulation of photosynthetic electron transport across plant evolution. *New Phytologist* 221(1), 105–109.

Allahverdiyeva, Y., Isojärvi, J., Zhang, P. and Aro, E. M. 2015. Cyanobacterial oxygenic photosynthesis is protected by flavodiiron proteins. *Life* 5(1), 716–743.

Araus, J. L., Sanchez-Bragado, R. and Vicente, R. 2021. Improving crop yield and resilience through optimization of photosynthesis: panacea or pipe dream? *Journal of Experimental Botany* 72(11), 3936–3955.

Aro, E. M., Virgin, I. and Andersson, B. 1993. Photoinhibition of photosystem II. Inactivation, protein damage and turnover. *Biochimica et Biophysica Acta* 1143(2), 113–134.

Badger, M. R., Sharkey, T. D. and von Caemmerer, S. 1984. The relationship between steady-state gas exchange of bean leaves and the levels of carbon-reduction-cycle intermediates. *Planta* 160(4), 305–313.

Björkman, O. and Demmig, B. 1987. Photon yield of O_2 evolution and chlorophyll fluorescence characteristics at 77 K among vascular plants of diverse origins. *Planta* 170(4), 489–504.

Buchanan, B. B. 2016. The carbon (formerly dark) reactions of photosynthesis. *Photosynthesis Research* 128(2), 215–217.

Bukhov, N. G., Wiese, C., Neimanis, S. and Heber, U. 1999. Heat sensitivity of chloroplasts and leaves: leakage of protons from thylakoids and reversible activation of cyclic electron transport. *Photosynthesis Research* 59(1), 81–93.

Clasquin, M. F., Melamud, E., Singer, A., Gooding, J. R., Xu, X., Dong, A., Cui, H., Campagna, S. R., Savchenko, A., Yakunin, A. F., Rabinowitz, J. D. and Caudy, A. A. 2011. Riboneogenesis in yeast. *Cell* 145(6), 969–980.

Colombo, M., Suorsa, M., Rossi, F., Ferrari, R., Tadini, L., Barbato, R. and Pesaresi, P. 2016. Photosynthesis Control: an underrated short-term regulatory mechanism essential for plant viability. *Plant Signaling and Behavior* 11(4), e1165382.

Dahal, K., Martyn, G. D. and Vanlerberghe, G. C. 2015. Improved photosynthetic performance during severe drought in Nicotiana tabacum overexpressing a nonenergy conserving respiratory electron sink. *New Phytologist* 208(2), 382–395.

Dingkuhn, M., Luquet, D., Fabre, D., Muller, B., Yin, X. and Paul, M. J. 2020. The case for improving crop carbon sink strength or plasticity for a CO2-rich future. *Current Opinion in Plant Biology* 56, 259–272.

Driever, S. M., Simkin, A. J., Alotaibi, S., Fisk, S. J., Madgwick, P. J., Sparks, C. A., Jones, H. D., Lawson, T., Parry, M. A. J. and Raines, C. A. 2017. Increased SBPase activity improves photosynthesis and grain yield in wheat grown in greenhouse conditions.

Philosophical Transactions of the Royal Society of London. Series B, Biological Sciences 372(1730), 20160384.

Eckardt, N. A., Snyder, G. W., Portis Jr., A. R. and Ogren, W. L. 1997. Growth and photosynthesis under high and low irradiance of *Arabidopsis thaliana* antisense mutants with reduced ribulose-1,5-bisphosphate carboxylase/oxygenase activase content. *Plant Physiology* 113(2), 575–586.

Elias, E., Liguori, N., Saga, Y., Schäfers, J. and Croce, R. 2021. Harvesting far-red light with plant antenna complexes incorporating chlorophyll d. *Biomacromolecules* 22(8), 3313–3322.

Essemine, J., Xiao, Y., Qu, M., Mi, H. and Zhu, X. G. 2017. Cyclic electron flow may provide some protection against PSII photoinhibition in rice (*Oryza sativa* L.) leaves under heat stress. *Journal of Plant Physiology* 211, 138–146.

Evans, J. R. 1989. Photosynthesis and nitrogen relationships in leaves of C_3 plants. *Oecologia* 78(1), 9–19.

Evans, J. R. and Clarke, V. C. 2018. The nitrogen cost of photosynthesis. *Journal of Experimental Botany* 70(1), 7–15.

Evans, J. R. and Poorter, H. 2001. Photosynthetic acclimation of plants to growth irradiance: the relative importance of specific leaf area and nitrogen partitioning in maximizing carbon gain. *Plant, Cell and Environment* 24(8), 755–767.

Evans, J. R. and Seemann, J. R. 1984. Differences between wheat genotypes in specific activity of ribulose-1,5-bisphosphate carboxylase and the relationship to photosynthesis. *Plant Physiology* 74(4), 759–765.

Fabre, D., Yin, X., Dingkuhn, M., Clément-Vidal, A., Roques, S., Rouan, L., Soutiras, A. and Luquet, D. 2019. Role of triose phosphate Utilization in photosynthetic response of rice to variable carbon dioxide levels and plant source-sink relations. *bioRxiv*, 633016.

Faralli, M. and Lawson, T. 2020. Natural genetic variation in photosynthesis: an untapped resource to increase crop yield potential? *Plant Journal: For Cell and Molecular Biology* 101(3), 518–528.

Farquhar, G. D. 1979. Models describing the kinetics of ribulose bisphosphate carboxylase-oxygenase. *Archives of Biochemistry and Biophysics* 193(2), 456–468.

Farquhar, G. D. and Sharkey, T. D. 1994. Photosynthesis and carbon assimilation. In: Boote, K. J., Bennett, J. M., Sinclair, T. R. and Paulsen, G. M. (Eds) *Physiology and Determination of Crop Yield*. Madison: Australian Society of Anaesthetists, CSSA, SSSA.

Farquhar, G. D., von Caemmerer, S. and Berry, J. A. 1980. A biochemical model of photosynthetic CO_2 assimilation in leaves of C_3 species. *Planta* 149(1), 78–90.

Field, C. and Mooney, H. A. 1986. The photosynthesis-nitrogen relationship in wild plants. In: Givnish, T. J. (Ed.) *On the Economy of Plant Form and Function*. Cambridge, London: Cambridge University Press.

Fisher, N., Bricker, T. M. and Kramer, D. M. 2019. Regulation of photosynthetic cyclic electron flow pathways by adenylate status in higher plant chloroplasts. *Biochimica et Biophysica Acta. Bioenergetics* 1860(11), 148081.

Fukayama, H., Miyagawa, F., Shibatani, N., Koudou, A., Sasayama, D., Hatanaka, T., Azuma, T., Yamauchi, Y., Matsuoka, D. and Morita, R. 2021. CO_2-responsive CCT protein interacts with 14-3-3 proteins and controls the expression of starch synthesis-related genes. *Plant, Cell and Environment* 44(8), 2480–2493.

Griffiths, C. A., Paul, M. J. and Foyer, C. H. 2016. Metabolite transport and associated sugar signalling systems underpinning source/sink interactions. *Biochimica et biophysica acta* 1857(10), 1715–1725.

Harrison, E. P., Olcer, H., Lloyd, J. C., Long, S. P. and Raines, C. A. 2001. Small decreases in SBPase cause a linear decline in the apparent RuBP regeneration rate, but do not affect RuBisCO carboxylation capacity. *Journal of Experimental Botany* 52(362), 1779–1784.

Johnson, J. E. and Berry, J. A. 2021. The role of cytochrome b_6f in the control of steady-state photosynthesis: a conceptual and quantitative model. *Photosynthesis Research* 148(3), 101–136.

Kacser, H. and Burns, J. A. 1973. The control of flux. *Symposia of the Society for Experimental Biology* 27, 65–104.

Kiirats, O., Cruz, J. A., Edwards, G. E. and Kramer, D. M. 2009. Feedback limitation of photosynthesis at high CO_2 acts by modulating the activity of the chloroplast ATP synthase. *Functional Plant Biology: FPB* 36(11), 893–901.

Körner, C. 2015. Paradigm shift in plant growth control. *Current Opinion in Plant Biology* 25, 107–114.

Kramer, D. M. and Evans, J. R. 2011. The importance of energy balance in improving photosynthetic productivity. *Plant Physiology* 155(1), 70–78.

Kubis, A. and Bar-Even, A. 2019. Synthetic biology approaches for improving photosynthesis. *Journal of Experimental Botany* 70(5), 1425–1433.

Kumarathunge, D. P., Medlyn, B. E., Drake, J. E., Rogers, A. and Tjoelker, M. G. 2019. No evidence for triose phosphate limitation of light saturated leaf photosynthesis under current atmospheric CO_2 concentration. *Plant, Cell and Environment* 42(12), 3241–3252.

Laporte, M. M., Galagan, J. A., Prasch, A. L., Vanderveer, P. J., Hanson, D. T., Shewmaker, C. K. and Sharkey, T. D. 2001. Promoter strength and tissue specificity effects on growth of tomato plants transformed with maize sucrose-phosphate synthase. *Planta* 212(5–6), 817–822.

Laporte, M. M., Galagan, J. A., Shapiro, J. A., Boersig, M. R., Shewmaker, C. K. and Sharkey, T. D. 1997. Sucrose-phosphate synthase activity and yield analysis of tomato plants transformed with maize sucrose-phosphate synthase. *Planta* 203(2), 253–259.

Lastdrager, J., Hanson, J. and Smeekens, S. 2014. Sugar signals and the control of plant growth and development. *Journal of Experimental Botany* 65(3), 799–807.

Lawson, T., Bryant, B., Lefebvre, S., Lloyd, J. C. and Raines, C. A. 2006. Decreased SBPase activity alters growth and development in transgenic tobacco plants. *Plant, Cell and Environment* 29(1), 48–58.

Leegood, R. C., Kobayashi, Y., Neimanis, S., Walker, D. A. and Heber, U. 1982. Co-operative activation of chloroplast fructose-1,6-bisphosphatase by reductant, pH, and substrate. *Biochimica et Biophysica Acta* 682(1), 168–178.

Lefebvre, S., Lawson, T., Zakhleniuk, O. V., Lloyd, J. C., Raines, C. A. and Fryer, M. 2005. Increased sedoheptulose-1,7-bisphosphatase activity in transgenic tobacco plants stimulates photosynthesis and growth from an early stage in development. *Plant Physiology* 138(1), 451–460.

Lima-Melo, Y., Gollan, P. J., Tikkanen, M., Silveira, J. A. G. and Aro, E.-M. 2019. Consequences of photosystem I damage and repair on photosynthesis and carbon utilisation in *Arabidopsis thaliana*. *Plant Journal* 97(6), 1061–1072.

Long, S. P., Ainsworth, E. A., Leakey, A. D. B., Nösberger, J. and Ort, D. R. 2006. Food for thought: lower-than-expected crop yield stimulation with rising CO_2 concentrations. *Science* 312(5782), 1918–1921.

McClain, A. M. and Sharkey, T. D. 2020. Building a better equation for electron transport estimated from chlorophyll fluorescence: accounting for non-photosynthetic light absorption. *New Phytologist* 225(2), 604–608.

McCree, K. J. 1972. The action spectrum, absorptance and quantum yield of photosynthesis in crop plants. *Agricultural Meteorology* 9, 191–216.

Morrison, J. F. 1969. Kinetics of the reversible inhibition of enzyme-catalysed reactions by tight-binding inhibitors. *Biochimica et Biophysica Acta* 185(2), 269–286.

Mott, K. A., Jensen, R. G., O'Leary, J. W. and Berry, J. A. 1984. Photosynthesis and ribulose 1,5-bisphosphate concentrations in intact leaves of *Xanthium strumarium* L. *Plant Physiology* 76(4), 968–971.

Nilsen, E. T. and Forseth, I. N. 2018. The role of leaf movements for optimizing photosynthesis in relation to environmental variation. In: Adams III, W. W. and Terashima, I. (Eds) *The Leaf: A Platform for Performing Photosynthesis*. Cham: Springer International Publishing.

Nishio, J. N. and Whitmarsh, J. 1993. Dissipation of the proton electrochemical potential in intact chloroplasts (II. The pH Gradient Monitored by Cytochrome f Reduction Kinetics). *Plant Physiology* 101(1), 89–96 (The, I. I. pH Gradient Monitored by Cytochrome f Reduction Kinetics).

Niyogi, K. K., Li, X. P., Rosenberg, V. and Jung, H. S. 2005. Is PsbS the site of non-photochemical quenching in photosynthesis? *Journal of Experimental Botany* 56(411), 375–382.

Ogawa, T., Tamoi, M., Kimura, A., Mine, A., Sakuyama, H., Yoshida, E., Maruta, T., Suzuki, K., Ishikawa, T. and Shigeoka, S. 2015. Enhancement of photosynthetic capacity in *Euglena gracilis* by expression of cyanobacterial fructose-1,6-/sedoheptulo se-1,7-bisphosphatase leads to increases in biomass and wax ester production. *Biotechnology for Biofuels* 8, 80.

Ono, K. and Watanabe, A. 1997. Levels of endogenous sugars, transcripts of rbcS and rbcL, and of RuBisCO protein in senescing sunflower leaves. *Plant and Cell Physiology* 38(9), 1032–1038.

Pierce, J., Lorimer, G. H. and Reddy, G. S. 1986. Kinetic mechanism of ribulose bisphosphate carboxylase: evidence for an ordered, sequential reaction. *Biochemistry* 25(7), 1636–1644.

Poolman, M. G., Olçer, H., Lloyd, J. C., Raines, C. A. and Fell, D. A. 2001 Computer modelling and experimental evidence for two steady states in the photosynthetic Calvin cycle. *European Journal of Biochemistry* 268(10), 2810–2816.

Poorter, H., Anten, N. P. R. and Marcelis, L. F. M. 2013. Physiological mechanisms in plant growth models: do we need a supra-cellular systems biology approach? *Plant, Cell and Environment* 36(9), 1673–1690.

Raines, C. A. 2011. Increasing photosynthetic carbon assimilation in C_3 plants to improve crop yield: current and future strategies. *Plant Physiology* 155(1), 36–42.

Raines, C. A., Harrison, E. P., Ölçer, H. and Lloyd, J. C. 2000. Investigating the role of the thiol-regulated enzyme sedoheptulose-1,7-bisphosphatase in the control of photosynthesis. *Physiologia Plantarum* 110(3), 303–308.

Rosenthal, D. M., Locke, A. M., Khozaei, M., Raines, C. A., Long, S. P. and Ort, D. R. 2011. Over-expressing the C_3 photosynthesis cycle enzyme sedoheptulose-1-7 bisphosphatase improves photosynthetic carbon gain and yield under fully open air CO_2 fumigation (FACE). *BMC Plant Biology* 11, 123.

Sage, R. F. 1990. A model describing the regulation of ribulose-1,5-bisphosphate carboxylase, electron transport, and triose phosphate use in response to light intensity and CO_2 in C_3 plants. *Plant Physiology* 94(4), 1728–1734.

Sage, R. F., Sharkey, T. D. and Seemann, J. R. 1989. The acclimation of photosynthesis to elevated CO_2 in five C_3 species. *Plant Physiology* 89(2), 590–596.

Sharkey, T. D. 2019. Discovery of the canonical Calvin-Benson cycle. *Photosynthesis Research* 140(2), 235–252.

Sharkey, T. D. 2021. Pentose phosphate pathway reactions in photosynthesizing cells. *Cells* 10(6), 1547.

Sharkey, T. D., Berry, J. A. and Sage, R. F. 1988. Regulation of photosynthetic electron-transport as determined by room-temperature chlorophyll a fluorescence in *Phaseolus vulgaris* L. *Planta* 176(3), 415–424.

Sharkey, T. D., Preiser, A. L., Weraduwage, S. M. and Gog, L. 2020. Source of ^{12}C in Calvin-Benson cycle intermediates and isoprene emitted from plant leaves fed with $^{13}CO_2$. *Biochemical Journal* 477(17), 3237–3252.

Sharkey, T. D., Seemann, J. R. and Berry, J. A. 1986a. Regulation of ribulose-1,5-bisphosphate carboxylase activity in response to changing partial pressure of O_2 and light in *Phaseolus vulgaris*. *Plant Physiology* 81(3), 788–791.

Sharkey, T. D., Stitt, M., Heineke, D., Gerhardt, R., Raschke, K. and Heldt, H. W. 1986b. Limitation of photosynthesis by carbon metabolism. II O_2 insensitive CO_2 uptake results from limitation of triose phosphate utilization. *Plant Physiology* 81(4), 1123-1129.

Shikanai, T. and Yamamoto, H. 2017. Contribution of cyclic and pseudo-cyclic electron transport to the formation of proton motive force in chloroplasts. *Molecular Plant* 10(1), 20–29.

Shirao, M., Kuroki, S., Kaneko, K., Kinjo, Y., Tsuyama, M., Förster, B., Takahashi, S. and Badger, M. R. 2013. Gymnosperms have increased capacity for electron leakage to oxygen (Mehler and PTOX reactions) in photosynthesis compared with angiosperms. *Plant and Cell Physiology* 54(7), 1152–1163.

Simkin, A. J., López-Calcagno, P. E. and Raines, C. A. 2019. Feeding the world: improving photosynthetic efficiency for sustainable crop production. *Journal of Experimental Botany* 70(4), 1119–1140.

Simkin, A. J., McAusland, L., Lawson, T. and Raines, C. A. 2017. Overexpression of the RieskeFeS protein increases electron transport rates and biomass yield. *Plant Physiology* 175(1), 134–145.

Sinclair, T. R., Rufty, T. W. and Lewis, R. S. 2019. Increasing photosynthesis: unlikely solution for world food problem. *Trends in Plant Science* 24(11), 1032–1039.

Storti, M., Alboresi, A., Gerotto, C., Aro, E. M., Finazzi, G. and Morosinotto, T. 2019. Role of cyclic and pseudo-cyclic electron transport in response to dynamic light changes in Physcomitrella patens. *Plant, Cell and Environment* 42(5), 1590–1602.

Strand, D. D., Fisher, N. and Kramer, D. M. 2017. The higher plant plastid NAD(P)H dehydrogenase-like complex (NDH) is a high efficiency proton pump that increases ATP production by cyclic electron flow. *The Journal of Biological Chemistry* 292(28), 11850-11860.

Strand, D. D., Livingston, A. K., Satoh-Cruz, M., Froehlich, J. E., Maurino, V. G. and Kramer, D. M. 2015. Activation of cyclic electron flow by hydrogen peroxide in vivo. *Proceedings of the National Academy of Sciences of the United States of America* 112(17), 5539-5544.

Takabayashi, A., Kishine, M., Asada, K., Endo, T. and Sato, F. 2005. Differential use of two cyclic electron flows around photosystem I for driving CO_2-concentration mechanism in C_4 photosynthesis. *Proceedings of the National Academy of Sciences of the United States of America* 102(46), 16898–16903.

Takizawa, K., Cruz, J. A., Kanazawa, A. and Kramer, D. M. 2007. The thylakoid proton motive force in vivo. Quantitative, non-invasive probes, energetics, and regulatory consequences of light-induced pmf. *Biochimica et Biophysica Acta* 1767(10), 1233–1244.

Takizawa, K., Kanazawa, A. and Kramer, D. M. 2008. Depletion of stromal P_i induces high 'energy-dependent' antenna exciton quenching (q_E) by decreasing proton conductivity at CF_O-CF_1 ATP synthase. *Plant, Cell and Environment* 31(2), 235–243.

Terashima, I., Fujita, T., Inoue, T., Chow, W. S. and Oguchi, R. 2009. Green light drives leaf photosynthesis more efficiently than red light in strong white light: revisiting the enigmatic question of why leaves are green. *Plant and Cell Physiology* 50(4), 684–697.

Tietz, S., Hall, C. C., Cruz, J. A. and Kramer, D. M. 2017. $NPQ_{(T)}$: a chlorophyll fluorescence parameter for rapid estimation and imaging of non-photochemical quenching of excitons in photosystem-II-associated antenna complexes. *Plant, Cell and Environment* 40(8), 1243–1255.

Vanlerberghe, G. C., Martyn, G. D. and Dahal, K. 2016. Alternative oxidase: a respiratory electron transport chain pathway essential for maintaining photosynthetic performance during drought stress. *Physiologia Plantarum* 157(3), 322–337.

von Caemmerer, S. 2020. RuBisCO carboxylase/oxygenase: from the enzyme to the globe: a gas exchange perspective. *Journal of Plant Physiology* 252, 153240.

von Caemmerer, S. and Farquhar, G. D. 1981. Some relationships between the biochemistry of photosynthesis and the gas exchange of leaves. *Planta* 153(4), 376–387.

Wada, M. and Kong, S. G. 2018. Actin-mediated movement of chloroplasts. *Journal of Cell Science* 131(2), jcs210310.

Weraduwage, S. M., Chen, J., Anozie, F. C., Morales, A., Weise, S. E. and Sharkey, T. D. 2015. The relationship between leaf area growth and biomass accumulation in *Arabidopsis thaliana*. *Frontiers in Plant Science* 6, 167.

Woodrow, I. E. and Walker, D. A. 1983. Regulation of stromal sedoheptulose-1,7-bisphosphatase activity and its role in controlling the reductive pentose phosphate pathway of photosynthesis. *Biochimica et Biophysica Acta* 722(3), 508–516.

Worrell, A. C., Bruneau, J. M., Summerfelt, K., Boersig, M. and Voelker, T. A. 1991. Expression of a maize sucrose phosphate synthase in tomato alters leaf carbohydrate partitioning. *The Plant Cell* 3(10), 1121–1130.

Yamori, W. and Shikanai, T. 2016. Physiological functions of cyclic electron transport around photosystem I in sustaining photosynthesis and plant growth. *Annual Review of Plant Biology* 67, 81–106.

Yang, J. T., Preiser, A. L., Li, Z., Weise, S. E. and Sharkey, T. D. 2016. Triose phosphate use limitation of photosynthesis: short-term and long-term effects. *Planta* 243(3), 687–698.

Yin, X., Busch, F. A., Struik, P. C. and Sharkey, T. D. 2021. Evolution of a biochemical model of steady-state photosynthesis. *Plant, Cell and Environment* 44(9), 2811–2837.

Yin, X., Struik, P. C., Romero, P., Harbinson, J., Evers, J. B., Van Der Putten, P. E. L. and Vos, J. A. N. 2009. Using combined measurements of gas exchange and chlorophyll fluorescence to estimate parameters of a biochemical C3 photosynthesis model:

a critical appraisal and a new integrated approach applied to leaves in a wheat (Triticum aestivum) canopy. *Plant, Cell and Environment* 32(5), 448–464.

Zhen, S. and Bugbee, B. 2020. Far-red photons have equivalent efficiency to traditional photosynthetic photons: implications for re-defining photosynthetically active radiation. *Plant, Cell and Environment* 43(5), 1259-1272.

Zhu, X. G., de Sturler, E. and Long, S. P. 2007. Optimizing the distribution of resources between enzymes of carbon metabolism can dramatically increase photosynthetic rate: a numerical simulation using an evolutionary algorithm. *Plant Physiology* 145(2), 513-526.

Zhu, X. G., Long, S. P. and Ort, D. R. 2010. Improving photosynthetic efficiency for greater yield. *Annual Review of Plant Biology* 61, 235–261.

Chapter 10

Improving proteins to optimize photosynthesis

James V. Moroney, Ashwani K. Rai, Hiruni Weerasooriya and Remmy Kasili, Louisiana State University, USA; and Marylou Machingura, Georgia Southern University, USA

1 Introduction

In 2013 it was estimated that crop yields need to increase 50% by the year 2050 to feed the increasing world population (Ray et al., 2013; Jaggard et al., 2010). Most of these increases will be realized through traditional plant breeding, but some improvements will be accomplished by introducing beneficial genes, such as those that increase herbicide resistance. However, in an additional effort to boost yields, scientists are also working to improve photosynthetic efficiency in terrestrial plants. Free air CO_2 enrichment experiments (FACE) have demonstrated that crop plants respond favorably to increases in CO_2, which enhances photosynthesis and yields (Zhu et al., 2010; Ainsworth and Long, 2021).

The observation that increases in CO_2 concentration lead to enhancements in growth and yield can be explained by the enzymatic properties of ribulose

http://dx.doi.org/10.19103/AS.2022.0119.14

1,5 bisphosphate carboxylase/oxygenase (RubisCO). RubisCO is the enzyme that catalyzes the addition of CO_2 to the sugar ribulose 1,5 bisphosphate and it is responsible for 'fixing' gaseous CO_2 into the organic molecules that make up plants. RubisCO has a low affinity for CO_2 and current atmospheric CO_2 levels are below levels that saturate RubisCO. This means that when CO_2 levels go up, the enzyme can catalyze more reactions and photosynthesis can increase. RubisCO also catalyzes the addition of O_2 to ribulose 1,5 bisphosphate, hence the 'oxygenase' in its name. This oxygenation reaction is considered wasteful because O_2 directly competes with CO_2 which reduces CO_2 fixation, photosynthesis and ultimately growth. Thus, by raising CO_2 in the field during the FACE experiments, CO_2 fixation was increased and the deleterious oxygenation reaction was reduced (Ainsworth and Long, 2021). The observed increases in yield led scientists to investigate other ways that CO_2 fixation could be increased in crop plants.

One potential way to increase photosynthetic efficiency is to introduce some or all of the components that make up algal CO_2 concentrating mechanisms (CCMs) into plant chloroplasts (Price et al., 2011; Atkinson et al., 2016; Nölke et al., 2019). Algae have adapted to the low CO_2 levels in the atmosphere by developing ways to increase the CO_2 concentration around RubisCO. This essentially does what the FACE experiments did in the field: increases CO_2 fixation and decreases the competing oxygenase activity.

Algal CCMs are extremely common in nature. They are found in prokaryotic cyanobacteria as well as almost all eukaryotic algae including green algae, red algae, diatoms and coccolithophores. It is thought that the overwhelming majority of these aquatic organisms employ a CCM to increase photosynthetic efficiency. The CCMs used by these organisms incorporate inorganic carbon transporters and carbonic anhydrase (CA) enzymes to increase CO_2 fixation. Inorganic carbon (C_i) refers to the inorganic carbon-containing molecules found in water, namely carbon dioxide (CO_2), bicarbonate (HCO_3^-) and carbonate (CO_3^{-2}). The proteins that help these molecules cross biological membranes are referred to as C_i transporters because it is often quite difficult to determine which form of carbon they transport.

The other protein that plays an essential role in algal CCMs is carbonic anhydrase or CA. CA catalyzes the interconversion of CO_2 and HCO_3^-, as seen in Equation 1. CAs are found in almost all organisms since a number of metabolic reactions require either CO_2 or HCO_3^- (DiMario et al., 2017). These two forms of carbon can interconvert in the absence of CA, but the process is slow. When CA is present, the interconversion is very fast, allowing metabolic reactions to take place efficiently.

$$CO_2 + H_2O \rightleftharpoons H_2CO_3 \rightleftharpoons H^+ + HCO_3^- \quad (1)$$

In contrast, most terrestrial plants do not have a CCM. One proposed explanation for this fact is that CO_2 diffusion in air is 10 000x faster than CO_2

diffusion in water. So, while CO_2 that enters a leaf is rapidly replenished by CO_2 diffusing in from the air, CO_2 that enters an aquatic organism is much more slowly replaced. Thus, algae have been forced to adapt to low CO_2 levels more than terrestrial plants. For many important crop plants such as soybean, rice and wheat, there is no evidence that they have an active CCM. A simple model of a leaf mesophyll cell chloroplast from one of these plants is shown in Fig. 1.

In aquatic environments, organisms often experience wide swings in C_i concentration in the environment and almost every photosynthetic organism has a CCM. In addition, algae represent an extremely diverse group of organisms, with many of the eukaryotic algae having different chloroplast origins. As a result, the molecular details of CCMs from divergent species vary, since each algal lineage evolved a CCM separately (Griffiths et al., 2017). However, even though the details vary, the basic structure of the algal CCM appears to be conserved. This basic CCM structure will be represented in the next section along with specific examples from cyanobacteria and green algae.

Before discussing the CCM in general terms, it is necessary to mention two important differences between CO_2 and HCO_3^- molecules. The first important point is that CO_2, a small uncharged molecule, can rapidly pass through biological membranes, while HCO_3^-, a charged molecule, crosses membranes

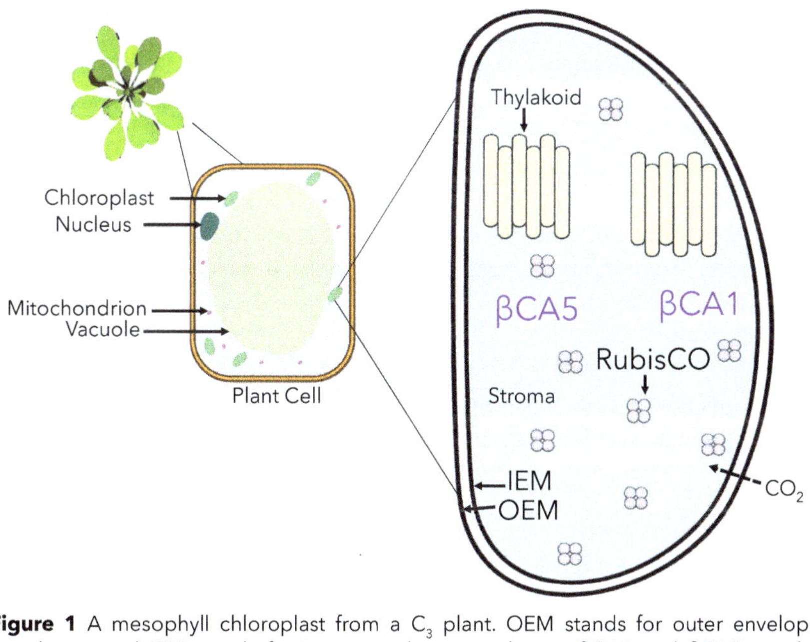

Figure 1 A mesophyll chloroplast from a C_3 plant. OEM stands for outer envelope membrane and IEM stands for inner envelope membrane. βCA1 and βCA5 are the carbonic anhydrases found in the chloroplast stroma. For C_3 plants, it is thought that CO_2 diffuses into the chloroplast where it is fixed by RubisCO.

over 1000 times more slowly by recent estimates (Stillwell, 2016; Tolleter et al., 2017). The second is that CO_2 is the substrate for RubisCO, not HCO_3^-. Therefore, a successful CCM must increase the concentration of CO_2 around RubisCO. But how can a unicellular alga concentrate a molecule like CO_2, which can readily leak out of the cell? The answer is that the alga concentrates HCO_3^-, which does not readily cross membranes, and then converts that HCO_3^- to CO_2 near RubisCO using a CA. This increases the CO_2 concentration and also dictates how CCMs are set up in algae - making it so that organisms as different as green algae and diatoms have CCMs that are similar in many aspects.

2 A general carbon dioxide concentrating mechanism model

To date, all CCMs that have been investigated have three similar properties. The first property is that RubisCO is packaged into a very specific location within the cell. For cyanobacteria, this location is the carboxysome. For eukaryotic algae, the location is the chloroplast pyrenoid. The second common feature of all CCMs is C_i transporters (Price et al., 2008). These transporters usually move C_i across a biological membrane, effectively increasing the HCO_3^- inside the cell or chloroplast. A wide variety of C_i transporters have been discovered in algae, but all of them share the common function of increasing HCO_3^- inside the cell or chloroplast. The third common component of all CCMs is a CA located close to the packaged RubisCO. In cyanobacteria, a CA is always present inside the carboxysome along with RubisCO (So et al., 2004; Peña et al., 2010). The function of this CA is to convert the HCO_3^- accumulated by the C_i transporters to CO_2 for RubisCO. This creates a microenvironment inside the carboxysome where the CO_2 concentration is higher, thus increasing the carboxylase activity of RubisCO. In green algae, there is a CA inside the thylakoid tubules that permeate the pyrenoid. Like the carboxysome, the CA in the pyrenoid is in close proximity to RubisCO, so it can increase CO_2 for that enzyme and enhance photosynthesis. These three CCM components - packaged RubisCO, C_i transporters and CA enzymes - work together to increase CO_2, allowing the algae to grow well even under limiting CO_2 conditions.

2.1 The cyanobacterial carbon dioxide concentrating mechanism

A general model of the cyanobacterial CCM is shown in Fig. 2. Cyanobacteria are prokaryotes and therefore have relatively simple architecture. Cyanobacteria have a plasma membrane that controls which molecules enter and exit the cell. This is where the C_i transporters are located. There are different types of C_i transporters which can have different properties. Some, like SbtA, are

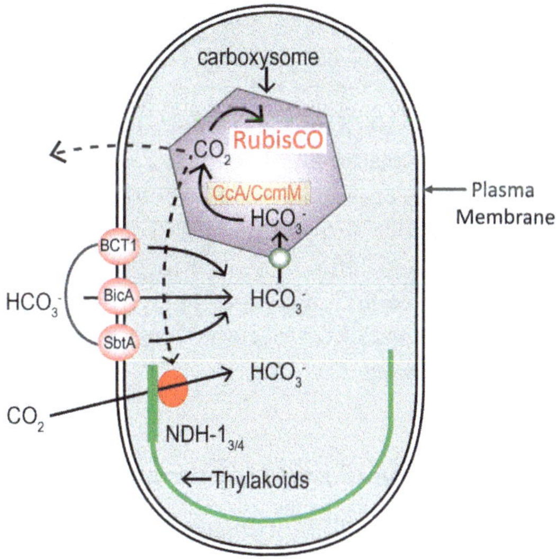

Figure 2 General model of the cyanobacterial CCM. BCT1, BicA and SbtA are C_i transporters located on the plasma membrane. The electron transport complex NDH-$1_{3/4}$ contains the CupA and CupB proteins that convert CO_2 to HCO_3^-. The HCO_3^- concentration in the cytoplasm can be over 1000x the external concentration. HCO_3^- enters the carboxysome through pores in the carboxysomal shell, where it is then converted to CO_2 by CAs. The CO_2 generated is fixed by carboxysomal RubisCO.

cotransporters that simultaneously move both Na^+ ions and HCO_3^-. SbtA is thought to utilize the Na^+ gradient across the plasma membrane (high Na^+ outside the cell, low Na^+ inside the cell) to 'pump' the HCO_3^- inside the cytoplasm of the cyanobacteria. Another protein, BCT1, is an ABC-type transporter which likely utilizes the energy in the form of ATP to 'pump' HCO_3^-, which increases the concentration of HCO_3^- inside the cell versus outside the cell (Omata et al., 1999; Shibata et al., 2002; Price, 2011). These transporters can increase the C_i concentration inside the cell by over 1000 fold (Price et al., 2008).

A second prominent feature of the cyanobacterial CCM is the carboxysome (Long et al., 2018). The carboxysome is a proteinaceous structure that contains all the cell's RubisCO. It is composed of shell proteins and is filled with RubisCO. In addition to RubisCO, the carboxysome contains a CA which converts the HCO_3^- that enters the carboxysome to CO_2. This creates a microenvironment high in CO_2 for RubisCO.

Two other features of the cyanobacterial CCM should be mentioned. The first is the role of the proteins CupA and CupB (Schuller et al., 2020). These proteins are part of the NDH complex of the cyanobacterial electron transport chain and appear to convert CO_2 to HCO_3^- inside the cell (Han et al., 2017). Interestingly, they appear to convert CO_2 to HCO_3^- but not in the reverse direction. The mechanism of these proteins is poorly understood, but it likely involves transport of H^+ across the thylakoid membrane. Their role is to convert any CO_2 leaking out of the carboxysome back to HCO_3^- before it can escape from the cell. The other notable feature of the cyanobacterial CCM is the lack of CA outside of the carboxysome. If an active CA was present in the cytoplasm, it would convert the accumulated HCO_3^- to CO_2, which would rapidly leak from the cell. This would set up a futile cycle, with the cell pumping in HCO_3^- only to lose the C_i as CO_2. Price and Badger (1989) confirmed this by inserting mammalian CA into cyanobacteria. This introduced CA short-circuited the CCM and created cells that could not grow at air levels of CO_2. As will be seen later, the intracellular locations of CA become very important when trying to construct a CCM in a crop plant.

2.2 The green algal carbon dioxide concentrating mechanism

The other well-studied CCM is the CCM of *Chlamydomonas reinhardtii* (hereafter referred to as Chlamydomonas), a unicellular green alga. Chlamydomonas is a eukaryote, so the architecture of the cell is more complicated than the cyanobacterial cell. In this case, C_i must cross three sets of membranes to reach the key CA: the plasma membrane, the chloroplast envelope and the chloroplast thylakoid membrane. However, the basic components of the CCM are the same as seen in the cyanobacterial CCM - namely packaged RubisCO, C_i transporters and CAs.

In Chlamydomonas, RubisCO is packaged in a structure known as a pyrenoid (Fig. 3). This arrangement of RubisCO is different than what is seen in terrestrial plants, where the RubisCO is evenly distributed throughout the

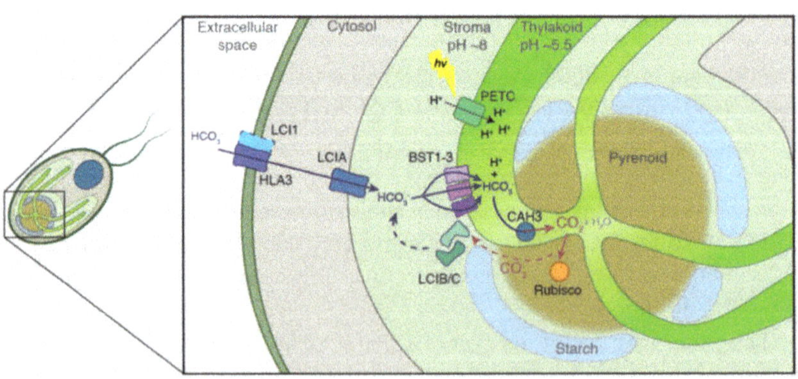

Figure 3 Model for the CCM of Chlamydomonas. C_i transporters bring the HCO_3^- across the plasma membrane (PM), the chloroplast envelope (CE) and the thylakoid membrane (TM) to the thylakoid lumen. Then the HCO_3^- is converted to CO_2 by CAH3. The thylakoid lumen is acidic, as the photosynthetic electron transport chain (PETC) pumps H^+ into the thylakoid lumen. The low pH drives the HCO_3^- toward CO_2.

chloroplast stroma (Fig. 1). There are two other important features of the pyrenoid. First, the pyrenoid is penetrated by thylakoid tubules (Meyer et al., 2017). These tubules are extensions of the thylakoid membranes and contain the carbonic anhydrase CAH3 (Mitra et al., 2005). CAH3 is the CA that converts HCO_3^- to CO_2 for RubisCO. Since it is present within the thylakoid tubules inside the pyrenoid, CAH3 is positioned to generate CO_2 in the middle of the packaged RubisCO and its role is similar to the CA in the carboxysome in cyanobacteria. The other notable feature is the prominent starch sheath that surrounds the pyrenoid. This starch sheath helps organize the pyrenoid structure and probably provides a barrier to CO_2 diffusion, reducing the leak of CO_2 from the pyrenoid.

The CCM of Chlamydomonas also has a number of C_i transporters. C_i transporters are found on the plasma membrane, chloroplast envelope and thylakoid membrane (Fig. 3). In the current model of the Chlamydomonas CCM, these transporters work together to bring HCO_3^- from outside of the cell, across three membranes, into the thylakoid lumen (Mukherjee et al., 2019). Finally, once inside the thylakoid, CAH3 converts HCO_3^- to CO_2 for RubisCO.

2.3 Differences between the cyanobacterial carbon dioxide concentrating mechanism and the Chlamydomonas carbon dioxide concentrating mechanism

There is a fundamental difference in how the Chlamydomonas CCM operates compared to the cyanobacterial system. The difference is in how CO_2 accumulation takes place. In cyanobacteria, the CO_2 concentration is increased by pumping up the cytoplasmic HCO_3^- concentration to very high levels using transport proteins. As mentioned earlier, HCO_3^- can accumulate inside the cell up to 1000 times the external HCO_3^- concentration. Some of this pumping is done by C_i transporters that take advantage of the Na^+ gradient across the cyanobacterial plasma membrane. Many of the best-studied cyanobacterial CCMs are from either marine organisms or cells that live in brackish water, where the external HCO_3^- and Na^+ concentrations are high. Cyanobacteria take advantage of these naturally occurring HCO_3^- and Na^+ concentrations to increase HCO_3^- inside the cell. Once the HCO_3^- is pumped into the cell it is converted to CO_2 in the carboxysome. Since the CO_2 and HCO_3^- are in equilibrium, this raises the CO_2 concentration as well. Therefore, the increase in CO_2 is mostly accomplished by the plasmalemma C_i transporters in cyanobacteria.

In contrast, Chlamydomonas appears to depend on the pH gradient across the thylakoid membrane to increase the CO_2 concentration. This gradient is set up because of the proton pumping that occurs during the light reactions of photosynthesis. When light energy is absorbed by the two photosystems,

H^+ ions are pumped from the chloroplast stroma into the thylakoid lumen. Because of that H^+ pumping, the chloroplast stroma has a basic pH of about 8.0, while the thylakoid lumen has an acidic pH between 5.5 and 6.0. The pKa of the CO_2 to HCO_3^- interconversion is ~6.4, which makes it so that the favored C_i species in the basic stroma is HCO_3^-, while in the acidic lumen CO_2 is the more abundant C_i species.

The CO_2 concentration in water that is in equilibrium with air is approximately 15 μM, and this value is independent of pH. At pH 8.0 the HCO_3^- is about 50 times the CO_2 concentration at equilibrium, while at pH 5.8 the HCO_3^- would be about one-third of the CO_2 concentration. Table 1 shows what the CO_2 and HCO_3^- concentrations would be in the cytoplasm, chloroplast stroma and thylakoid lumen if those organelles were in equilibrium with air. These concentrations represent what we would expect in a C_3 crop plant or a Chlamydomonas cell without a CCM. The CO_2 concentration is constant at 15 μM but the HCO_3^- concentration varies because of the different pH of each compartment. Table 2 shows what can happen if a CCM is introduced. In this case, the HCO_3^- concentration is held constant at the value found in the cytoplasm of Table 1. In an organism with a CCM, there are transport proteins that can move HCO_3^- across the different membranes (Fig. 4). What this means is that if HCO_3^- is moved from the chloroplast stroma to the thylakoid lumen, it will go from an environment that strongly favors HCO_3^- formation to one that favors CO_2 formation. So if the HCO_3^- concentration was 75 μM in the stroma, moving the HCO_3^- into the lumen could easily raise the CO_2 concentration

Table 1 CO_2 and HCO_3^- concentrations in different cell compartments with CO_2 held constant at 15 μM. We assume the CO_2 concentration is in equilibrium with air. At standard temperature and pressure, the concentration of CO_2 dissolved in water is about 15 μM and this concentration does not vary with pH. However, since H_2CO_3 is a weak acid, the concentration of HCO_3^- varies with pH

Cell compartment	pH	Air level CO_2	[HCO_3^-]
Cytoplasm	7.0	15 μM	75 μM
Chloroplast stroma	8.0	15 μM	750 μM
Thylakoid lumen	5.8	15 μM	5 μM

Table 2 CO_2 and HCO_3^- concentrations in different cell compartments when the HCO_3^- concentration is held constant at 75 μM. When the HCO_3^- is allowed to enter the acidic thylakoid lumen, a high concentration of CO_2 can be generated

Cell compartment	pH	[CO_2] at equilibrium	[HCO_3^-]
Cytoplasm	7.0	15 μM	75 μM
Chloroplast stroma	8.0	1.5 μM	75 μM
Thylakoid lumen	5.8	225 μM	75 μM

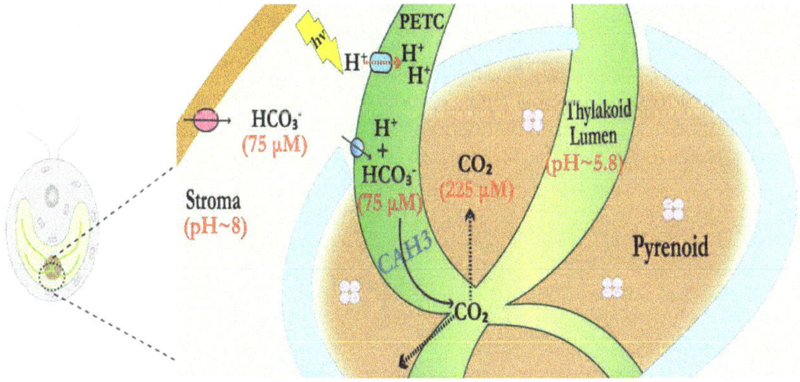

Figure 4 How a pH gradient can increase [CO_2]. The photosynthetic electron transport chain (PETC) pumps H^+ into the thylakoid lumen. This causes the thylakoid lumen to become more acidic, while the chloroplast stroma becomes more basic. As HCO_3^- enters the thylakoid, it goes from a basic environment that favors HCO_3^- formation to an acidic environment that favors CO_2 formation. This results in a large increase in the CO_2 concentration. The CO_2 then diffuses out of the thylakoid, creating a microenvironment high in CO_2 for RubisCO.

inside the lumen to 75 µM or higher as it is converted to CO_2 by CAH3 (Fig. 4 and Table 2). Since CO_2 can easily cross membranes, it will diffuse out of the thylakoid tubules to RubisCO. This means that photosynthetic electron transport is the driver of CO_2 accumulation in Chlamydomonas.

2.4 Putting algal carbon dioxide concentrating mechanism components into crop plants

As our understanding of algal CCMs increased, scientists began to model whether adding a CCM to the chloroplast of a C_3 plant could increase photosynthetic efficiency. The models predicted that CO_2 fixation could potentially be increased between 9% and 60%, depending on the number of CCM components put into the crop plants and the efficiency of the system (Price et al., 2011; McGrath and Long, 2014). For this article, we will consider five different strategies (Table 3) that could be used to introduce a CCM into terrestrial crop plants. They range from the simple envelope transporter model to full-blown pyrenoid-based or carboxysome-based models.

3 The envelope transporter-based strategy

The simplest way to introduce a partial CCM into a crop plant is to add a C_i transporter to the inner envelope membrane of the chloroplast (Fig. 5). In theory, the addition of a C_i transporter at the envelope would increase the HCO_3^- in the

Table 3 Five different strategies to incorporate CCM components into terrestrial plants

Model name	Brief description
Envelope transport	The simplest model with only a chloroplast envelope C_i transporter added.
Thylakoid-based model	A model based on the thylakoid pH-driven CO_2 generation seen in Chlamydomonas.
Protopyrenoid-based	A model where RubisCO is packaged along with a CA. The C_i would be increased using an envelope transporter.
Pyrenoid-based	A model where close to the entire Chlamydomonas CCM is transferred to the plant.
Carboxysome-based	A model where the components of the cyanobacterial carboxysome are transferred to the plant along with envelope C_i transporters.

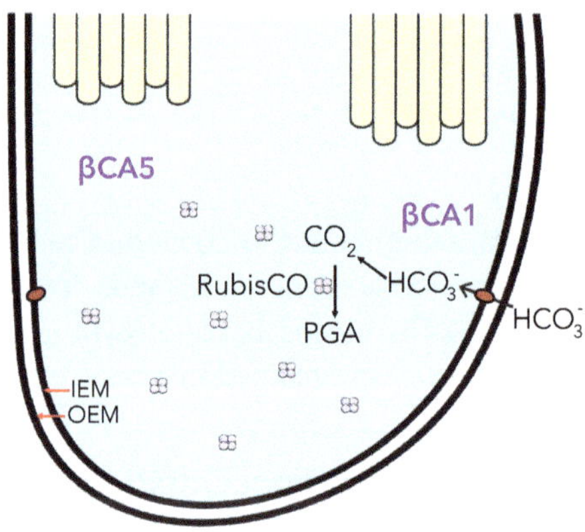

Figure 5 The envelope transporter strategy. For this modification, C_i transporters are introduced (red ovals) to the IEM of the plant chloroplast. This would increase both the HCO_3^- and CO_2 concentrations inside the chloroplast stroma. In this depiction, the endogenous CAs, βCA1 and βCA5, are left in place.

chloroplast stroma. Since higher plants have a lot of CA activity in the stroma, any increase in HCO_3^- would also cause an increase in CO_2. Since the concentration of CO_2 inside the chloroplast stroma is below the $K_m(CO_2)$ of RubisCO, any increase in CO_2 will lead to an increase in carboxylase activity. Various groups have estimated this potential increase in carboxylation activity to be between 5% and 10%. While this increase in fixation is modest, it is still significant. Multiple groups have attempted to introduce a C_i transporter into terrestrial plants. Later in this chapter we will summarize these attempts and some of the technical difficulties associated with the introduction of these transporters.

The advantage of this approach is that it is the simplest. It would require the addition of only one or a few genes encoding chloroplast envelope C_i

transporters. This would result in an increase in HCO_3^- and thus an increase in CO_2. The disadvantage of this approach is that it is predicted to produce the smallest increase in photosynthesis of any of the strategies.

4 The thylakoid-based strategy

The goal of the thylakoid-based approach is to insert the thylakoid components of the Chlamydomonas CCM into a crop plant. To accomplish this, at least two proteins must be added to the terrestrial plant. One is a C_i transporter that needs to be localized to the thylakoid membrane in the plant chloroplast (Fig. 6). This transporter is a bestrophin-like protein normally found in the thylakoid membrane of Chlamydomonas. These bestrophin-like proteins were first reported in Chlamydomonas in 2019 and they are widely distributed in eukaryotic organisms (Mukherjee et al., 2019). Bestrophins carry a number of ions, but Cl^- and HCO_3^- are the most common. Chlamydomonas has three closely related bestrophin-like proteins found in the thylakoid, and reducing their expression severely reduces C_i uptake by the alga.

The second component that would have to be added in this approach is a carbonic anhydrase called CAH3 (Moroney et al., 1989). CAH3 is found in the thylakoid lumen and, more importantly, the pyrenoid thylakoid tubules of Chlamydomonas (Karlsson et al., 1998; Blanco-Rivero et al., 2012). The physiological role of CAH3 is to convert the HCO_3^- that enters the thylakoid to CO_2 for fixation by RubisCO. As mentioned earlier, the pH gradient across the thylakoid membrane is what drives the HCO_3^- to CO_2 (Fig. 6 and Table 2).

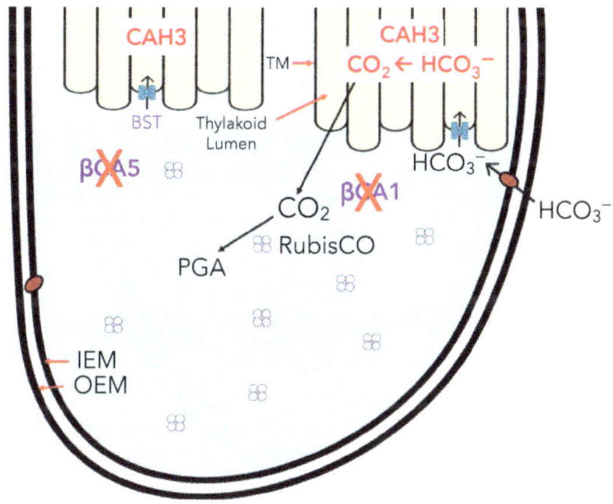

Figure 6 Thylakoid-based strategy. For this approach, C_i transporters are put into the IEM and the thyalkoid membrane (TM). The two endogenous CAs are eliminated and a new thylakoid CA, CAH3, is added. The method would take advantage of the pH gradient across the TM to generate increased concentrations of CO_2.

However, putting in bestrophins and CAH3 is probably not enough to increase CO_2 fixation. The existing chloroplast CAs, βCA1 and βCA5, will have to be eliminated for this approach to work. Our current thinking is that if these CAs were present, they would create a futile cycle of HCO_3^- going to CO_2 and then back to HCO_3^- (Fig. 7). In this cycle, HCO_3^- would be brought into the thylakoid and converted to CO_2 by CAH3. The resulting CO_2 would then leak out of the thylakoid and cause the CO_2 and HCO_3^- in the chloroplast stroma to be in disequilibrium because the CO_2 concentration is now too high. If βCA1 and/or βCA5 were still present in the chloroplast stroma, the CO_2 would simply be converted back to HCO_3^- and there would be no increase in CO_2 for RubisCO (Fig. 7). If the two CAs are knocked out, however, the high amount of CO_2 coming in from the thylakoid would persist and CO_2 fixation by RubisCO would increase. The elimination of these two CAs is required for most of the strategies discussed in this section. Later, there will be an additional section about these chloroplast CAs and efforts being made to eliminate their expression in terrestrial plants.

An advantage of this strategy is that bestrophins and CAH3 are normally targeted to the thylakoid in Chlamydomonas, so they should go to their correct locations in a plant chloroplast. In addition, no change needs to be made to RubisCO or its localization. The endogenous CAs will need to be removed, but most of the approaches require the same CAs to be removed.

The disadvantage of this approach is it could lead to a futile cycle if the leak of CO_2 from the plant chloroplast is too high. This strategy might also uncouple the thylakoids if the conversion of HCO_3^- to CO_2 in the thylakoid creates a

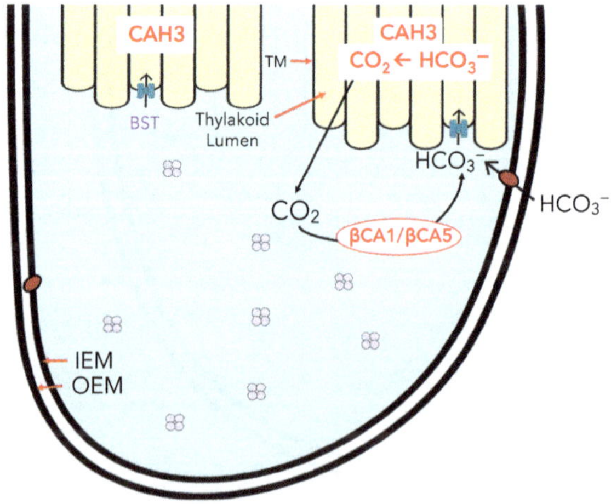

Figure 7 The possible futile cycle created by a thylakoid C_i transporter and CAH3. In this scheme, HCO_3^- is brought into the thylakoid and converted to CO_2 by CAH3. The CO_2 generated then leaks out of the thylakoid and, instead of being fixed by RubisCO, is converted back to HCO_3^- by the endogenous CAs, βCA1 and βCA5.

reduction in the pH gradient across the membrane. In Chlamydomonas, it is likely that CAH3 is active only in the thylakoid tubules within the pyrenoid (Blanco-Rivero et al., 2012). If that is the case, then the pH gradient in the rest of the thylakoid would be normal, leading to normal ATP production. Overexpression of either the thylakoid C_i transporter or CAH3 in crop plants might lead to lower ATP production and cripple photosynthesis. The reason for this is that the plant uses the H^+ gradient for ATP production and the generation of CO_2 from HCO_3^- consumes an H^+ ion (Equation 1).

5 The protopyrenoid-based strategy

Like the thylakoid-based approach, the protopyrenoid-based strategy will partially recreate the Chlamydomonas CCM in terrestrial plants (Fig. 8). The name 'protopyrenoid' arises from a recent paper (Atkinson et al., 2020) in which the authors report on a partially rebuilt pyrenoid in tobacco. The researchers gathered most of the RubisCO together by inserting the protein EPYC1 into Arabidopsis and substituting the Chlamydomonas RubisCO small subunit RbcS2 for the endogenous Arabidopsis RbcS genes. In Chlamydomonas, EPYC1 has five RbcS binding sites and binds to the small subunit of RubisCO. As a result, EPYC1 binds to multiple RubisCO holoenzymes and gathers them together. Jonikas and his colleagues have identified other proteins that help further define the mature pyrenoid structure (Mackinder et al., 2016).

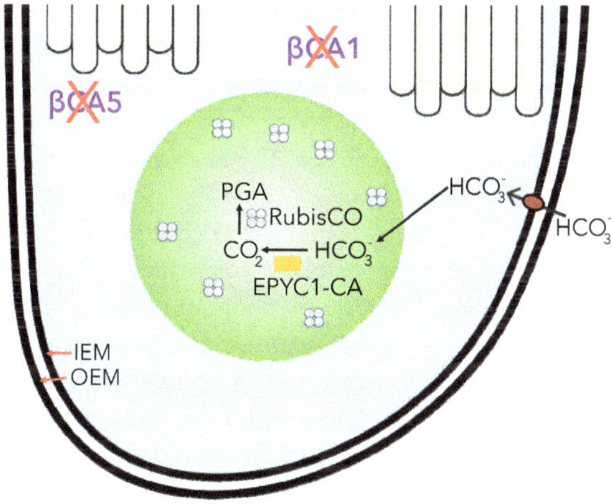

Figure 8 The protopyrenoid strategy. For this approach, Chlamydomonas EPYC1 and Chlamydomonas RbcS are introduced into the plant. This gathers RubisCO together into a protopyrenoid structure. A C_i transporter is also included to increase HCO_3^- and CO_2 concentrations inside the chloroplast stroma. In addition, the two endogenous CAs are eliminated and a new protopyrenoid-localized CA is added. In this model, the CA is linked to EPYC1 to bind it to the protopyrenoid.

In previous work, it was shown that RbcS from Arabidopsis did not bind EPYC1, thus necessitating the substitution of RbcS2 from Chlamydomonas (Meyer et al., 2012). It is impressive that the hybrid RubisCO containing large subunits from Arabidopsis and small subunits from Chlamydomonas shows catalytic properties similar to the Arabidopsis or Chlamydomonas holoenzymes.

So far, part of the protopyrenoid strategy, namely the assembly of RubisCO, has been accomplished. However, at least three other tasks must be accomplished for this approach to increase CO_2 for RubisCO. First, a C_i transporter would need to be included on the chloroplast envelope (Fig. 8). The addition of the transporter would resemble the first strategy. Second, the endogenous CAs, βCA1 and βCA5, would need to be eliminated as described in the thylakoid-based strategy. If the endogenous CAs were left in the chloroplast, any HCO_3^- that was pumped into the chloroplast by the C_i transporter would be converted to CO_2 and leak out before reaching RubisCO. Finally, a CA would need to be present inside the protopyrenoid (Fig. 8). This way, the HCO_3^- pumped into the chloroplast would be converted to CO_2 inside the protopyrenoid. This would raise the CO_2 concentration in the protopyrenoid where RubisCO is assembled. It is not clear how a CA would be brought into the protopyrenoid, but in theory, there are multiple ways this might be accomplished. First, a CA could be linked to EPYC1 so that it would be part of the protopyrenoid as it assembles (Mackinder et al., 2016). A CA could also be linked to an EPYC1 binding domain, similar to the domain on RbcS, so the CA also binds to EPYC1 and becomes part of the protopyrenoid. The CA could also be designed to have two or three RbcS2 binding domains, mimicking EPYC1. In this model, the added CA would somewhat resemble the cyanobacterial CcmM protein, which has a catalytic CA domain, plus three RbcS-like domains on the C-terminus (Long et al., 2007). CcmM has both catalytic and structural functions that not only help produce CO_2 for RubisCO but also aid in the assembly of the carboxysome.

The advantage of this strategy is that it looks quite feasible. Protopyrenoid assembly has already been demonstrated and knocking out the chloroplast CAs has been accomplished. The protopyrenoid-containing Arabidopsis plants are quite viable, although their photosynthetic capacity appears to be slightly reduced (Atkinson et al., 2020). Similarly, the double CA knockout plants are viable, as long as a C_i transporter is present on the chloroplast envelope.

What has not been done yet is successfully increase the HCO_3^- concentration inside the chloroplast or localize CA specifically to the protopyrenoid. Another disadvantage is the hybrid RubisCO consisting of Arabidopsis large subunits and Chlamydomonas small subunits has kinetic properties similar to higher plant RubisCO, meaning that the enzyme is still slow. In addition, there are no models predicting the increase in photosynthesis that this strategy would generate.

6 Pyrenoid-based strategy

In the pyrenoid-based strategy, a complete or nearly complete pyrenoid would be assembled in the plant chloroplast (Fig. 9). This strategy would combine the thylakoid-based approach with the protopyrenoid-based approach, while including the other alterations. Like the thylakoid-based strategy, this approach would require putting CAH3 into the thylakoid lumen, adding bestrophins to the thylakoid membrane and adding a C_i transporter to the chloroplast envelope (Figs. 6 and 9). In addition, as seen in both the thylakoid- and protopyrenoid-based strategies, the endogenous CAs would need to be removed. Another similarity to the protopyrenoid-based approach is the assembly of RubisCO in the pyrenoid. This can be accomplished by adding EPYC1 along with the Chlamydomonas RbcS2 gene (Atkinson et al., 2020). One important difference between this strategy and the protopyrenoid approach is the location of the added CA. In the protopyrenoid approach, the CA was part of the assembled protopyrenoid. For the pyrenoid-based approach, CAH3 is the added CA and it needs to be localized to the thylakoid tubules that transect the pyrenoid. Having CAH3 inside the pyrenoid tubules would allow the plant to use the pH gradient across the thylakoid to raise the CO_2 concentration inside the pyrenoid.

The biggest difference between this approach and earlier strategies is the inclusion of thylakoid tubules that penetrate the pyrenoid. This would make the overall structure resemble the Chlamydomonas pyrenoid. The activity of CAH3

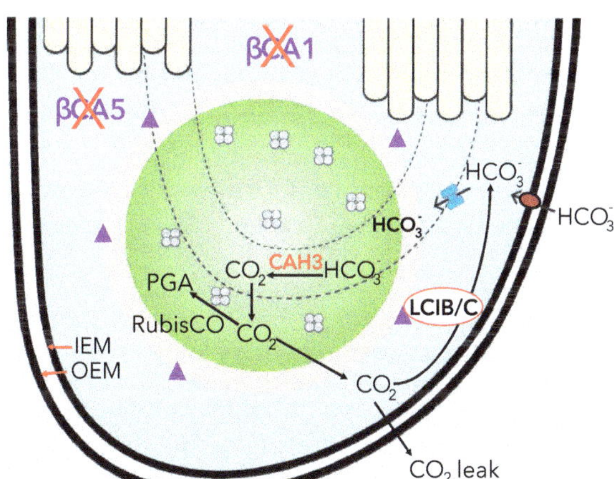

Figure 9 The pyrenoid-strategy. The modifications needed for the protopyrenoid-based approach and the thylakoid-based approach are both required here. In addition, modified thylakoid tubules must enter the pyrenoid to be effective. The CA that will be used is the thylakoid lumen CAH3, as in the thylakoid-based approach. Two more changes would be made to reduce the leak of CO_2 out of the pyrenoid: adding a starch sheath around the pyrenoid and inserting the proteins LCIB and LCIC. LCIB and LCIC normally coat the starch sheath and are thought to 'recapture' any CO_2 leaking from the pyrenoid.

would also have to be tightly controlled. Any CAH3 inside the thylakoid but away from the pyrenoid would have to be inactivated. The reason for this is that any CAH3 that is active but away from the pyrenoid would set up a futile cycle as depicted in Fig. 7. For this approach to work, only CAH3 that is physically inside tubules in the pyrenoid should be active.

Finally, to create a fully developed pyrenoid, a starch sheath and LCIB would need to be added. The starch sheath would slow the leak of CO_2 from the pyrenoid, while LCIB would 'recapture' CO_2 leaking from the pyrenoid. In Chlamydomonas, LCIB is associated with the starch sheath - converting CO_2 back to HCO_3^- if the CO_2 exits the pyrenoid.

The advantage of this strategy is that it uses a two-pronged approach to raising CO_2. It uses a C_i transporter on the envelope to increase the stromal HCO_3^- concentration but also allows the HCO_3^- to enter the thylakoid to be converted to CO_2. The more acidic pH inside the thylakoid would assure that the majority of C_i entering the thylakoids is converted to CO_2. This strategy should also minimize the leakage of CO_2. Therefore, this method uses concepts introduced in the first three strategies and further minimizes CO_2 leakage. Theoretically, increases in CO_2 fixation of 40% or more could be realized if the full implementation of this strategy were successful.

The disadvantage of this approach is its complexity. It would require an assembly of a nearly complete pyrenoid in a C_3 plant. Recent studies have shown that the assembly of RubisCO into a protopyrenoid or pyrenoid is relatively simple. It involves EPYC1 and the correct EPYC1-binding site on the RubisCO small subunit. However, how to organize the pyrenoid thylakoid tubules is unknown right now. In addition, how CAH3 is regulated in Chlamydomonas is not understood, meaning it will be hard to design CAH3 so that it is only active in the tubules that penetrate the pyrenoid (Blanco-Rivero et al., 2012). Those are the most challenging aspects of the project; it should also be pointed out that many foreign genes would have to be introduced into the plant for this strategy to work.

7 The carboxysome-based strategy

In the carboxysome-based approach, a partial cyanobacterial CCM will be constructed inside the plant chloroplasts (Fig. 10). Similar to the envelope-based strategy, this approach will put C_i transporters on the chloroplast envelope. Also, like the thylakoid- and protopyrenoid-based strategies, the endogenous CAs βCA1 and βCA5 will have to be either removed or inactivated. What makes this approach so different is that a carboxysome will be assembled in the chloroplast. A carboxysome is a large structure, but it is composed of a fairly small number of proteins. In cyanobacteria, there is often an operon encoding most of the proteins that make up the carboxysome. These include

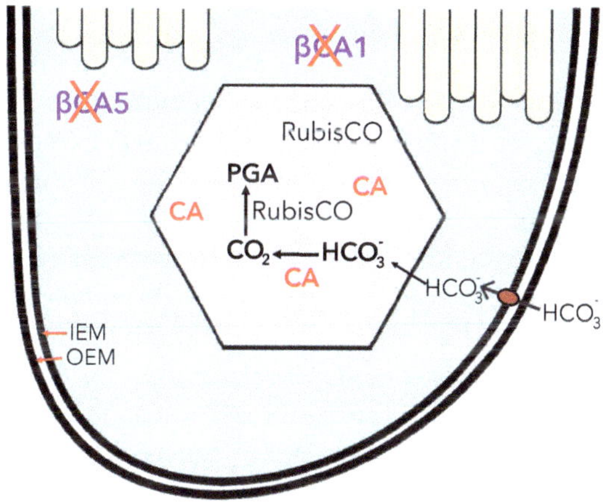

Figure 10 The carboxysome-based strategy. The genes encoding the carboxysome shell, cyanobacterial RubisCO and cyanobacterial CA would be transformed into the plant. Cyanobacterial RubisCO replaces the plant RubisCO. Envelope C_i transporters are added and the endogenous chloroplast CAs, βCA1 and βCA5 are removed.

shell proteins, the RubisCO large and small subunits, a CA and one or more protein(s) that organize RubisCO.

Partial carboxysomes have already been assembled in plant chloroplasts (Long et al., 2018; Orr et al., 2019). To date, the construction of a complete carboxysome has not been achieved, but it is encouraging that these partial carboxysomes have been observed in tobacco plants transformed with the carboxysome genes. For this method to work, the assembled carboxysomes must contain a high percentage of the cell's RubisCO and the carboxysome CA must be in the carboxysome as well. Overall, this CCM would operate similarly to the cyanobacterial CCM, with the C_i transporters on the chloroplast envelope increasing the HCO_3^- concentration in the chloroplast. The accumulated HCO_3^- would then enter the carboxysome, where it is converted to CO_2 by the carboxysomal CA. This would create a microenvironment with elevated CO_2 for RubisCO.

For this method to work, cyanobacterial RubisCO needs to be used. The use of cyanobacterial RubisCO is potentially one of the strongest aspects of this strategy, since cyanobacterial RubisCO has a higher catalytic rate than terrestrial plant RubisCO (Hanson et al., 2016). This means, in theory, that crop plants containing the cyanobacterial RubisCO would need less RubisCO to achieve the same rate of CO_2 fixation. This could potentially reduce the plant's nitrogen consumption, since proteins require a lot of nitrogen and it would need to make less RubisCO. Since RubisCO can make up close to 50% of the soluble protein in leaves, the nitrogen savings could be significant. The other strategies tested in this article would all employ either the endogenous plant

RubisCO or Chlamydomonas RubisCO and these cases would not result in nitrogen savings.

The substitution of the cyanobacterial RubisCO could have its disadvantages. First, the endogenous RubisCO would need to be significantly reduced for this strategy to work. Arabidopsis plants with reduced RubisCO have been made, but methods to reduce RubisCO have to be developed for each crop plant. In addition, while the cyanobacterial RubisCO is a faster enzyme, it also has a higher $K_m(CO_2)$ of about 110 μM compared to the $K_m(CO_2)$ of 15–20 μM measured for the RubisCO of terrestrial plants (Jordan and Ogren, 1981). This means that higher levels of CO_2 must be maintained for the RubisCO to remain active. Therefore, the C_i transporters in the chloroplast envelope would need to increase the HCO_3^- concentration to an even higher concentration than is needed for the other strategies.

Since the CO_2 is being generated inside the carboxysome, it is thought that CO_2 leakage will be minimized using the carboxysome-based strategy. At this point, there have been no efforts to insert the CupA or CupB protein, although in theory their presence would reduce CO_2 leakage even further. Like the pyrenoid-based strategy, this approach is complex. It requires all of the proteins that comprise the carboxysome, C_i transporters, cyanobacterial RubisCO and the removal of the endogenous CAs. However, it is encouraging that the assembly of partial carboxysomes has already been accomplished.

8 Technical challenges: targeting inorganic carbon transporters

All the CCM-based strategies described in this chapter involve the introduction of an algal C_i transporter into a plant chloroplast. Most of the time this transporter is to be placed on the chloroplast envelope of the plant cell. There are two major technical challenges to this approach. The first is how to correctly target the transport protein to the chloroplast envelope. Many of the transport proteins being used are bacterial transporters from the plasma membrane – the cyanobacterial transporters are in this category. Other transporters, like HLA3, are from the plasma membrane of Chlamydomonas. For these CCM-based strategies to work, these proteins must be relocated to the chloroplast envelope. This will involve reengineering the protein.

The second challenge is how to determine whether the protein is correctly working once it is correctly localized. The intracellular targeting of the introduced protein is usually monitored by 'tagging' that protein with a fluorescent probe, usually green fluorescent protein (GFP) or some modified version of GFP. The researcher can then visualize where the protein is expressed in the target plant cell. If the chloroplast envelope was targeted, one would expect to observe the GFP fluorescence around the periphery of the chloroplasts. This information tells you the protein is in the correct organelle but does not tell you if the protein is in the correct orientation in the membrane or whether the protein is functional. In addition, when GFP is added to a protein, the protein is modified because an entire second protein has been attached. This modification could potentially alter the transport protein's activity.

Focusing first on the challenges of relocating the protein, it should be noted that the protein needs to be re-engineered to be correctly targeted. If a researcher is trying to target a bacterial or cyanobacterial protein to the chloroplast envelope, one can approach this from two directions. If the chloroplast genome is transformed by adding the gene encoding the protein, then the protein would enter the envelope after being made in the chloroplast stroma. The problem with this approach is that very few envelope proteins are encoded by the chloroplast genome, so it is unknown how a protein would be directed to the envelope after being made by chloroplast ribosomes. A second problem is that methods for efficient transformation of chloroplasts have not been developed for most crop plants. For this reason, most research has focused on introducing the transgene into the nuclear DNA and having the protein made in the cytoplasm and subsequently targeted to the chloroplast.

In plants, almost every envelope protein is encoded by the nuclear genome. The protein is made as a preprotein, which is the protein with an N-terminal extension. This N-terminal extension or target sequence directs the protein to the chloroplast. If the protein is soluble and localized to the chloroplast stroma, the target sequence is usually cleaved off as the protein enters the chloroplast. After cleavage, the mature protein then folds and becomes active. Some chloroplast target sequences, like the sequence directing the RubisCO small subunit (RbcS) to the stroma, are well known (Rolland et al., 2016). By attaching the RbcS target sequence to a protein, a soluble protein normally found in the cytoplasm can be redirected to the chloroplast stroma. However, less is known about the signals that direct a protein to the chloroplast envelope.

It should be noted that researchers plan to target proteins to the inner envelope membrane (IEM). The chloroplast envelope consists of two membranes, an IEM and an outer envelope membrane (OEM). The OEM has porin proteins and is not the osmotic barrier for small molecules and ions. The IEM contains almost all of the carrier proteins for metabolites and ions. Therefore, any C_i transporter should be targeted to the IEM.

IEM proteins also have a target sequence, but so far researchers have not settled on a specific target sequence like RbcS for targeting the envelope. In fact, there is evidence that parts of the mature protein might also be needed to correctly direct a protein to the IEM (Rolland et al., 2016). If part of the mature protein is also needed, that will make the targeting of a foreign protein more difficult. The exact mechanism or mechanisms for the insertion of a protein into the IEM is not completely understood. Some of the IEM proteins are pushed all the way through both envelope membranes into the stroma and then get inserted into the IEM (Rolland et al., 2017). Other IEM proteins appear to have stop-transfer sequences that lock them into the IEM as they enter the chloroplast.

Any bacterial protein used in a CCM-based strategy will need to be re-engineered to include a sequence targeting the protein to the IEM. So far, a number of schemes have been tried for directing cyanobacterial transporters. These methods use the targeting sequences from known transport proteins to direct to the IEM. In some cases, parts of the mature IEM protein have been added as well. Using GFP-tagged C_i transporters, these procedures have been somewhat successful, as there is enrichment of the fluorescence in the envelopes (Rolland et al., 2016; Uehara et al., 2016, 2020). However, these methods have also been plagued with protein degradation and so far, there is no physiological evidence that the introduced transporters are functional.

If a eukaryotic protein like the Chlamydomonas plasma membrane protein HLA3 is used, the existing target sequence needs to be removed before the IEM targeting sequence is added. At that point, the same problems exist for these eukaryotic proteins as the bacterial C_i transporters. There is one published report using HLA3 and it showed only slight relocalization to the IEM (Atkinson et al., 2016). However, if a Chlamydomonas IEM C_i transporter like LCIA is used, the protein appears to be correctly targeted to the IEM in tobacco plants. This indicates that the targeting sequences are still recognized between alga and terrestrial plants. If IEM proteins from other eukaryotes, like diatoms, are used, it is not clear that the plant would recognize the targeting sequence, as the diatom chloroplast arose from a distinctly different endosymbiotic event.

The second location for targeting a C_i transporter is the thylakoid membrane. This would be the target if the thylakoid-based or pyrenoid-based strategy is used. The proteins that would be used are the bestrophin-like proteins recently identified in Chlamydomonas (Mukherjee et al., 2019). An advantage of using these proteins is that they might be correctly targeted if put into terrestrial plants without any modifications since green algae and higher plants are related. While no one has reported a test of this approach yet, there will likely be papers on this topic in the near future.

The second difficulty facing researchers is the lack of available methods for directly assaying the activity of C_i transporters once they have been inserted into higher plants. To be active and useful, the protein needs to be in the correct orientation in the membrane and any needed post-translational modifications must take place. Also, some proteins might need cofactors or use an ion gradient (like a Na^+ gradient) to transport HCO_3^-. Ultimately, researchers will look at the photosynthetic properties of plants containing the C_i transport transgenes for evidence that they are functional. However, intermediate indications of success or failure will be needed for this process. For example, if C_i is being enriched in the chloroplast, then one would expect to observe a reduction in the CO_2

compensation point of the plant. Another way activity could be confirmed is by measuring $\Delta^{13}C$ discrimination. Two stable isotopes of carbon are found in nature, with ^{12}C making up 98% of the carbon atoms and ^{13}C making up 1.1%. One might therefore expect about 1.1% of a plant's carbon to be ^{13}C. However, plants without a CCM tend to have slightly less ^{13}C than expected because the diffusion of ^{13}C across membranes is slightly slower than ^{12}C. However, if a CCM is operating, as in a C_4 plant, the discrimination against ^{13}C is measurably less than it is in C_3 plants. C_3 plants often have a $\Delta^{13}C$ of −25%. In C_4 plants, this discrimination is reduced significantly to −12%. For transgenic plants with an active C_i transporter, one would expect the $\Delta^{13}C$ discrimination to be less than what is seen in the untransformed C_3 plant.

9 Technical challenges: replacing and relocating carbonic anhydrases

Most of the CCM-based strategies require the endogenous CAs to be removed and replaced with CAs expressed in other specific locations. Depending on the CCM-based strategy used, these replacement CAs would need to be either in the carboxysome, the pyrenoid or inside the thylakoid lumen. In the case of the pyrenoid-based strategy, the CA should only be active in the thylakoid tubules within the pyrenoid. All these strategies begin with the removal of the endogenous CAs in the plant's chloroplast stroma.

Higher plants have many genes encoding CA. For the plant genomes that have been sequenced, the total number of CA genes is fairly conserved (DiMario et al., 2017). Fortunately, only two of the CAs appear to be located in the leaf chloroplast stroma and none in the thylakoid. The two CAs in the stroma are βCA1 and βCA5 (Fabre et al., 2007). In terrestrial plants where these CAs have been studied, βCA1 appears to make up 90% or more of the CA present in leaf chloroplasts, while βCA5 makes up less than 10% of the CA. Surprisingly, if βCA1 is deleted in these plants, they show no obvious growth defect. In Arabidopsis, however, if βCA5 is knocked out, plants show a profound growth reduction and are sterile (Medina-Puche et al., 2017). In tobacco, the story is similar but not identical. In tobacco, if βCA1 and βCA5 are deleted individually, there is no obvious growth problem. However, if both CA genes are mutated, the plant shows severe growth and developmental defects.

So why are Arabidopsis plants missing βCA5 so sick? It turns out that the growth of these CA-deficient plants can be restored by high CO_2. This implies that some metabolic pathway(s) is/are starved for either CO_2 or HCO_3^-. The explanation may come from what has been learned in *E. coli* and yeast. In both microorganisms, if the CA genes are knocked out, the bacteria or yeast cannot grow in air (Merlin et al., 2003; Aguilera et al., 2005). They can, however, grow on elevated CO_2 – like what is seen in CA-defective plants. In yeast and *E. coli*, it turns out that a number of biosynthetic pathways require HCO_3^- and the carboxylases that catalyze essential steps in these pathways have a $K_m(HCO_3^-)$ of 1–2 mM. These pathways include fatty acid, purine and pyrimidine biosynthesis. Each of these pathways has a step where one carbon is added to a molecule and this reaction is catalyzed by a biotin-dependent carboxylase (DiMario et al., 2017). At ambient levels of CO_2 and pH 7.0, the CO_2 concentration is expected to be about 15 µM and the HCO_3^- concentration about 120 µM. Therefore, the HCO_3^- level in ambient air is well below the $K_m(HCO_3^-)$ of the carboxylases (DiMario et al., 2017). It has been hypothesized that the high CO_2 requirement of the CA-defective strains is due to low levels of HCO_3^-. Aguilera et al. (2005) restored growth to CA-deficient yeast by adding key nucleotides and lipids to the cells and therefore produced data that strongly supports this idea.

Will the removal of the chloroplast stromal CAs (βCA1 and βCA5) be a problem in implementing these CCM strategies? Since each strategy involves adding a CA back to a location in the chloroplast but not in the stroma, it would appear that the loss of βCA1 and βCA5 should not be a problem. In addition to adding CA to the chloroplast, each strategy will employ a C_i transporter on the IEM to increase HCO_3^- in the chloroplast. Since both CO_2 and HCO_3^- concentrations should be higher if the CCM-based strategies work, plants that have the CCM components but lack the endogenous CAs should be fine, even on ambient CO_2.

Once the endogenous CAs have been removed, a new CA must be added. In the case of the protopyrenoid, this CA would need to be localized in the protopyrenoid itself. That would require a CA tagged with either an EPYC1 or RubisCO binding site. Such modifications would tether the CA to the protopyrenoid. For the thylakoid- and pyrenoid-based approaches, CAH3 could be put into the plant. CAH3 normally goes to the thylakoid lumen in Chlamydomonas. So far, there are no reports in the literature of CAH3 being put into plants, but the protein should be directed to the thylakoid lumen if the plant protein targeting machinery recognizes all of the signals in the Chlamydomonas CAH3. CAH3 must cross two membranes to reach the correct location: the IEM as well as the thylakoid membrane. An additional technical problem of the pyrenoid-based strategy is that the CAH3 should only be active in the thylakoid tubules inside the pyrenoid. How CAH3 activity is controlled in Chlamydomonas is not well understood, although phosphorylation of the protein as well as its redox state both appear to influence its activity.

For the carboxysome-based approach, a CA must be present inside the carboxysome itself. This may not be a serious technical problem, as the cyanobacteria carboxysome normally has a CA embedded within it. In theory, the CA should be made correctly and assembled into the carboxysome if the gene encoding it is expressed along with the other carboxysome genes.

10 Conclusion and future trends

Zhu et al. (2010) outlined nine ways photosynthesis might be improved. When groups began to work on improving photosynthesis, the strategy employing algal CCM components was considered to be a high-risk, high-reward approach. Indeed, it was estimated that it will take between 15 and 25 years to implement this strategy. In the past 5–10 years, progress has been made on many of the technical challenges of this approach. For packaging of RubisCO, both protopyrenoids and partial carboxysomes have been successfully assembled in tobacco. While complete carboxysomes and fully functional pyrenoids still need to be constructed, the steps that have already been made are encouraging.

Progress has also been made on the incorporation of C_i transporters. Some have been successfully put into the IEM of higher plants, but active C_i transport has not been demonstrated. The somewhat slow progress in part reflects our poor understanding of how proteins get targeted to the IEM. Another frustrating aspect is the lack of good ways to measure the movement of C_i across biological membranes. In addition, it is unclear whether other transporters will have to be added to balance the C_i movement. Examples of other transporters might include Na^+ or H^+ ion transport to drive the C_i uptake or Cl^- carriers to balance the charge if HCO_3^- is the ion moved. Surprisingly, the introduction of the C_i transporters might be the most challenging aspect of the algal-mechanisms strategy.

Good progress is being made regarding the CAs. We now know which CAs are in the chloroplast and how to delete or disrupt the genes encoding these CAs. New CAs still need to be successfully introduced into higher plants. While introducing new proteins into a plant is always a challenge, these CAs are soluble proteins and may offer fewer technical problems compared to the membrane-bound C_i transporters.

Finally, the strategies presented in this chapter are all based on either the cyanobacterial CCM or the Chlamydomonas CCM. That is because these are the best-studied systems. However, there are many algae that have CCMs and, outside of diatoms, most other types of algae have not been studied at all. It is likely that each major algal lineage (green, red, diatoms, etc.) has adapted to low CO_2 independently. That means there are many other C_i transporters and CAs to be discovered. Indeed, whole new families of CAs have been discovered recently in diatoms. It is likely that one or more of these algal lineages are using CCM mechanisms that significantly differ from the cyanobacterial or Chlamydomonas CCM. It is also likely that some of these yet-to-be-discovered-CCM components might be more easily adapted for use in higher plants. In that regard, it is important to continue doing basic work to resolve these different CCMs, even as we work to introduce the components we understand into crop plants. Any increase in CO_2 that can be achieved using CCM components should improve both photosynthesis and yield in higher plants.

11 Where to look for further information

11.1 Key journals/conferences

- Gordon Research Conference - CO_2 Assimilation in Plants: from Genome to Biome - holds meetings every 2 years, attended by many researchers in this field.
- Inorganic Carbon Uptake by Aquatic Photosynthetic Organisms - meets every few years and many CCM researchers attend this meeting.
- International Photosynthesis Congress - probably the largest regularly scheduled photosynthesis conference, meets every 4 years.

11.2 Major international research projects

- RIPE - Realizing Increased Photosynthetic Efficiency is an international project funded by the Bill & Melinda Gates Foundation along with FFAR and FDCO. https://ripe.illinois.edu/.
- Engineering an algal pyrenoid into higher plants to enhance yields - is an international project funded by the US NSF and the UK BBSRC agencies. https://gtr.ukri.org/projects?ref=BB%2FS015337%2F1.

12 References

Aguilera, J., Van Dijken, J. P., De Winde, J. H. and Pronk, J. T. (2005). Carbonic anhydrase (Nce103p): an essential biosynthetic enzyme for growth of *Saccharomyces cerevisiae* at atmospheric carbon dioxide pressure. *Biochem. J.* 391(2): 311-316.

Ainsworth, E. A. and Long, S. P. (2021). 30 years of free-air carbon dioxide enrichment (FACE): what have we learned about future crop productivity and its potential for adaptation? *Glob. Change Biol.* 27(1): 27-49.

Atkinson, N., Feike, D., Mackinder, L. C. M., Meyer, M. T., Griffiths, H., Jonikas, M. C., Smith, A. M. and McCormick, A. J. (2016). Introducing an algal carbon-concentrating mechanism into higher plants: location and incorporation of key components. *Plant Biotechnol. J.* 14(5): 1302-1315.

Atkinson, N., Mao, Y., Chan, K. X. and McCormick, A. J. (2020). Condensation of RuBisCO into a proto-pyrenoid in higher plant chloroplasts. *Nat. Commun.* 11(1): 6303.

Blanco-Rivero, A., Shutova, T., Román, M. J., Villarejo, A. and Martinez, F. (2012). Phosphorylation controls the localization and activation of the lumenal carbonic anhydrase in *Chlamydomonas reinhardtii*. *PLoS ONE* 7(11): e49063.

DiMario, R. J., Clayton, H., Mukherjee, A., Ludwig, M. and Moroney, J. V. (2017). Plant carbonic anhydrases: structures, locations, evolution, and physiological roles. *Mol. Plant* 10(1): 30-46.

Fabre, N., Reiter, I. M., Becuwe-Linka, N., Genty, B. and Rumeau, D. (2007). Characterization and expression analysis of genes encoding α and β carbonic anhydrases in Arabidopsis. *Plant Cell Environ.* 30(5): 617–629.

Griffiths, H., Meyer, M. T. and Rickaby, R. E. M. (2017). Overcoming adversity through diversity: aquatic carbon concentrating mechanisms. *J. Exp. Bot.* 68(14): 3689–3695.

Han, X., Sun, N., Xu, M. and Mi, H. (2017). Co-ordination of NDH and Cup proteins in CO_2 uptake in cyanobacterium *Synechocystis* sp. PCC 6803. *J. Exp. Bot.* 68(14): 3869–3877.

Hanson, M. R., Lin, M. T., Carmo-Silva, A. E. and Parry, M. A. (2016). Towards engineering carboxysomes into C_3 plants. *Plant J.* 87(1): 38–50.

Jaggard, K. W., Qi, A. and Ober, E. S. (2010). Possible changes to arable crop yields by 2050. *Phil. Trans. R. Soc. B Biol. Sci.* 365(1554): 2835–2851.

Jordan, D. B. and Ogren, W. L. (1981). Species variation in the specificity of ribulose biphosphate carboxylase/oxygenase. *Nature* 291(5815): 513–515.

Karlsson, J., Clarke, A. K., Chen, Z. Y., Hugghins, S. Y., Park, Y. I., Husic, H. D., Moroney, J. V. and Samuelsson, G. (1998). A novel alpha-type carbonic anhydrase associated with the thylakoid membrane in *Chlamydomonas reinhardtii* is required for growth at ambient CO_2. *EMBO J.* 17(5): 1208–1216.

Long, B., Badger, M., Whitney, S. and Price, D. (2007). A structural role for CcmM in beta-carboxysome shell formation. *Photosynth. Res.* 91: 223–223.

Long, B. M., Hee, W. Y., Sharwood, R. E., Rae, B. D., Kaines, S., Lim, Y. L., Nguyen, N. D., Massey, B., Bala, S., von Caemmerer, S., Badger, M. R. and Price, G. D. (2018). Carboxysome encapsulation of the CO_2-fixing enzyme RuBisCO in tobacco chloroplasts. *Nat. Commun.* 9(1): 3570.

Mackinder, L. C. M., Meyer, M. T., Mettler-Altmann, T., Chen, V. K., Mitchell, M. C., Caspari, O., Freeman Rosenzweig, E. S., Pallesen, L., Reeves, G., Itakura, A., Roth, R., Sommer, F., Geimer, S., Mühlhaus, T., Schroda, M., Goodenough, U., Stitt, M., Griffiths, H. and Jonikas, M. C. (2016). A repeat protein links RuBisCO to form the eukaryotic carbon-concentrating organelle. *Proc. Natl Acad. Sci. U. S. A.* 113(21): 5958–5963.

McGrath, J. M. and Long, S. P. (2014). Can the cyanobacterial carbon-concentrating mechanism increase photosynthesis in crop species? A theoretical analysis. *Plant Physiol.* 164(4): 2247–2261.

Medina-Puche, L., Castelló, M. J., Canet, J. V., Lamilla, J., Colombo, M. L. and Tornero, P. (2017). β-carbonic anhydrases play a role in salicylic acid perception in Arabidopsis. *PLoS ONE* 12(7): e0181820.

Merlin, C., Masters, M., McAteer, S. and Coulson, A. (2003). Why is carbonic anhydrase essential to *Escherichia coli*? *J. Bacteriol.* 185(21): 6415–6424.

Meyer, M. T., Genkov, T., Skepper, J. N., Jouhet, J., Mitchell, M. C., Spreitzer, R. J. and Griffiths, H. (2012). RuBisCO small-subunit α-helices control pyrenoid formation in *Chlamydomonas*. *Proc. Natl Acad. Sci. U. S. A.* 109(47): 19474–19479.

Meyer, M. T., Whittaker, C. and Griffiths, H. (2017). The algal pyrenoid: key unanswered questions. *J. Exp. Bot.* 68(14): 3739–3749.

Mitra, M., Mason, C. B., Xiao, Y., Ynalvez, R. A., Lato, S. M. and Moroney, J. V. (2005). The carbonic anhydrase gene families of *Chlamydomonas reinhardtii*. *Can. J. Bot.* 83(7): 780–795.

Moroney, J. V., Husic, H. D., Tolbert, N. E., Kitayama, M., Manuel, L. J. and Togasaki, R. K. (1989). Isolation and characterization of a mutant of *Chlamydomonas reinhardtii* deficient in the CO_2 concentrating mechanism. *Plant Physiol.* 89(3): 897–903.

Mukherjee, A., Lau, C. S., Walker, C. E., Rai, A. K., Prejean, C. I., Yates, G., Emrich-Mills, T., Lemoine, S. G., Vinyard, D. J., Mackinder, L. C. M. and Moroney, J. V. (2019). Thylakoid localized bestrophin-like proteins are essential for the CO_2 concentrating mechanism of *Chlamydomonas reinhardtii*. *Proc. Natl Acad. Sci. U. S. A.* 116(34): 16915–16920.

Nölke, G., Barsoum, M., Houdelet, M., Arcalís, E., Kreuzaler, F., Fischer, R. and Schillberg, S. (2019). The integration of algal carbon concentration mechanism components into tobacco chloroplasts increases photosynthetic efficiency and biomass. *Biotechnol. J.* 14(3): e1800170.

Omata, T., Price, G. D., Badger, M. R., Okamura, M., Gohta, S. and Ogawa, T. (1999). Identification of an ATP-binding cassette transporter involved in bicarbonate uptake in the cyanobacterium *Synechococcus* sp. strain PCC 7942. *Proc. Natl Acad. Sci. U. S. A.* 96(23): 13571–13576.

Orr, D. J., Worrall, D., Lin, M. T., Carmo-Silva, E., Hanson, M. R. and Parry, M. A. J. (2019). Hybrid cyanobacterial-tobacco RuBisCO supports autotrophic growth and Procarboxysomal aggregation. *Plant Physiol.* 182(2): 807–818.

Peña, K. L., Castel, S. E., de Araujo, C., Espie, G. S. and Kimber, M. S. (2010). Structural basis of the oxidative activation of the carboxysomal γ-carbonic anhydrase, CcmM. *Proc. Natl Acad. Sci. U. S. A.* 107(6): 2455–2460.

Price, G. D. (2011). Inorganic carbon transporters of the cyanobacterial CO_2 concentrating mechanism. *Photosynth. Res.* 109(1–3): 47–57.

Price, G. D. and Badger, M. R. (1989). Expression of human carbonic anhydrase in the cyanobacterium *Synechococcus* PCC7942 creates a high CO_2-requiring phenotype: evidence for a central role for carboxysomes in the CO_2 concentrating mechanism. *Plant Physiol.* 91(2): 505–513.

Price, G. D., Badger, M. R. and von Caemmerer, S. (2011). The prospect of using cyanobacterial bicarbonate transporters to improve leaf photosynthesis in C_3 crop plants. *Plant Physiol.* 155(1): 20–26.

Price, G. D., Badger, M. R., Woodger, F. J. and Long, B. M. (2008). Advances in understanding the cyanobacterial CO_2-concentrating-mechanism (CCM): functional components, C_i transporters, diversity, genetic regulation and prospects for engineering into plants. *J. Exp. Bot.* 59(7): 1441–1461.

Ray, D. K., Mueller, N. D., West, P. C. and Foley, J. A. (2013). Yield trends are insufficient to double global crop production by 2050. *PLoS ONE* 8(6): e66428.

Rolland, V., Badger, M. R. and Price, G. D. (2016). Redirecting the cyanobacterial bicarbonate transporters BicA and SbtA to the chloroplast envelope: soluble and membrane cargos need different chloroplast targeting signals in plants. *Front. Plant Sci.* 7: 185–185.

Rolland, V., Rae, B. D. and Long, B. M. (2017). Setting sub-organellar sights: accurate targeting of multi-transmembrane-domain proteins to specific chloroplast membranes. *J. Exp. Bot.* 68(18): 5013–5016.

Schuller, J. M., Saura, P., Thiemann, J., Schuller, S. K., Gamiz-Hernandez, A. P., Kurisu, G., Nowaczyk, M. M. and Kaila, V. R. I. (2020). Redox-coupled proton pumping drives carbon concentration in the photosynthetic complex I. *Nat. Commun.* 11(1): 494.

Shibata, M., Katoh, H., Sonoda, M., Ohkawa, H., Shimoyama, M., Fukuzawa, H., Kaplan, A. and Ogawa, T. (2002). Genes essential to sodium-dependent bicarbonate transport in cyanobacteria: function and phylogenetic analysis. *J. Biol. Chem.* 277(21): 18658–18664.

So, A. K. C., Espie, G. S., Williams, E. B., Shively, J. M., Heinhorst, S. and Cannon, G. C. (2004). A novel evolutionary lineage of carbonic anhydrase (epsilon class) is a component of the carboxysome shell. *J. Bacteriol.* 186(3): 623-630.

Stillwell, W. (2016). Membrane transport. *An Introduction to Biological Membranes*. Elsevier, 423-451.doi: 10.1016/B978-0-444-63772-7.00019-1.

Tolleter, D., Chochois, V., Poiré, R., Price, G. D. and Badger, M. R. (2017). Measuring CO_2 and HCO_3^- permeabilities of isolated chloroplasts using a MIMS-^{18}O approach. *J. Exp. Bot.* 68(14): 3915-3924.

Uehara, S., Adachi, F., Ito-Inaba, Y. and Inaba, T. (2016). Specific and efficient targeting of cyanobacterial bicarbonate transporters to the inner envelope membrane of chloroplasts in *Arabidopsis*. *Front. Plant Sci.* 7: 16-16.

Uehara, S., Sei, A., Sada, M., Ito-Inaba, Y. and Inaba, T. (2020). Installation of authentic BicA and SbtA proteins to the chloroplast envelope membrane is achieved by the proteolytic cleavage of chimeric proteins in Arabidopsis. *Sci. Rep.* 10(1): 2353.

Zhu, X. G., Long, S. P. and Ort, D. R. (2010). Improving photosynthetic efficiency for greater yield. *Annu. Rev. Plant Biol.* 61: 235-261.

Index

Lightning Source UK Ltd.
Milton Keynes UK
UKHW022103150123
415352UK00001B/7

9 781801 461290